BUILDING SUPPLY CHAIN EXCELLENCE IN EMERGING ECONOMIES

BUILDING SUPPLY CHAIN EXCELLENCE IN EMERGING ECONOMIES

Edited by

HAU LEE
Graduate School of Business
Stanford University
Stanford, CA

CHUNG-YEE LEE
Department of Industrial Engineering & Logistics Management
The Hong Kong University of Science and Technology
Hong Kong

 Springer

Hau L. Lee
Stanford University
California, USA

Chung-Yee Lee
The Hong Kong Univ. of Science & Tech.
Kowloon, China

Library of Congress Control Number:

ISBN-10: 0-387-38428-6 (HB)
ISBN-13: 978-0387-38428-3 (HB)

ISBN-10: 0-387-38429-4 (e-book)
ISBN-13: 978-0387-38429-0 (e-book)

Printed on acid-free paper.

9 8 7 6 5 4 3 2 1

springer.com

Contents

Chapter 19

Transforming an Indian Manufacturing Company: The Rane Brake Linings Case

Ananth V. Iyer and Sridhar Seshadri

PREFACE

Hau L. Lee and Chung-Yee Lee

In the last decade, we have seen major progresses in the development of the theories and practice of supply chain management in many industries. The most notable advances started in the apparel industry, dubbed "Quick Response (QR)," and shortly after, followed by the grocery industry, dubbed "Efficient Consumer Response (ECR)." Both industry-led movements have awaken major companies in the US and Europe on the need to integrate their supply chains. Academic research has followed, and courses on supply chain management are also standard at many business and engineering schools.

These movements, however, have involved companies that are in developed countries. Many of the leading edge supply chain examples that one can find in the literature are about powerful companies such as Dell Computer, Cisco, Seven-Eleven Japan, Wal-Mart, and Zara, etc. But the bulk of the focus has been on their excellent processes and systems in developed economies.

Yet the forces of globalization have resulted in a significant part of the supply chain of almost every industry being located in emerging economies such as China and India. Increasingly, these emerging economies also form the end-markets of a lot of industrial and consumer products. Our knowledge and experience of operating and managing a supply chain that involves emerging economies, however, is still very limited.

Supply chains are definitely increasingly global, as a result of the unprecedented growth of global trade. In 2004, global trade has grown by more than 10%, constituting 10% of the world's GDP. In fact, between 1973 and 1999, global trade has grown annually at three times the rate of worldwide GDP growth. In 1970, global foreign exchange transactions occurred at a rate of $10 billion a day. Today, that exchange is occurring at a rate of $10 billion a second. A recent Accenture study[1] showed that, in 2005, major companies had 35% of their revenues generated outside their home market, and 31% sourced raw materials, semi-finished goods, or finished goods from outside their home

---

[1] Accenture Global Operations Survey, 2005.

market. Such figures were expected to grow to 42% and 38% respectively in 2008. Companies can no longer afford to ignore managing their supply chain outside of their home country effectively.

With the rapidly developing new economies such as China, India, Hungary, Vietnam, Costa Rica, Mexico, Brazil, etc., emerging economies are at the crossroads of almost all major supply chains. The BRIC countries (Brazil, Russia, India and China) constitute 15% of the economic size of today's G6, but they are expected to grow to surpass today's G6 in less than 40 years' time. The special cultural and organizational barriers, infrastructure development, technological advances, logistics challenges, and public/private collaborations, all play central roles in the evolution of supply chains in these parts of the world. Increasingly, these parts of the world are playing the roles of suppliers, design centers, final assembly, and markets. Both forward and reverse logistics are critical for successful supply chain management.

Given the physical, social and cultural characteristics of the emerging economies, managing supply chains there could be even more challenging than in developed economies. How can we manage supply chains well in emerging economies, coordinate information flows with multiple partners, tackle challenges such as unexpected disruptions, diversify the risks and increase flexibilities, be efficient but at the same time contribute to the social and environmental developments of these economies, and use supply chain concepts and practices to improve the economic welfare of these countries, such as basic infrastructure developments and disaster relief, are topics of heightened interest.

This book seeks to provide some insights on the answers to the above questions. It is our hope that the collection of articles will enable practitioners to gain insights on the developments, challenges and opportunities when operating supply chains in emerging economies; and learn about some innovative approaches and experiences by some progressive companies and thought leaders. We also hope that this collection will stimulate researchers to gain deeper understanding and develop methods in operating supply chains in emerging economies.

We have organized the book in three key sections. The first section develops the overall framework in managing global supply chains and developing strategies. The second section describes the challenges and opportunities in supply chain management of emerging economies – the infrastructure constraints, the logistics inefficiencies, and limitations in service operations; and discusses how to create opportunities in such adverse conditions. The third section is devoted to a number of industrial cases that showcase innovative approaches to gain excellence, and share insights and lessons from such experiences.

Global Supply Chain: General Strategies and Framework

There are five chapters on the general strategies and framework. In "On the Globalization of Operations and Supply Chain Strategies – A Conceptual Framework and its Applications," Panos Kouvelis and Julie Niederhoff describe the forces that shape globalization and a framework to develop strategies. The authors illustrate how the framework can be used with the case of Acrilan, an acrylic fiber manufacturer.

As the emerging economies begin to mature, and the supply and demand points in a supply chain begin to shift, companies need to re-optimize the design of their global supply chain, so as to make the best use of their global resources. This is the subject of "Globalization and Emerging Markets: The Challenge of Continuous Global Network Optimization," by Peter Koudal and Douglas A. Engel.

Many emerging economies have lower direct labor costs, and are so attractive offshoring locations. But there could be many hidden costs. The decision to offshore has to be based on a sound comprehensive analysis of the total landed costs, the tradeoffs of associated risks, and the business strategies of the company. David Pyke provides us with the approach to tackle this decision, and share with us his personal experience in helping companies to make such decisions, in his chapter titled "Shanghai or Charlotte? The Decision to Outsource to China and Other Low Cost Countries."

It is not just commercial goods that would be of concern to operating supply chains in emerging economies. Such economies are also prone to natural disasters. In "Life-Saving Supply Chains: Challenges and the Path Forward," Anisya Thomas and Laura Kopczak show how humanitarian relief organizations could make use of information technologies and supply chain principles to improve the effectiveness of relief operations. The private sector can probably learn from humanitarian disaster relief operations, as supply chain disruptions in emerging economies are not uncommon and we have to be just as responsive and efficient.

Eric Johnson describes how Mattel developed its capacity expansion strategies in "Dual Sourcing Strategies: Operational Hedging and Outsourcing to Reducing Risk in Low-Cost Countries." The Mattel story can be used to develop better risk-hedging strategies, which is crucial given the often higher risk exposures in supply chains of emerging economies. Eric Johnson gives us a comprehensive treatment of all the risks in managing a supply chain.

Supply Chain Management in Emerging Economies: Challenges and Opportunities

There are six chapters in this section. India is one of the major emerging economies. Most of us think of India as a growing source of labor in software and computing technologies. But in fact, the supply chain of India encompasses many more products and services. In "Managing Supply Chain Operations in India – Pitfalls and Opportunities," Jayashankar Swaminathan gives a thorough overview of the state of supply chain management in India, as both a source and market, and outlines the necessary steps in order to gain control of the supply chain.

Another major emerging economy is China. Given the size of the country and the fact that the logistics infrastructure of most of the inland of China is still not well developed, order fulfillment is a major challenge. In "Integrated Fulfillment in Today's China," Jamie M. Bolton and Wenbo Liu discuss what these challenges are, and how these challenges are changing as a result of China entering WTO. They also give some lessons, based on their experience, on what companies need to do as they increasingly make use of China as a part of the supply chain. In "Logistics Management in China: Challenges, Opportunities and Strategies," Gengzhong Feng, Gang Yu, and Wei Jiang discuss in detail the transportation and logistics problems in China, but shows how the trends are changing and that opportunities can be created.

Hong Kong has long been a major logistics hub connecting the East and the West. But the economic growth of Southern China and its reliance on Hong Kong as a major outbound port has resulted in significant congestion and potential productivity degradation. In "Connectivity at Inter-Modal Hub Cities: the Case of Hong Kong," Raymond Cheung and Allen Lee describe such problems and discuss ways to improve logistics flow.

Besides the forward supply chain, the reverse supply chain is equally important as companies start to develop emerging economies as their markets. Part of the reverse supply chain is the provision of after-sales customer service. In "Service Parts Management in China," Steven Aschkenase and Keith Nash articulate the importance of managing service parts in China, and show that, by managing the service chain well, great values can be created.

Finally, supply chain flows in emerging economies are complicated by the existence of regulations, trade agreements, and other governance rules. In "DHL in China: The Role of Logistics Governance," Kevin Leung and Paul Forster use the DHL experience to discuss how companies need to be cognizant of logistics governance factors so that they could overcome barriers and gain control.

Building Supply Chain Excellence: Innovations and Success Cases

There are eight chapters in this section. The first one deals with the use of information technology to support supply chains of agri-business in India. Despite the poor infrastructure and the highly inefficient supply chain, smart use of information technology can help to transform supply chains and make a difference, leading to benefits to all parties in the supply chain. In "Supply Chain Reengineering in Agri-Business – A Case Study of ITC's e-Choupal," Ravi Anupindi and S. Sivakumar give us the case of e-Choupal, in which such transformation had been successfully implemented.

Going against the trend of outsourcing, Esquel, an apparel manufacturer, developed a vertically integrated supply chain going all the way from cotton farms in Xinjiang, China, to fabric weaving and dyeing, garment manufacturing, and even retailing. In "Esquel Group: Going Beyond the Traditional Approach in the Apparel Industry," Barchi Peleg-Gillai describes how Esquel can run such a vertically integrated supply chain with efficiency, social responsibility, environmental sensitivity, and sound business results.

Most global companies would develop supply chain processes using developed economies as the test-bed, and then localize the processes in emerging economies. CEMEX, a major cement manufacturer, did it the other way round. It used Mexico, a country with very difficult physical logistics infrastructure, diverse customer needs, and very demanding customers service requirement, as its test-bed for innovative approaches and methods; and then extend such processes to the rest of the world. In "End-To-End Transformation in the CEMEX Supply Chain," David Hoyt and Hau L. Lee describe how CEMEX was able to use such an approach to become the world's leader in cement manufacturing.

As the emerging economies grow and mature, the increasing middle class enables such economies to become major markets for consumer goods as well. Distributing in these economies is not easy. But IDS, a Li and Fung company (which has often been dubbed as *the Supply Chain Architect of Apparel and Consumer Goods*), created an innovative approach to distribution. It first unbundled the multiple distribution services, and then re-integrate them to give the greatest values to customers. This is recorded in "The IDS Story – Reinventing Distribution," by Ben Chang and Joseph Phi.

As companies source materials from emerging economies that are in underdeveloped countries, the risk of supply disruptions is increased. Starbucks Corporation has recognized that sustainability of the supply bases is an important objective of a supply chain. In "Building a Sustainable Supply Chain – Starbucks' Coffee and Farm Equity Program," Hau L. Lee, Stacy Duda, LaShawn

James, Zeryn Mackwani, Raul Munoz, and David Volk describe the Starbucks initiative to help farmers in Africa, East Asia and Central America. A sustainable supply chain is also a socially-responsible supply chain.

Since emerging economies are just beginning to be growing markets for industrial and consumer goods, multi-national giants have not penetrated in many such economies. This provided a window of opportunity for smaller players to build its business. The value proposition can be based on sound customer service. In "Building a Distribution System in Eastern Europe: Organic Growth in the Czech Republic," Eric Johnson describes how this can be done, and draw learning lessons from this successful case.

Emerging economies often made use of highly focused industrial and logistics parks as a way to attract foreign investments to develop its manufacturing sector. In "A Path to Low Cost Manufacturing for Integrated Global Supply Chain Solutions." Wesley Chen describes the experience of Solectron in making use of the Suzhou Industrial and Logistics Park to create its manufacturing center of excellence. It requires a lot of collaborative work with the local government, but the payoffs are huge.

Finally, in "Transforming an India Manufacturing Company: the Rane Brake Lining Case," Ananth Iyer and Sridhar Seshadri describe the journey of Rane Brake Lining, an Indian manufacturing company. Emerging economies are often not known for high quality standards, but Rane Brake Lining ran against the conventional wisdom. Its persistent pursuit of quality management is a lesson for others in emerging economies.

Acknowledgment

Some of the chapters of this book were based on a conference in Shanghai in 2004, co-sponsored by Hong Kong University of Science and Technology, Stanford University, Eindhoven University and the China-Europe International Business School, with financial support by the World Bank. The theme of the conference was "China at the Crossroad of the Global Supply Chain." We are grateful to our collaborators of the event, which gave rise to the idea of developing the current book. We are also grateful to many of our colleagues, who encouraged us to expand our focus from China to many other emerging economies. Some of these colleagues also contributed chapters to this book.

Part I

GLOBAL SUPPLY CHAIN: GENERAL STRATEGIES AND FRAMEWORK

Chapter 1

ON THE GLOBALIZATION OF OPERATIONS AND SUPPLY CHAIN STRATEGIES
A Conceptual Framework and Its Application

Panos Kouvelis and Julie Niederhoff
Olin School of Business
Washington University, USA

Abstract: In a global market, companies do not compete solely as individuals but as part of an entire supply chain, and strategic managers must consider the whole supply chain and fully understand global forces and relevant trends when making operational decisions. We present a conceptual framework through which managers can evaluate the many forces affecting global operating strategies. This framework is composed of four types of forces: global market forces, technology forces, global cost forces, and political or macroeconomical forces. Global market forces, in general, motivate a company to seek a larger market for its goods or services. Technological forces generally ease barriers to globalization or require global operations strategies in order to access cutting edge technology. Global cost forces seek reduced or shared costs through high quality, lower total cost global production sites. Finally, political or macroeconomical forces cause a firm to seek competitive advantages through the careful utilization of exchange rates, regional trade agreements, or nontariff barriers. By using the global forces framework presented here, managers can clearly analyze the various important factors that shape global operations decisions and understand the implications of recent global events and trends on their supply chain strategies. We clearly illustrate the application of our framework in shaping supply chain strategy via a real case study from our consulting practice.

1.1.   Introduction

Over the past two decades, a new global business environment has evolved. If current trends continue, world exports of goods and services will reach $12 trillion dollars by 2006, or close to 30% of world gross domestic product, up from a mere 10% two years ago. The vast majority of businesses now have some form of global presence – through exports, strategic alliances, joint ventures, or as part of a committed strategy to sell in foreign markets or locate production plants or business process services abroad. American manufacturers have 8,000 units overseas employing almost 5 million workers, equal to nearly 25% of US manufacturing employment. Previously, large multinational corporations dominated the international marketplace, which domestic firms generally ignored. However, according to the 2001 census, 97 percent of manufacturers who exported were considered small- or medium-sized. In today's marketplace, most companies realize that it is essential to be aware of and participate in international markets. The main goal of this chapter is to examine the factors and forces that are driving the increasing globalization of operations activities and to organize them into a framework through which managers can analyze globalization decisions.

In a global market, companies do not compete solely as individuals, but as part of an entire supply chain, and strategic management must consider the whole supply chain and the global forces and trends shaping the new competitive environment when making operational decisions. Using the case of an acrylic fiber producer in the apparel supply chain, we illustrate how a viable company can lose its competitive edge when a less competitive member of the forward supply chain acts as a bottleneck to the market. In the global market, competition is at a supply chain level and this case, along with many other examples, illustrates many of the driving forces for globalization in the supply chain, analyzed through the "globalization forces" framework. Effective management in a global operations setting requires a sophisticated understanding of these global forces and the new manufacturing order of dispersed manufacturing with supply chain and logistics as core management capabilities.

1.1.1   Globalization is Increasing

Trends among U.S. businesses illustrate the growing size and importance of global operations:

- According to the U.S. Census, from 2003 to October 2005, the value of goods imported rose nearly 45 percent while the value of goods exported rose only 31.3 percent. Overall the trade deficit for goods rose 62 percent in that time.
- When other characteristics of companies are held constant, exporting firms perform much better than nonexporters. Worker productivity in exporting firms is 20 percent higher than that of non-exporting firms. Export jobs are better jobs: production workers in exporting firms earn 6.5

percent more. They are also more stable jobs: exporting firms are 9 percent less likely to go out of business than comparable nonexporting firms. (Bergsten, 2001)

- According to U.S. Census information from 2001 to 2002, nearly 20 percent of U.S. manufacturers participate in exporting, accounting for over 71 percent of the total known value for exports.
- Manufacturing assets held by multinational enterprises in foreign countries are substantial and rapidly increasing.
- Much of the US trade deficit represents what US corporations buy from their overseas units. In the late 1990's such foreign affiliates of US corporations generated close to $3 trillion dollars in sales, with 65% of that sold to local markets and the remaining 35% brought back to US.
- In 2004 about 8 percent of administrative office work had been outsourced, much of it to India. In IT services, 16 percent of all IT work is done by an outsourced party. Much like manufacturing in the 1980's and 1990's, Business Process Outsourcing (BPO) is a growing trend, estimated to grow from $3.6 billion in 2003 to $21-$24 billion by 2008.

Two factors underlie the dramatic rise in globalization. First, global reach is important to a firm's survival. Second, multinational firms are more profitable and grow faster. Among twenty major U.S. manufacturing industries, multinational firms grew faster than domestics in nineteen of the cases and were more profitable in seventeen cases (Bergsten, 2001).

Given this environment, the goal of this chapter is to understand the development of successful global operations strategies when competing at a supply chain level. These strategies achieve business objectives through a dynamic process of leveraging and managing manufacturing, logistics, and research and development (R & D) activities. Using the framework presented here (Figure 1.1), managers can organize the decision process for globalization strategy. The essence of our framework is that there are four major driving forces for globalization: global market forces, technological forces, global cost forces, and political and macroeconomical forces. The power of this chapter is the usefulness of this framework to aid managers in analyzing how these forces are impacting a specific industry or product and to illuminate potential advantages or pitfalls to various globalization options.

1.2. The Driving Forces of the Globalization Process

The conceptual framework that follows classifies the major factors and driving forces behind the globalization process. Each of these factors affects different industries, even different products, to varying degrees. The affects

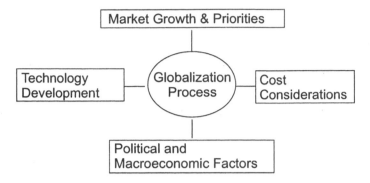

Figure 1.1. Four Forces Globalization Framework.

of each on the globalization efforts of Acrilan are detailed in Section 7. Even though generalizations are impossible and "recipes" undesirable in strategy development, the framework will allow an operations manager to structure his or her thinking process in understanding changes in the global environment, prioritizing the importance of various factors, and developing strategic alternatives.

The motivations for globalization of operations can be grouped into four categories, as shown in Figure 1.1. These forces, and specifically how they emerge in the Acrilan case, are detailed in the following sections.

1.2.1   Global Market Forces

Market forces for globalization include the need to establish a presence in foreign markets to capitalize on foreign demand and recoup domestic demand lost to foreign imports, as well as to build a global presence to minimize foreign threats into the domestic market through a competitive balance. As the existence of secondary markets for end-of-life products begins to fade, speed to market sometimes dictates a global production and distribution network. Finally, as state-of-the-art niche markets develop in specific countries, companies must consider establishing a presence in these markets to stay competitive and abreast of the latest technology developments and demands.

1.2.2   Technological Forces

Many companies seek a global market in order to achieve economies of scale as they simultaneously narrow the scope of their product to a niche market by differentiating themselves in a commodity market. This, in effect, is the mass customization of a good at a global-level.

As technological production skills develop globally, multinational firms need to tap into the technological knowledge of various countries and integrate the new technology into their own product as necessary to stay competitive. In order to stay current and access these new developments in real time, global companies may benefit by forming close relationships with dominant foreign suppliers instead of investing in an in-house capability. Additionally, by placing production facilities close to the state-of-the-art suppliers and competitors, costly delays due to breakdowns are minimized. Globally located firms can also engage in technology sharing and interfirm collaborations, and take advantage of state-of-the-art or lower priced R&D workers in countries such as Taiwan, China, and India.

1.2.3 Global Cost Forces

Perhaps the most commonly recognized force for globalization, experiences from previous global expansion projects, and some of their failures, indicates that costs to consider for offshore sourcing should go beyond just direct labor and definitely include quality, differential productivity, and design costs, and carefully account for added logistical and transportation costs. Often, direct labor is such a small component of total product cost that it is misleading to offshore based solely on reduced labor costs. In industries with capital-intense production facilities, globalization can be a natural result of multiple firms seeking a joint production facility or a single firm seeking economies-of-scale through high utilization of a private production facility.

1.2.4 Political or Macroeconomical Forces

Currency rate fluctuations can help or hurt global operations and require careful analysis and, preferably, a portfolio of options to balance unfavorable fluctuations. Regional trade agreements, such as NAFTA, also influence the decision to globalize operations. Finally, the imposition of trade protection mechanisms, such as tariffs, trigger price mechanisms, local content requirements, technical standards, health regulations, and procurement policies, influence where a company may choose to locate global operations in order to make the most of its supply chain.

1.3. Global Market Forces

Manufacturers cannot afford to ignore the tremendous growth potential of foreign markets. First they need to attack competitors abroad to develop a competitive balance and protect domestic market share. They also may need to acquire market knowledge in markets other than the home country, to respond quickly to customer orders, and customize products for various local markets. For many products and industries, penetration of global markets depends on having global facilities and/or distribution and supply networks to respond to customer demand in all the relevant competitive dimensions of cost, quality, service, and flexibility.

The nature of global market forces, how they contribute to the globalization of operations activities, and the strategies manufacturers pursue in response to them, are characterized by five main themes.

- Intensified foreign competition in local markets
- Growth in foreign demand
- Global market presence as a competitive threat
- Changing competitive priorities in product markets
- Establishing a presence in state-of-the-art markets

1.3.1   Intensified Foreign Competition in Local Markets

Among U.S. manufacturers, penetration of foreign goods in the U.S. consumer goods markets doubled during the 1980s and the value of foreign goods imports more than tripled between 1990 and 2004. The numbers are even more exaggerated in capital goods markets, where foreign penetration rose from about 14 percent to over 40 percent in the 1980s and the value of capital goods imports in 2004 almost tripled that of 1990. Manufactured goods such as machines and transport equipment increased from a mere 4 percent of China's total exports in 1985 to nearly 35 percent in 2000, many of which are exported to the United States. Clearly, foreign competition has intensified in virtually every industry and affects all companies to some degree.

As a result, even small- or medium- sized firms that have never marketed or produced products abroad need to understand developments in the global environment. The openness of most international markets today allows foreign firms to compete directly with domestic firms in previously protected local markets. These competitors frequently are large multinationals with integrated global operations. They are adept practitioners of world-class manufacturing and logistics standards, which forces many small- and medium-sized firms to upgrade their operations, to keep abreast of product and process innovations, and to adopt the latest in just-in-time and total quality management techniques. In many cases, small- and medium-sized firms must even consider expanding into international operations – either through exporting, outsourcing some production, or entering into alliances and licensing agreements with foreign partners.

1.3.2   Growth in Foreign Demand

In recent years, the world market has grown disproportionately larger relative to the U.S. market, and most of the growth is coming from developing country markets. It therefore makes sense for operations executives to target those markets for future growth potential.

Growth in foreign-market demand necessitates the development of a global network of factories, as well as an expanded sales and distribution network. If *economies of scale* are important in an industry, the global network probably will consist of a few centralized production facilities in countries that offer comparative advantages in the critical production process inputs (i.e., labor, resources). The multinational firm then uses its global economies of scale to attack local markets.

If *customization and fast response* drive the industry, and economies of scale are less important, then the resulting global network will contain multiple facilities, each dedicated to serving a specific local or regional market.

Having the opportunity to operate in, and meet the demand of global markets complicates the production-planning task of the global operations manager. It requires attention to complicating factors such as currency movements and coordination of dispersed production facilities. On the other hand, operation on a global scale allows more efficient utilization of resources and a more stable production plan. Companies can take advantage of regionally different demand fluctuations to smooth production.

1.3.3   Global Market Presence as a Competitive Threat

Global presence can be used as a defensive tool to stop aggressive moves by foreign competitors toward penetrating a firm's home market. For example, in the ready-to-eat cereal market, the U.S.-based Kellogg company and its large European competitor Nestlé have large market share in their home markets, but limited presence in their competitor's home market. The two companies maintain a gentleman's agreement of nonaggressive penetration of each other's home markets following unsuccessful past efforts and heavy revenue bleeding from subsequent retaliations. In general, a company that is unable to retaliate against aggressive foreign competitors attempting to penetrate its home market is in a vulnerable competitive position.

1.3.4   Changing Competitive Priorities in Product Markets

For many years the dominant theory in international production was based on the concept of an *international product life cycle*. Under this theory, a company introduced a product in one or several developed-country markets. When the product entered the decline stage of its life cycle in these markets, the company simply began shipping it to developing-country markets. The strategy regenerated or extended the product's total life cycle by sequentially cultivating markets that lagged behind in customer needs and knowledge on the latest product and process technology developments.

Unfortunately, few industries remain today in which this theory still applies. Product markets, particularly in technologically intensive industries, are changing rapidly. Product life cycles are shrinking as customers demand new products faster. In addition, the advances in communication and transportation technology give customers around the world immediate access to the latest in available products and technologies. Thus, manufacturers hoping to capture global demand must introduce their new products simultaneously to all major markets. Furthermore, the integration of product design and development of related manufacturing processes have become the key success factors in many high-technology industries, where fast product introduction and extensive customization determine market success. As a result, companies must maintain or source from production facilities, pilot production plants, engineering resources, and even R & D facilities all over the world. For example, in the 1980s, Apple Computer built a global manufacturing and engineering infrastructure with facilities in California, Ireland, and Singapore. This network allowed Apple to introduce new products simultaneously in the American, European, and Asian markets. With the improvements to logistics technology in the 1990s, they moved to entirely outsourced production based largely in Asia, while maintaining design in California.

1.3.5   Presence in State-of-the-Art Markets

In certain industries, particularly hypercompetitive high-technology segments, certain country markets demand state-of-the-art products to meet their consumer needs. For these industries and product markets, customer preferences drive the next generation of product and process innovations. Firms that intend to remain product/process leaders in these state-of-the-art markets must set up production, and in some cases, product development facilities there. Examples of state-of-the-art markets are:

- Japan: Semiconductor process equipment, consumer electronics and machine tools
- Germany: Machine tools
- United States: Aerospace, computers, software

Companies use the state-of-the-art markets as learning grounds for product development and effective production management, and then transfer this knowledge to their other production facilities worldwide. This rationale explains why Mercedes-Benz decided to locate a huge manufacturing plant in Vance, Alabama. The company recognized that the United States is the state-of-the-art market for sport utility vehicles.

1.4. Technological Forces

In recent years, transportation and communication costs have fallen dramatically, and international operating activities have become much easier to organize. At the same time, the sources of creation and dissemination of knowledge have globalized. Competitive success depends more and more on how quickly and effectively a firm incorporates new product and process technology into the design and production of its products and services. This need for speed has prompted companies to locate more production, R & D, and business processes services abroad, closer to the suppliers of advanced technological knowledge in component production or of crucial process equipment or skill bases. Companies have formed joint ventures to share technological knowledge in exchange for market presence. They have located R & D facilities in countries with the most cost-effective technological resources and scientific infrastructure.

Technological forces have shaped the global operations strategies of multinationals in four ways.

- Technological advancements and effective mass customization in global markets
- Diffusion of technological knowledge and global location
- Technology sharing and interfirm collaborations
- Global location of R & D facilities

1.4.1 Technological Advancements and Effective Mass Customization in Global Markets

Another major trend in operations strategy is *fewness*. Fewness characterizes markets that have a limited number of producers. The concept is not new, but is a strategic quality firms should seek by segmenting existing markets or creating new product niches from scratch. This tradeoff reduces competition to just those within the niche market, but also reduces the market, leading to the necessary pursuit of a larger global market.

The average number of all competitors in specific industry segments dropped by 45 percent in the late 1980 and early 1990s, according to The Conference Board manufacturing database. The same trends intensified in the

90s and the new millennium. *Fewness* generally is associated with increasing profitability. A firm that achieves market leadership and has few competitors in its market segment is likely to be more profitable than one that typically has many competitors or small market share. Indeed, in this era of globalization, it is easier for companies to expand within a market segment across borders than to expand across diverse product lines within a country.

How are firms able to pursue fewness and simultaneous profitability in the global marketplace? The answer is fairly straightforward, and relies on exploiting a unique synergy of global market forces and recent technological advancements in manufacturing and logistics. Fewness is the result, in part, of two forces: diversity among products and uniformity across national markets. Product diversity has increased as products have grown more complex and differentiated and as product life cycles have shortened. At the same time, national markets have become increasingly similar, especially in the industrial countries, and particularly for intermediate goods. Companies have been able to expand their global market presence and simultaneously realize economies of scale, as a result of more flexible manufacturing and distribution methods and better communications and transportation technologies. Falling transportation costs are driven in part by dramatic improvements in containerization and supply chain management.

1.4.2   Diffusion of Technological Knowledge and Global Location

Advanced technological/production knowledge is no longer the preserve of large American or European multinationals. The U.S. only produces about 7 percent of the world's engineers, less than China, Japan, and Russia. The share of the U.S. market for high-technology goods supplied by imports from foreign-based companies rose from a negligible 5 percent to more than 20 percent in the early 1990s. Moreover, the sources of such imports expanded beyond Europe to include Japan and the newly industrialized countries of Hong Kong, Singapore, South Korea, Taiwan and more recently China. China's market share of the semiconductor market is expected to grow from 5 percent in 2000 to 15 percent in 2010 due to the influx of Taiwanese semiconductor technology into China and the tremendous growth in local demand. By 2010, China is expected to dominate the low-end design chips market. China's abundance of engineers makes it a popular destination for R & D facilities. Companies such as GM, Siemens, and Nokia have research centers in China.

The abundance of engineers and MBAs in India has made it the most common choice for business process offshoring. Adding to India's appeal are the 2.5 million graduates each year who have the necessary skills to do BPO work, such as accounting, telesales, technical support, and insurance. Along with their technical knowledge and language skills, technical infrastructure, such as telecommunications, has enabled immediate service at minimal cost to companies seeking to offshore business processes in India.

In response to this diffusion of technological capability, multinational firms need to improve their ability to tap multiple sources of technology located in various countries. They also must be able to absorb quickly, and commercialize effectively, new technologies that, in many cases, were invented outside the firm – thus overcoming the destructive and pervasive "not-invented-here" attitude and resulting inertia.

The need to have access to critical technological components for their products forces firms to develop close relationships with dominant foreign suppliers in certain product/technology areas. Supplier involvement in new product design efforts becomes critically important, thereby causing some companies to locate production facilities close to their suppliers. Two examples of dominant suppliers are Canon and Fanuc, many of whose industrial customers have located close to their plants. Canon is the leader in global production of motors for fax machines and laser printers, with more than 80 percent of the world market. Fanuc controls more than 70 percent of the world market for machine-tool controllers.

The alternative to dependence on dominant foreign suppliers is a *deep-pocket* investment strategy of developing component production capabilities in-house. This strategy is not only expensive but also risky, especially if the required know-how lies outside the firm's core capabilities. However, some firms pursue it, with IBM's manufacturing of engines for its own laser printers being one such example.

Another clear trend is for companies to locate production facilities close to foreign suppliers of critical process equipment because of the sophisticated nature of this equipment, the devastating effect of prolonged breakdowns or production slowdowns, and the need for process technology know-how in the accelerated cycle of new product-manufacturing process development. In a high-technology environment, for instance, buying process equipment long distance is not usually a viable solution. For example, both IBM and Xerox chose to produce video displays in Japan, which has the best process capability of this technology. A large U.S. paper company has built operations in Europe, not only to penetrate the European market, but also to gain fast and easy access to process development emerging from the major European equipment suppliers, as well as from smaller European competitors. These smaller paper firms are open to licensing agreements for their process innovations, an attractive alternative for U.S. firms.

In services, technology has reduced the need for proximity between markets and service providers. Dramatically lower telecommunication costs, significant increases in telecommunication capacity, and radical improvements in digitization of services have redefined the range of services that may be provided remotely. Fast food orders at some McDonald's restaurants in Missouri are routed through call centers a few states away, and there is no reason why that can not be a continent away.

1.4.3   Technology Sharing and Interfirm Collaborations

In many cases, the main motivation behind interfirm collaborative agreements such as joint ventures, participation in international consortia, or technology licensing, for example, and a variety of other alternatives, is the need to gain access to technological development. In the partnership between Texas Instruments (TI) and Hitachi, two of the world's top-ten chip-makers, the firms began collaborating in 1988 with the intent of sharing basic technologies. As the joint venture succeeded, however, TI and Hitachi decided to capitalize on their success by expanding the agreement to include manufacturing as well. The two partners have now built a joint facility in Richardson, Texas, in order to create the next generation of Dynamic Random Access Memory (DRAM) chips.

The main motivation of joint ventures in the steel industry between U.S. and Japanese firms (for example, LTV and Sumitomo, Inland Steel and Nippon Steel, National Steel and Marubeni) was the desire of the U.S. firms to gain access to advanced process technology, combined with the need for financial backing from Japanese producers. In return, the Japanese obtained broader access to the U.S. market and a production base to supply the American plants of Japanese automakers.

The well-known joint ventures in the auto industry between U.S. and Japanese firms (GM-Toyota, Chrysler-Mitsubishi, Ford-Mazda) followed a similar pattern. U.S. firms needed to obtain first-hand knowledge of Japanese production methods and accelerated product development cycles, while the Japanese producers were seeking ways to overcome U.S. trade barriers and gain access to the vast American auto market.

1.4.4   Global Location of R & D Facilities

As competitive priorities in global products markets shift more toward product customization and fast new product development, firms are realizing the importance of co-location of manufacturing and product design facilities abroad. In certain product categories, such as Application Specific Integrated Circuits (ASICs), this was the main motivation for establishing design centers in foreign countries. Other industries such as pharmaceuticals and consumer electronics also have taken this approach.

The availability of low-cost, high-quality engineers in some developing countries has been a major factor contributing to the location of R & D facilities abroad, as the U.S. produces only 7 percent of engineers globally. Taiwan has been a primary location for firms looking for highly trained mechanical and electrical engineers; India is a rich source of software engineering talent. Access to creative and highly trained technical workers also seems to be the main motivation behind the recent overseas expansion of Japanese R & D. Since 2001, Western companies have established more than 130 R & D facilities in China, the largest producer of engineers in the world, almost three times that of America.

As different, more demanding technical standards are increasingly set both by customers and regulators, companies are forced to locate design and R & D facilities in foreign countries to support their global manufacturing and marketing networks. Firms that keep up with developing demands and changing standards in many national markets are more likely to be at the forefront of innovation. Additionally, they may be able to influence the development of new industrial standards (for example, a firm with operation in Europe can influence industrial standards set by the European Union), and thus gain and advantage over firms just exporting to that market.

1.5. Global Cost Forces

The comparative cost advantage of some countries in various inputs to the manufacturing process or business processes has always driven the expansion of multinational operations to new "lower cost" paradises. But as technology diminishes the importance of various cost components (e.g., direct labor cost), while accentuating the magnitude of others (e.g., capital costs), the expansion and location of the various firm activities start to be driven by new categories of total costs (e.g., taxes and total quality costs) and the economic relationship between a company and the local government (e.g., government subsidies). As the order-winners in the product markets shift away from production-cost considerations toward quality, delivery speed, and customization, factors such as transportation, telecommunications, and supplier infrastructure assume increasing importance in determining the location of production and business processes activities. For business processes that are characterized as labor intense with low commodity skill sets, offshoring can be a cost savings provided the necessary training and telecommunications infrastructure are in place. Without question, the cost of business process inputs affects the global location of activities. However, cost priorities are not simply that of direct labor cost, but also of new cost categories such as poor quality and product design costs. In fact, cost priorities have become much more dynamic, and, as a result, are reshaping global strategies more frequently. In this section we discuss the three factors that have influenced this evolution.

- Diminishing importance of direct labor cost in offshore sourcing strategies
- Emergence of new cost priorities in the location of global operating activities
- Increasing capital intensity of production facilities

1.5.1   Diminishing Importance of Direct Labor Cost in Offshore Sourcing Strategies

Sourcing from foreign suppliers can be a legitimate tactic for staying even with or gaining advantage over the competition. The 1970s and 1980s witnessed an explosive growth of offshore sourcing by multinational manufacturers, driven by an obsessive search for the lowest labor costs. This strategy was particularly popular with U.S. manufacturing firms that perceived offshore sourcing as the only alternative to the aggressive invasion of their local markets by low-priced, high-quality imports. In some cases, this approach was the correct one. In the assembly of electronic devices, for example, the typical decision involves choosing between setting up highly automated, very expensive assembly technologies or opting for an offshore sourcing arrangement. The right answer depends on the product and the characteristics of its market. For products with very short model lives, for example, low labor cost locations may be preferable to automation to minimize capital investment.

In many other cases, however, choosing offshore sourcing arrangements was an incorrect decision, based on a misunderstanding of the firm's cost structure. Companies misunderstand their cost structures for a variety of reasons. In many cases, manufacturing managers overstate the importance of direct labor cost because it is the easiest to quantify and the most readily apparent cost element. For many years, standard accounting practices seriously inflated the importance of direct labor costs. Overhead cost allocations were typically based on direct labor costs, thus prompting companies to allocate a large percentage of fixed costs, many of them unaffected by the location decision, to products with slightly higher direct labor cost components. This form of overhead cost allocation cast offshore sourcing as the panacea for reducing overhead expenses, particularly for companies in mature or declining markets. After pursuing such offshore strategies, many companies were unpleasantly surprised to find that offshore sourcing can lead to fragmented production processes, fewer but less productive facilities, and in many cases, higher total overhead.

Advancements in technology and production methods (e.g. implementation of just-in-time manufacturing methods and set-up time reduction) have reduced direct labor cost to less than 15 percent of total production costs for most manufacturing industries. In high technology industries it is often less than 5 percent. These facts make an offshore sourcing strategy obsolete in many cases. It is irrational to allow the cost category that represents the smallest percentage in overall product costs to drive the location of operations activities.

The exception to this for many firms is products or processes that are labor intensive and require low or commoditized skills. Low skill processes such as call-centers, forms processing, and data entry, or commoditized skills such as accounting and IT are all labor intensive tasks which could be performed by trained foreign workers for a lower cost. Whether as a captive offshore facility or an independent outsourcing company, many firms seek cost savings in business process costs by performing these tasks offshore.

Many companies have aggressively pursued offshore sourcing as the backbone of their manufacturing strategy. The choice of offshore locations is typically based purely on labor cost consideration. But the comparative advantages of countries with respect to low wages change over time, affected by shifts in exchange rates and in labor demands. Korea and Taiwan were cheap labor wage countries in the 1970s and Thailand in the early 1980s, but China became the favorite offshore sourcing ground of the late 1980s and early 1990s.

The continuous pursuit of the lowest labor cost forces companies to shift operations from one country to another, thus giving rise to the appropriately named *island hopping strategy*. However, the labor cost savings of such a strategy in many cases can be easily consumed by the increased capital, logistical and operating start-up costs in a new country. Many firms have found that the hidden costs of island hopping strategies are hard to estimate when making the decision to pursue the strategy. Shoe manufacturer Nike learned this lesson the hard way with start-up problems at a new production facility in China in the early 1980s. Hidden costs can include additional training due to lack of skilled workers, high quality costs due to poor workmanship and a lack of quality culture among workers, increased lead-times and associated inventory costs due to poor transportation and communication infrastructure, and unexpected logistics complications due to multilevel and bureaucratic government structures. As of 2002, China's logistics costs as a percentage of nominal GDP were 18.5 percent compared to the U.S. at 10.1 percent and by 2010 were predicted to improve only slightly to 14.5 percent while the U.S. decreased to 9.2 percent.

Business process outsourcing also has hidden costs. Many companies that seek lower labor costs for low skill, labor-intensive work such as call centers and application processing do not consider the process fully. In order to work effectively, service operations must already be standardized and digitized. Also, seeking purely the lowest labor cost for business process outsourcing may result in a choice that requires more extensive and costly training, making the labor savings an illusion. Furthermore, offshoring initiatives that have cost savings as their main reason often do not allow companies to capture greater revenues from the market. Such companies do not commit themselves to the organizational changes that are necessary for offshoring to help them, such as customizing products or services, and compressing new product development cycles. Also often ignoring cultural incompatibilities between the firm's management and local workers can easily erode cost savings due to turnover and failures in productivity and quality. Many US firms brave their way to China, Indonesia and Malaysia, often ignoring such issues, while most European companies are slower to embrace offshoring in such exotic locations concerned about cultural incompatibility issues. Those that do, find it advantageous to outsource in the Czech Republic or Poland for cultural compatibility with the company and its customers rather than the potentially lower cost of India or China.

1.5.2   Emergence of New Cost Priorities in the Location of Global Operating Strategies

New competitive priorities in manufacturing industries – that is product and process conformance quality, delivery reliability and speed, customization, and responsiveness to customers – have forced companies to reprioritize the cost factors that drive their global operations strategies. The total quality management (TQM) revolution brought with it a focus on total quality costs, rather than just direct labor costs. Companies realized that early activities such as product design and worker training substantially impact production costs. They began to emphasize prevention rather than inspection. In addition, they quantified the costs of poor design, low input quality, and poor workmanship by calculating internal and external failure costs. All these realizations placed access to skilled workers and quality suppliers high on the priority list for firms competing on quality. Similarly, just-in-time (JIT) manufacturing methods, which companies widely adopted for the management of mass production systems, emphasized the importance of frequent deliveries by nearby suppliers. Wages are not the only cost to consider in BPO, either. Considerations such as the reliability of the telecommunications infrastructure and the language skills of workers affect the quality and reliability of an offshore call center or other administrative process.

Obviously, successful implementation of such techniques requires the presence of an adequate supplier infrastructure in the location chosen for manufacturing facilities and necessary telecommunications infrastructure for BPO. Lack of worker skills, inadequate transportation and communication infrastructure, and low-quality supply typically prove extremely disruptive for the implementation of any modern production management methods. They also can easily lead to deterioration of productivity standards. These considerations explain why most U.S. companies have chosen to make their overseas manufacturing investments in developed economies (more than 60 percent in developed countries).

1.5.3   Increasing Capital Intensity of Production Facilities

A number of high-technology industries have experienced dramatic growth in the capital intensity of production facilities. A state-of-the-art semiconductor factory, for instance, costs close to half a billion dollars. When R & D costs are included, the cost of production facilities for a new generation of electronic products can easily exceed $1 billion. Similarly huge numbers apply for the development and production of new drugs in the pharmaceutical industry. Such high costs drive firms to adopt an economies-of-scale strategy that concentrates production in a single location – typically in a developed country that has the required labor and supplier infrastructure. They then achieve high capacity utilization of the capital-intensive facility by aggressively pursuing the global market.

Sometimes in capital-intensive industries, competing manufacturers find it beneficial to enter joint venture agreements in order to share capital costs and associated risks. In the semiconductor industry, Texas Instruments and Hitachi, Motorola and Toshiba, and IBM and Siemens share production facilities for DRAM chips. Such partnerships in fast-changing industries tend to be successful, freeing up needed resources for product development.

The location of capital-intensive production processes is strongly influenced by the availability of local government subsidies in terms of reduced interest rates, favorable price breaks for price-controlled industries, tax holidays for low tax rates, and cost sharing on plants and equipment. In some cases, countries offer deals for special access to raw materials, local financing, grants, employee training programs, and priority access to foreign exchange. Firms searching for foreign assembly or production locations often negotiate directly with the government the terms of investment.

Taxation factors can dominate the location decisions of capital-intensive industries, such as pharmaceuticals and semiconductors. These industries have a significant presence in low-tax countries. Tax benefits have driven pharmaceutical companies to locate in Puerto Rico and Ireland, for instance.

1.6.   Political and Macroeconomical Forces

The international economic and political environment can be best characterized as turbulent and increasingly complex. A variety of political and macroeconomic factors – exchange rates volatility, regional trade agreements, open markets, and managed trade, for example – continuously shape the global manufacturing and business process environment. As a company develops markets globally, it faces a powerful reason to invest abroad to try to match the currencies of its costs to the currencies of its revenues. Such matching reduces the risk of losses due solely to currency movements. Additionally, multigovernment initiatives to open markets and managed trade, such as NAFTA, dramatically change corporate plans to meet changing regulations and requirements.

Political and macroeconomic forces have always shaped global operations strategies. This section provides examples of such forces that have shaped the strategies of multinational companies.

- Exchange rate fluctuations and the value of operating flexibility in global manufacturing and sourcing networks
- Emergence of regional trade agreements and their impact on the structuring of global manufacturing and logistics networks
- Effects of trade protection mechanisms on global operations strategies

1.6.1   Exchange Rate Fluctuations and the Value of Operating Flexibility in Global Manufacturing and Sourcing Networks

Getting hit with unexpected or unreasonable currency devaluations in the foreign countries in which they operate is a nightmare for global operations managers. Managing exposure to changes in nominal and real exchange rates is a task the global operations manager must master.

A wide array of financial mechanisms is available for hedging against currency fluctuations, but they are most effective for short-term variations. It is important for the global operations manager to recognize and respond to the wide-ranging strategic problems and opportunities that long-term currency shifts present. Since disequilibria in exchange rates may last for several months or even years, the firm should strive to maximize its operational flexibility by diversifying production geographically and effectively using global sourcing networks. As a firm's local currency moves out of equilibrium the company should be in a position to shift its production to those facilities or suppliers that, by virtue of their location, can produce the product or provide input at the lowest cost in the local currency. In the automobile manufacturing industry,

European and Japanese carmakers have expanded manufacturing capacity in dollar-dominated markets. Toyota plans to build a new plant in Texas by 2006 to increase U.S. exports to Europe and Japan and increase purchases of U.S. supplies. Volkswagen moved production from Germany and Slovakia to Mexico where it has also increased its share of North American components. Firms that have global facility networks or established relationships with vendors in a variety of countries can implement such a strategy most effectively. Saab and Volvo have both increased production in their U.S. based partner companies, GM and Ford, respectively.

If the economics are favorable, the firm may even go so far as to establish a supplier in a foreign country where one does not yet exist. For example, if the local currency is chronically undervalued, it is to the firm's advantage to shift most of its sourcing to local vendors. In any case, the firm may still want to source a limited amount of its inputs from less favorable suppliers in other countries if it feels that maintaining an ongoing relationship may help in the future when strategies need to be reversed.

The need to exploit operational flexibility can play a determining role in the location of the firm's activities. Partial concentration of production and sourcing activities allows companies to respond effectively to exchange rate movements and bargain effectively with suppliers, labor unions, and host governments.

Building operational flexibility is a challenging and sometimes costly proposition, however. It requires establishing a global facility and sourcing network with excess production capacity, if economically feasible. It also requires developing a global product line with a high degree of cross-country standardization; designing and implementing production/supplier switching strategies; and creating incentive systems for managers that reward good decisions in this area.

1.6.2 Emergence of Regional Trade Agreements and Their Impact on the Structuring of Global Manufacturing and Logistics Networks

The emergence of trading blocks in Europe (Europe 1992), North America (NAFTA), and the Pacific Rim has serious implications for the way firms will structure or rationalize their global manufacturing/sourcing networks. These trends are apparent in many industries. Implementing the strategic changes required for a rationalized factory and distribution network can be painful, and firms can make mistakes.

NAFTA called for the elimination of nearly 7,000 individual tariffs, duties, and nontariff barriers to trade over a ten to fifteen year timeframe. Although

NAFTA is more trade oriented than production oriented, the easier access to a larger market and greater freedom for sourcing opened up multiple new opportunities for multinational firms. Naturally, these opportunities impact the development of operations strategies.

For example, the role of American owned maquiladoras had been to allow for cheap labor on production which would incur tariffs only on the low-cost value-added portion of the product. NAFTA made this logic invalid and changed the role of maquiladoras for American companies. Similarly, Mexico's historically low national content requirement of 36 percent was raised to a NAFTA standard of 62.5 percent. This encouraged European and Japanese firms to invest in North American-based production in order to qualify for preferential tariff treatment.

1.6.3   Effects of Trade Protection Mechanisms on Global Operations Strategies

The two broad types of barriers to international trade are tariff and nontariff barriers. Tariff barriers are types of direct price protection and are imposed as taxes (duties) on imported goods. They are assessed either as a percentage of the value of the imported good or as a flat tax. The General Agreement on Tariffs and Trade (GATT) has been very successful in drastically reducing the tariffs for most industrial goods traded among developed countries, thereby significantly contributing to the expansion of operations globally.

With the success of GATT and its successor the World Trade Organization (WTO) with its wide membership, tariff barriers have become less important as a form of trade protection than nontariff barriers, or forms of indirect, non-price protection of exports and imports. The most common form of nontariff barrier is the quota, a quantitative restriction on the volume of imports. Textile quotas imposed by industrialized countries against textile imports from developing countries are a good example of a nontariff barrier. For Asian producers, quotas are not effective deterrents to exporting and are traded throughout the Asian community like any other commodity. Changes in the structure of quotas could strongly affect the nature of competition in the garment industry. Import quotas on apparel and textiles from 45 countries, including China, were set to expire in 2005. Upon expiration, China negotiated self-imposed quotas that were far from adequate and effective and thus resulted in a massive influx of garment imports of all types, drastically cutting the market for domestic garment producers.

The 1970s and 1980s also saw the increasing application of another form of nontariff barrier – trigger price mechanisms (i.e., establishment of a minimum price for sales by an exporter from a foreign country), particularly by the United States and the European Union. The U.S. government used trigger price mechanisms in both semiconductors and steel as a way to eliminate dumping by foreign manufacturers. Unfortunately, in these cases the trigger price mechanisms misfired. They caused the exporters' prices to increase in the United States, thus generating additional profits for low-cost foreign producers. These profits enabled the companies to make greater investments in technologies that reduce production costs and further expand manufacturing facilities in North America.

Governments also use other forms of nontariff barriers, including local content requirements, technical standards and health regulations, and procurement policies. As discussed under NAFTA, local content requirements are designed to promote import substitution by specifying that a certain portion of the value added must be produced inside the country. For example, both Texas Instruments and Intel built semiconductor facilities in Europe (Italy and Ireland, respectively) in response to increases in the amount of semiconductor processing required by local content rules in Europe. Similarly, the European Union has established strict local content rules (with specific time tables) for Japanese car manufacturers in Europe.

Response to local content requirements has been an important factor in the globalization of operations strategies. In some cases, local content requirements may preclude penetration of a specific market. The firm may deem the sales potential in the market too small to justify a manufacturing investment, or may view the local supplier base as inadequate to fulfill quality specifications. In other cases, market presence may be essential to tap sales potential or to gain access to specific resources, but the local content rules force decisions on the amount and form of manufacturing investment that are suboptimal from the firm's perspective.

Technical standards and health regulations relate to matters such as consumer safety, health, the environment, labeling, packaging, and quality standards. The United States, for instance, prohibits imports of many types of agricultural products on such grounds; Japan refused to import U.S. skis for a number of years on grounds that Japanese snow was different from U.S. snow. The Philips manufacturing plant in Brugge, Belgium, used to employ seventy engineers just to adjust the seven different types of television sets rolling off the assembly lines to meet widely differing reception and technical standards throughout the European Union. These adjustments cost Philips about $20 million annually.

Governments also adopt procurement policies that favor domestic producers. For instance, in the 1980s and 1990s, the U.S. government attempted "Buy American" regulations to give U.S. producers up to 50 percent price advantage on Defense Department contracts. However the campaign was not successful and quickly abandoned in most industries, and especially so in apparels and cars.

Fortunately, thanks to harmonization of standards and regulations in the EU, these situations are changing rapidly today, and operating across borders is becoming much easier.

1.7.   Applying the Framework: The Acrilan Case

In September of 1999, Acrilan, a division of the Monsanto spin-off company Solutia, faced an eroding market for acrylic fiber, which typically comprised about 45 percent of its total sales. Traditionally, Acrilan faced competition primarily from within the North American supply chain. The Asian Crisis of 1997, in which many Asian currencies devalued drastically and Asian local demand for goods collapsed, combined with the presence of overcapacity led to a sharp decrease in the Chinese prices for Acrylic fibers (see Figure 1.2). The serious drop in demand for Chinese fibers and apparels from nearby Asian markets left Chinese producers with surplus goods and the incentives to take advantage of their low costs via heavy exports. This stimulated an increase in North American imports of Chinese acrylic fibers and finished goods, which reduced the market share for North American based knitters, sewers, and spinners. Facing this market competition, Acrilan was forced to examine its own supply chain and realized it needed to fix the weak links. Acrilan's manufacturing process was designed for flexible custom orders and ran at a cost disadvantage to less flexible acrylic fiber producers who were driven by economies of scale. Thus, it could not compete as a supplier for Asian or European spinners, with the added substantial transportation costs further eliminating the option as a consideration, and needed to focus on making its North American supply chain more competitive. Meanwhile, technological forces in the form of trained employee clusters in Mexico and political forces such as the passing of the North American Free Trade Agreement (NAFTA) had shaped new global market incentives, making Mexican and Caribbean Basin options more attractive. Acrilan had to find a way to strengthen its entire supply chain, especially its weak link of cut and sew operators in Brooklyn, New York (see Figure 1.3), in order to compete with the forcefully emerging Asian supply chain threats.

1.7.1   Global Market Forces and Acrilan

Intensified competition in local markets.   Foreign competition was a major motivation in Acrilan's management decisions. While the U.S. sweater market is large and growing with acrylic fibers having the largest share of the market at 36 percent, most of the acrylic sweaters are imported from foreign

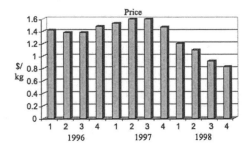

Figure 1.2. Chinese pricing in acrylic fibers $/KG CIF(Cost, Insurance & Freight), which decline sharply following the Asian Crisis in 1997.

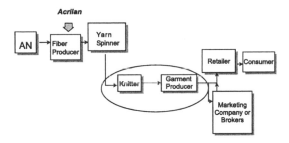

Figure 1.3. Acrylic supply chain. The weak links in Acrilan's forward supply chain were the knitters and garment producers (sewers) which faced strong Asian competition.

garment makers (70 percent)(Figure 1.4) with about 99 percent of those being from Asia (Figure 1.5). Retailers view acrylic sweaters as cheap or low-end products with low to mid price points and rank price and quality of the good over just-in-time delivery, making Asia, with its reputation for low cost and high quality, a strong supplier despite a longer leadtime. The Asian Crisis of 1997, as explained previously, ultimately strengthened China's position as the low-cost provider of many products, including acrylic sweaters. This direct competition with members of Acrilan's forward supply chain, namely the cut and sewers, was affecting the entire backward supply chain. American retailers and apparel manufacturers preferred to place orders with Asian firms, who of course used Asian supply chains, thus limiting the demand for yarn and fibers in the corresponding American supply chains. To stay viable, Acrilan needed to find a way to make the forward supply chain more competitive against foreign competition in the domestic market. As we will discuss later, this was far from an easy challenge as the "weak link" itself was composed of undercapitalized firms that were short in management talent and global experience, who

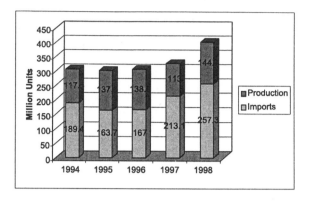

Figure 1.4. Imports Share of U.S. Sweater Market (all fibers).

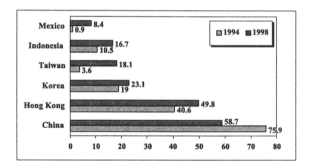

Figure 1.5. U.S. Sweater Imports: Major Suppliers.

had neither the capabilities nor the needed investments to reverse the competitive trends.

Growth in Foreign Demand.   Globally, acrylic consumption is about 80 percent of available capacity (Figure 1.6) with an oversupply in North America (Figure 1.7) and an expected undersupply in Latin America (Figure 1.8). In order to recoup the reduction of domestic demand, Acrilan considered establishing a customer base with spinners, knitters, and sewers in the undersupplied Latin America. This would simultaneously work to make the forward supply chain more competitive and allow Acrilan to directly tap into the unmet acrylic fiber demand of foreign knitters and sewers.

Changing Competitive Priorities in Product Markets.   Acrilan saw that Asian producers, especially South Korean firms, were quickly moving into the Mexican and Caribbean Basin for cheaper production closer to the U.S. mar-

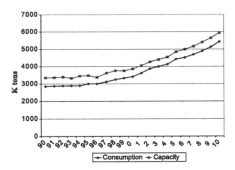

Figure 1.6. Global acrylic demand/supply balance shows an overall overcapacity.

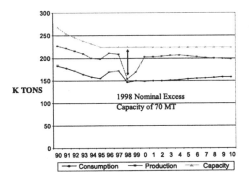

Figure 1.7. The North American demand/supply balance shows excess in capacity and an over-supply in production.

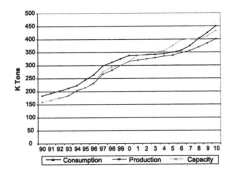

Figure 1.8. The Latin American supply/demand balance predicts an undersupply in production and capacity.

ket. While cost and quotas were the major motivations for these investments for some Hong Kong producers, quick delivery to the U.S. market was also a factor. Since Acrilan could not compete effectively as a low cost producer, they sought to fill a niche as a quick-response full packet supplier, with a competitive lead-time for the entire production process from order and design through delivery. The Korean producers had accomplished a similar full packet service, quite remarkable in its ability to turn around quotations for new orders in less than two weeks, but were no longer competitive in costs. Acrilan had an opportunity to meet this new competitive priority for a full packet, high service, and quick response solution at a cost and lead-time advantage by developing the right North American supply chain. But as we will see, that was a challenge insurmountable for a slow moving company entrenched in the culture of an old chemical company and operating within a slow clock speed industry as such.

Presence in Sate-of-the-Art Markets.   Acrilan realized that if the globalization strategy called for a move to foreign-based spinners, knitters, and sewers, they must retain the design and sales teams of the Brooklyn-based knitters in Brooklyn, where it had traditionally been done in innovative and effective ways. The Brooklyn design team had and it will continue to have a presence in the fashion industry. Time to market and product innovation were considered two of the keys to a successful strategy, and Brooklyn had a long history of a strong design team which was embedded in the fashion industry and could serve as a sensor to emerging design trends. If the Brooklyn design clusters could be paired with adequate cut-and-sew capability clusters in Mexico, an effective solution could emerge for Acrilan.

1.7.2   Global Technology Forces and Acrilan

Mass Customization.   Acrilan had more than two-thirds of the American sourced fiber market, having increased its market share to 70 percent after DuPont divested its acrylic business in the late 1980's. However, the American acrylic fiber business in segments other than sweaters was too small of a niche market and accounted for only one-third of the U.S. acrylic fiber production – the rest is exported (Figure 1.9). The undifferentiated, non-branded, narrow profit-margins nature of acrylic fibers, and the involved transportation costs in accessing foreign markets, made it difficult to achieve global economies of scale via pursuit of small niche markets in various countries. A global niche strategy was not an available option to Acrilan.

Diffusion of Technological Knowledge.   Foreign sourcing in Mexico offered Acrilan two regions in which skilled garment producers were available to work. Both regions were small and unsophisticated. Moroleon is located in

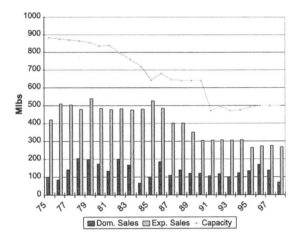

Figure 1.9. Domestic Utilization Rates of Acrylic fiber.

Guanajuato and has about 250 small factories specializing in acrylic sweaters. Cuautepec, near Tulancingo in Hedalgo, has about 200 factories. Due to the migrant nature of workers, production only runs half of the year and labor is hard to find in Moroleon. However, in general the quality of the Mexican work was considered lower quality than the American or Asian production facilities. A giant cotton apparel complex called Nustart, in the Mexican state of More-los, was also an interesting option for Acrilan. These facilities were owned by various cotton garment manufacturers, each with its own cut and sew opera-tion in its own building. Acrilan considered exploiting other existing clusters of skilled workers to make a similar all-in-one facility to group acrylic spin-ners, knitters, cutters, and sewers into an integrated production facility through a joint venture with U.S. based yarn spinners. However, in doing so the com-panies must carefully consider where these clusters of skilled workers existed. They needed to keep the Brooklyn-based design cluster in Brooklyn where it had traditionally been done and move only the labor-intensive production work to the skilled garment producing clusters located in Mexico.

Technology Sharing and Interfirm Collaborations.   If Acrilan were in-terested in Mexico, the technology would most likely have to be developed from scratch through a heavy investment. An Asian-based competitor, Crys-tal Kobe, had already done a similar investment in a cut and sew operation in Mexico, but was not interested in partnering with Acrilan. Better skilled work-ers with high production capacity already existed in other countries, but freight costs and transportation lead-times made them less desirable. If Acrilan were to invest in a spinning, cutting, or sewing facility in Mexico, it would likely be through a joint venture with another domestic member of its forward supply chain. However, this would be complicated as competing members were each seeking an exclusive alliance with Acrilan. Acrilan was not seeking to vertically integrate into the labor-intensive textile operations, but at the same time existing domestic textile companies were hesitant to take on the globalization process alone. The level of commitment was subject to debate, however, ranging from moving the forward member's company to Mexico with the assistance of Acrilan to simply helping them find an outsource provider.

1.7.3   Global Cost Forces and Acrilan

Diminishing Importance of Direct Labor Costs.   In Acrilan's supply chain, the knitters and sewers were losing competitive edge due to high labor costs. However, by moving to a global operation, the lower labor costs must be measured against a longer lead time, lower productivity, and duties for some of the options. See Figures 1.10, 1.11, and 1.12 for some of these cost factors. As seen in Figure 1.10, the total cost of labor must consider the productivity of workers in a given location. The full cost of labor must account for values such as wages, benefits, vacation days, and bonuses. Figure 1.11 then compares the total supply chain costs of a garment considering productivity, transportation and duties, both before and after the impending WTO (World Trade Organization) negotiations. After all of this analysis, there is a cost advantage to locate in Indonesia or Mexico (Figure 1.12).

Emergence of New Cost Priorities.   Acrilan recognized a niche market for serving the priority of a quick response, low cost full package solution from order and design through delivery. In order to provide such a service, Acrilan knew it needed to find a low cost location for some members of the forward supply chain that was geographically close enough to the North American market to provide fast service without compromising quality.

1.7.4   Political Factors and Acrilan

Exchange Rate Fluctuations and the Value of Operating Flexibility in Global Manufacturing and Sourcing Networks.   Exchange rates were also an important factor for Acrilan when choosing where to locate a potential global operation. The Asian Crisis of 1997 had triggered many of the problems the industry was now facing. Future rate fluctuations weighed on the globalization decisions. Exchange rates for Indonesia, for example, were highly volatile over a two-year period (Figure 1.13), exposing the company to financial risks if they chose to pursue the post-WTO lowest cost alternative in Indonesia (Figure 1.14). Currently Indonesia was a strong competitor for a global operation site,

Year 1999	Guatemala	Costa Rica	El Salvador	Honduras	Domin. Rep.
Min. Wage-$/hr.	0.34	0.81 - 1.11	0.69	0.34	0.71
Fringe Benefits	0.24	0.19	0.16	0.14	n/a
Total - $/hr.	0.58	1.00 - 1.30	0.85	0.48	n/a
Fringe Benefits- as % of salary	69.6%	23 - 27%	23.2%	41%	n/a
Vacation Days	10	10	15	10	12
Year End Bonus	2 mo.	1 mo.	½ mo.	2 mo.	1 mo.

Source: IDC Guatemala, SECO Francfort

Figure 1.10. Comparison of Wages and Benefits in Various Offshoring Location Alternatives.

Location of Activity	Fabric: Cutting: Assembly:	USA USA USA	Korea Korea Korea	USA Honduras Honduras	USA USA Mexico	USA Mexico Mexico	China China China	Indonesia Indonesia Indonesia	Mexico Mexico Mexico
Costs									
Direct Materials		2.10	1.30	2.10	2.10	2.10	1.30	1.20	1.50
Cut and Sew		2.20	1.40	0.70	0.90	0.80	0.40	0.40	0.80
Freight		0.02	0.30	0.15	0.10	0.10	0.30	0.30	0.10
Duty			0.92	0.95			0.58	0.54	
Landed Cost $/Garment		4.32	3.92	3.90	3.10	3.00	2.58	2.44	2.40
% Of US		1.00	0.91	0.90	0.72	0.69	60.00	56.50	55.50

1. DUTIES ARE CALCULATED AT THE RATE OF 34%
2. THIS PARTICULAR SWEATER HAS A RETAIL PRICE OF $30 IN DEPARTMENT STORES
3. STATED COSTS FULLY ACCOUNT FOR PRODUCTIVITY DIFFERENCES ACROSS COUNTRIES

Country	Mexico	China	Korea	Indonesia	Honduras
Productivity as % of US	85%	65%	85%	65%	80%

Figure 1.11. Acrylic sweater costs under current duty rates, accounting for productivity rates.

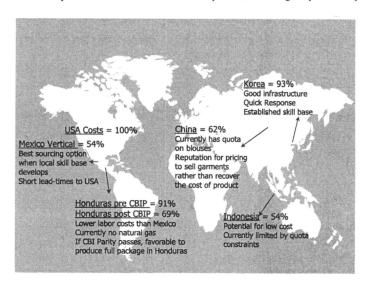

Figure 1.12. Acrylic Knit Blouse: Global Summary of Costs and Barriers.

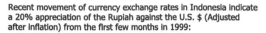

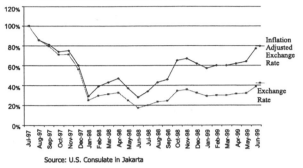

Source: U.S. Consulate in Jakarta

Figure 1.13. Exchange Rates Appreciation in Indonesia.

Location of activity								
Fabric	USA	Korea	USA	USA	USA	China	Indonesia	Mexico
Cutting	USA	Korea	Honduras	USA	Mexico	China	Indonesia	Mexico
Assembly	USA	Korea	Honduras	Mexico	Mexico	China	Indonesia	Mexico
Costs								
Direct Materials	2.1	1.3	2.1	2.1	2.1	1.3	1.2	1.5
Cut and Sew	2.2	1.4	0.7	0.9	0.8	0.4	0.4	0.8
Freight	0.02	0.3	0.15	0.1	0.1	0.3	0.3	0.1
Duty (.26%)		0.08	0.7			0.43	0.4	
Duty (.15%)		-0.41	-0.42			-0.25	-0.24	
Landed Cost	4.32	3.68	3.65	3.1	3	2.43	2.3	2.4
$/ Garment		-3.41	-3.37			-2.25	-2.14	
% of US		85%	84.50%	71.70%	69.40%	56.20%	53.20%	55.50%
		-79%	-78%			-52.10%	-49.50%	

Figure 1.14. Acrylic Sweater Costs: Post WTO.

but the exchange rate played a heavy role in this advantage. If the exchange rate rebounded, Acrilan would lose the competitive edge they had gained by locating there.

Emergence of Regional Trade Agreements and Their Impact on the Structuring of Global Manufacturing and Logistics Networks.   Political factors played a large role in Acrilan's decision process. NAFTA and the corresponding bilateral trade agreement with the Caribbean nations, the Caribbean Basin Initiative (CBI-Parity) agreement, would strongly affect future costs for duties if Latin America were chosen. Furthermore, NAFTA's "Fiber Forward" protection and local content requirements could influence decisions for Asian alternatives and became strategic deterrents to Asian producers. For example, Asian-owned maquiladoras in Mexico were largely established to circumvent quota and duties restrictions but with new regulations were being rendered ineffective for this purpose. Quotas imposed under the Multi-Fiber Agreement were scheduled to end in 2005 and while this would eliminate quotas on fiber, duties would remain in effect. The effect of quota elimination would be a massive influx of Asian produced fibers. An impending "sub-Saharan Initiative" could allow for higher production from Chinese fiber suppliers through African-based production facilities. Under this initiative, garments with 40 percent country content in certain African nations could enter the U.S. market free of duties. Strategic Chinese companies would take advantage of such an opportunity and build cut-and-sew capacity in Africa, with the Chinese fiber and yarn spinners supplying it. The resulting imports to the U.S. market would further diminish Acrilan's supply chain's market share. China's recent addition to the WTO allowed them the opportunity to negotiate for more favorable trade terms, which could strengthen their supply chain in the U.S. In fact, in 2005, with the expiration of the Multifiber Agreement and despite China's self-imposed quotas, a dramatic spike in imports for all types of clothing created trading tensions between China and its two major markets, Europe and US.

Effects of Trade Protection Mechanisms on Global Operations Strategies. For Asian producers, quotas are not effective deterrents to exporting and are traded throughout the Asian community like any other commodity. Changes in the structure of quotas could strongly affect the nature of competition in the garment industry. Import quotas on apparel and textiles from 45 countries, including China, were set to expire in 2005, and this posed a serious threat to Acrilan's entire supply chain. Upon expiration, China negotiated self-imposed quotas that were far from adequate and effective and thus resulted in a massive influx of garment imports of all types, drastically cutting the market for domestic garment producers.

Governments also use other forms of nontariff barriers, including local content requirements. Acrilan sought to capitalize on the local content requirements in the fiber forward supply chain in their search for a foreign-based knitting and sewing operation. The "Fiber Forward" protection in NAFTA, local content requirements, and quotas under the Multi-Fiber Agreement influenced decisions for globalization and were strategic deterrents to Asian producers. The "Fiber Forward" protection, specifically, would allow the entire forward supply chain of Acrilan to be produced in Mexico without incurring duties.

Acrilan's Decision. The right decision must be phased in with flexibility, due to the uncertainties. The option needs to offer value beyond fiber production and seek to manage the weak link in the forward supply chain to extract the maximum value from the chain. Too much dependence on the yarn spinners may put the company at a disadvantage and should be avoided. As knowledge and skill clusters develop in the U.S., Mexico, or Asia, they should be exploited. Always, the company must remain attractive to potential partners.

In the end, Solutia chose to leave the acrylic fiber business, and later left nylon altogether and invested in other businesses. Solutia preferred to sell Acrilan as a packet with both chemical and fiber capabilities, but finding a buyer proved difficult. The search for a joint venture partner was painful with no real synergies or benefits. The shutdown costs, both financial and emotional, were high and the eventual Solutia bankruptcy instigated the final closure of Acrilan.

In a flexible, fast-moving culture, Acrilan could have pursued globalization more aggressively. However, the slow-moving culture of a large chemical company lacked the necessary entrepreneurial mentality and support for such a strategy, limiting the options that they could pursue.

1.8.   Summary

Globalization affects every industry and every firm, both large and small. As competition evolves to include the entire supply chain, firms must consider

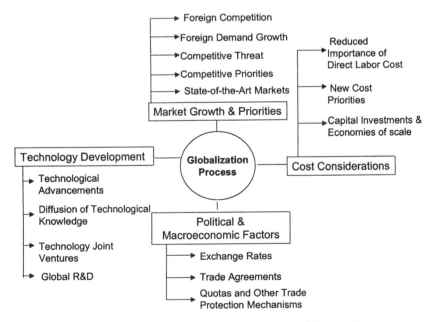

Figure 1.15. Four Forces of Globalization Conceptual Framework.

how globalization could be utilized to improve the competitive status of its forward and backward supply chain. Firms will compete based less on cost and more on time, with supply chain management and logistics being core management capabilities of successful firms. By using the conceptual global forces framework presented here, managers can clearly analyze the various important factors that shape global operations decisions in terms of market, cost, technology, and political forces (Figure 1.15). While these forces affect every industry and product in a unique way, the framework can be applied to any industry facing the pressures of a global marketplace.

References

Dornier, Philippe-Pierre, R. Ernst, M. Fender, and P. Kouvelis (1998). *Global Operations and Logistics: Text and Cases*. New York: John Wiley & Sons.

Kouvelis, Panos. *Global Supply Chain Competition in Acrylic Fibers in the New Millennium: Solutia, Acrilan Division & The Asian Crisis Case*.

www.census.gov

Bergsten, C. Fred (2001). *The US Export-Import Bank: Meeting the Challenges of the 21st Century*. Institute for International Economics Testimony before the Subcommittee on International Trade and Finance committee on Banking, Housing, and Urban Affairs United States Senate, Washington, DC, May 17.

Hau L. Lee and Chung-Yee Lee (Eds.)
*Building Supply Chain Excellence
in Emerging Economies*
©2007 Springer Science + Business Media, LLC

Chapter 2

GLOBALIZATION AND EMERGING MARKETS
The Challenge of Continuous Global Network Optimization

Peter Koudal
*Deloitte Research
Deloitte Services LP, USA*

Douglas A. Engel
Deloitte & Touche LLP, USA

Abstract: In pursuit of new revenue opportunities and lower-cost operations, manufacturers around the world are creating ever-more complex global networks of sourcing, manufacturing, marketing, sales, and service, and research and development activities. Over the last two years, we have monitored the development of such networks through our global benchmark studies of the global operations of nearly 800 manufacturers based in North America, Western Europe, Central and Eastern Europe, South Africa, and Asia-Pacific. These companies represent a broad range of industries, including consumer business, automotive, high-tech, diversified industrials, pharmaceuticals, and the chemical process sector, and together account for about $1 trillion in global sales.[1]

Our research finds that most companies have made little progress in optimizing their operations from a global perspective. Rather than take a holistic view in the expansion and optimization of their global networks – the complex web of suppliers, production and R&D facilities, distribution centers, sales subsidiaries, channel partners, and customers, and the flows of goods, services, information, and finance that link them – most global manufacturers focus on fixing individual pieces of the network. In spite of launching many improvement initiatives across their global operations, most are overwhelmed by increasing strategic

---

[1] For further details on this research, see Deloitte Research, *Unlocking the Value of Globalization: Profiting from Continuous Optimization* (New York and London, 2005) upon which study this chapter is based.

and operational complexity. And the complexity will only increase as companies continue their global expansion efforts – as our research indicates they will. The problem is that those who let their global footprint grow without continuously determining how the various pieces of their operations should be redesigned, rationalized and optimized unwittingly build in huge redundant costs while losing opportunities for higher growth and profits.

2.1.   The Optimization Paradox

Coordinating product development, supply chain, and sales and marketing activities that are oceans and time zones apart will become even more difficult in the years ahead as companies' operations become more fragmented with continued globalization.

This is just one of the key findings our global research on how companies can effectively optimize global networks. It is based partly on our comprehensive, in-depth global benchmark survey with executives at nearly 800 companies or business units around the world.

Over the next three years, more than 50 percent of North American manufacturers plan to enter or expand sourcing and marketing/sales operations in China. More than 40 percent say they will enter or expand into markets in Central and Eastern Europe. And more than 20 percent will initiate or expand sourcing and manufacturing operations in Mexico (Figure 2.1).

Western European manufacturers are not standing still either. With the eastern expansion of the European Union, more than 50 percent expect to increase their market activities in Central and Eastern Europe over the next three years and nearly 40 percent expect to enter or expand their sourcing and marketing/selling in China.

Such moves cannot but help introduce major inefficiencies into the value chains of global manufacturers. In addition, shrinking product cycles means less time for an increasingly dispersed workforce to collaborate and manage product transitions in each product cycle. This is especially important as companies come to rely more and more on new products to boost revenues and satisfy ever-more demanding customers. On average, companies expect new product share of total revenues to hit 35 percent in 2007, a 66 percent increase from 1998.[2]

---

[2] New products are defined as products introduced over the last three years. See also Deloitte Research, *Mastering Innovation: Exploiting Ideas for Profitable Growth* (New York and London, 2004).

Top 3 Future Growth Destinations		
Percentage of Companies Planning to Enter or Expand Operations over Next Three Years		
	North American Companies	Western European Companies
SOURCING		
China	55%	39.4%
Mexico	23.1%	
Other SE Asia	20.4%	19%
Eastern Europe		36.2%
MANUFACTURING		
China	35.8%	26.9%
Mexico	22.3%	
Western Europe	11.5%	
Central Europe		16.5%
Eastern Europe		27.4%
ENGINEERING		
China	23%	13%
India	13%	
Eastern Europe	10%	14%
Central Europe		13%
MARKETING/SELLING		
China	53.1%	38.7%
Central Europe	40%	51.3%
Eastern Europe	40.4%	60.3%

Source: Deloitte Research

Figure 2.1. Relentless Globalization: Emerging Markets Dominate Top Three Growth Destinations.

Despite the clear advantages that can be derived from optimizing the value chain, most manufacturers lack the capabilities to do so. Less than a third (30 percent) report an advantage in supply chain cost structure. In comparison, 70 percent say they had better product quality than their primary competitors (Figure 2.2). Perhaps not surprisingly, then, over the last three years, companies on average ranked initiatives to upgrade their supply chain network structures at

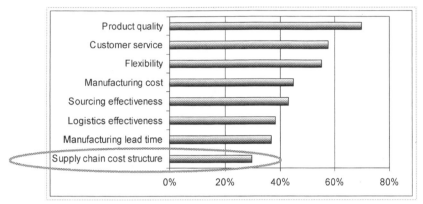

Percentage of Companies with Near-Strong or Strong Advantage Compared to Primary Competitors

Figure 2.2. Forget Global – Most Optimization Is Local.

Ranking of <u>Performance Improvement Initiatives</u> Undertaken Over Last Three Years

Rank	All Industries	Automotive	Consumer Products	Discrete Manufacturing	High Tech	Life Sciences	Process / Chemicals
1	Operations/ Manufacturing	Operations/ Manufacturing	Operations/ Manufacturing	Operations/ Manufacturing	Operations/ Manufacturing	Operations/ Manufacturing	Operations/ Manufacturing
2	Forecasting/ Planning	Engineering	Forecasting/ Planning	Sourcing/ Procurement	Order Management/ Customer Service	Forecasting/ Planning	Order Management/ Customer Service
3	Order Management/ Customer Service	Forecasting/ Planning	Distribution/ Logistics	Forecasting/ Planning	Sourcing/ Procurement	Order Management/ Customer Service	Forecasting/ Planning
4	Sourcing/Procurement	Order Management/Customer Service	Order Management/Customer Service	Order Management/Customer Service	Forecasting/Planning	Distribution/Logistics	Sales & Marketing
5	Distribution/ Logistics	Sourcing/ Procurement	Sourcing/ Procurement	Sales & Marketing	Engineering	Research & Development	Distribution/ Logistics
6	Sales & Marketing	Distribution/ Logistics	Sales & Marketing	Engineering	Distribution/ Logistics	Sourcing/ Procurement	Sourcing /Procurement
7	Engineering	Research & Development	Supply Chain Network Structure	Distribution/ Logistics	Research & Development	Sales & Marketing	Research & Development
8	Research & Development	Sales & Marketing	Research & Development	Research & Development	Sales & Marketing	Supply Chain Network Structure	Engineering
9	Supply Chain Network Structure	Supply Chain Network Structure	Engineering	Supply Chain Network Structure	Supply Chain Network Structure	Engineering	Supply Chain Network Structure

Figure 2.3. Supply Chain Network Structure Optimization Initiatives at Bottom of List in Nearly All Industries.

the bottom of their list of improvements (Figure 2.3). Less than a third had undertaken "extensive" or "near extensive" initiatives to improve supply chain network structure performance over that period.

Thus, while companies are globalizing just about everything, most optimization still remains "local." We refer to this as the "global optimization paradox," and it creates a number of problems in a number of areas.

Attempting to enter new markets with new or existing products is always fraught with challenges; manufacturers that underestimate the strain on the global network and have limited insight into the true cost of products sold in different markets can jeopardize their investments and growth plans. For example, some companies are pursuing opportunities in low-cost countries such as China without realizing that the gains from lower unit costs of products can be eaten up by delays and uncertainty, regulatory and tax issues, and huge logistics costs.

The case of one U.S.-based multinational highlights such pitfalls. After spinning off its manufacturing subsidiaries in Singapore, China, and other Asian countries, the firm set up a "commissionaire" structure to sell to European and U.S. markets. This meant, for example, that the company's European divisions would be paid in commissions rather than profiting from value-added manufacturing activities as had been the case in the previous network structure. For the purpose of determining duties, however, the cost of goods was calculated on the basis of ownership of the product as it entered the European markets. As simply an agent, the company never owned the goods. Therefore, it had to pay duties on the sales price rather than the manufacturing cost – an increase of 50 percent. After a lengthy customs audit, the company determined that the miscalculation cost millions of euros in current and back duties.

Similarly, the logistics department of a Dutch company thought it could save 5 percent in production costs by outsourcing assembly to China, where individual parts were already being produced. The finished assemblies would then be imported into the Netherlands. However, the company's tax department was paying the duties on the imported goods. With limited visibility and collaboration between the two departments, it took a year for the company to realize that its total costs had actually *increased* by almost 10 percent. While the company could import parts in this category duty-free, the final assembly came with a hefty 14 percent duty.

2.2.   Why Are Companies Falling Behind in Optimizing Their Global Networks?

Given the wide range of problems the optimization paradox often creates, why is it that most companies are not making significant efforts to resolve it?

Today, the pace of change in most industries is significantly higher than it was 10 or 20 years ago. Faster product cycles, new and more diverse sources of supply, and ever-more-complex global networks increase the need for companies to continually optimize their value chain networks.

Our research shows that the average time for manufacturers to bring a new product to market will be less than 13 months by 2007 – a more than 30 percent reduction from the 18-months it took in 2001. Putting more new products through the "development," "demand" and "supply" chains will further raise cost and complexity – particularly with the increased number of plants, warehouses, and R&D centers through which those products will likely pass.

Also, as the process of outsourcing major pieces of manufacturers' value chains continues unabated, companies will find it increasingly difficult to monitor and assess the total network cost and impact of new initiatives.

In addition, if not executed well, mergers and acquisitions can play havoc with existing networks. Financial markets increasingly penalize companies that make acquisitions without harvesting the fruits of consolidation and optimization of global networks. Some of the greatest benefits of acquisitions come from optimizing demand, supply, and product innovation networks and processes. Leaving supply chain, product development, sales and marketing, and other facilities intact after an acquisition ignores the benefits of optimization.

Changes in more complex economic and political matters – in regulations, environmental protection, international trade and investments, currency rate fluctuations, and taxation – compound the problem. This includes recent developments such as increased border controls and security concerns, the continued evolution of World Trade Organization (WTO) rules, the expansion of the European Union, new regulations on environmental safety and health, fluctuating currencies, and the emergence of new global players such as China and India, to name a few.[3]

In just one example, electronics makers in Europe will likely be forced to spend an estimated US$100 billion over the next decade to comply with new EU directives on hazardous materials that become operational in 2006.[4] Companies in a variety of industries that manufacture products with electronics content such as automobiles or lighting equipment will also be affected by these regulations. The impact will in fact be global: Every company importing relevant products into the European Union will have to comply.

To meet these new standards, companies must also prove that they comply at every stage of the value chain, from design and production to service and

---

[3] See also Deloitte Research, *Prospering in the Secure Economy* (New York, 2004).

[4] Based on cost estimates from a European trade group, ORGALIME, in reference to EU Directives on Waste from Electronics and Electrical Equipment (WEEE) and Restriction of Hazardous Substances (RoHS).

disposal. This means detailed product traceability across the entire global network. Expanding supply chains into new and emerging markets will only make this even harder to achieve. To comply, companies will need change many of their current business practices and spend a lot of money in the process. OR-GALIME, a European trade body, predicts European companies will spend up to €15 billion in up-front costs to redesign their processes. The biggest portion, however, will likely be for retiring products in circulation, a cost estimated to be €40 billion.

Keeping pace with change on a global scale is a challenge for even the best companies. Sony realized this when it had to recall 1.3 million Sony PlayStation 1 game systems and 800,000 accessories because of cadmium levels in peripheral coupling cables that did not meet environmental standards.[5]

The global automotive industry finds itself under similar pressure to address environmental issues throughout the product lifecycle. Consider the End-of-Life Vehicles Directive (ELV) that will be in effect in Europe by January 1, 2006. Cars will have to be 85 percent recyclable, a figure that increases to 95 percent by 2015; this is up from 75 percent today.[6] In addition, to comply with emissions legislation and new fuel efficiency requirements, it is estimated that Ford and GM will need to spend US$400 per vehicle. This could reduce margins between 10 percent and 15 percent by 2015.[7] BMW, for its part, would have to spend more than US$600 per vehicle, although this will impact BMW less than to other automakers because of the company's higher margins. Other companies would be less affected for other reasons. Honda, for example, is

---

[5] By some estimates, the company experienced a US$110 million loss in revenue due to the incident. See "Sony: Dutch authorities seized PlayStations; cadmium fears," *Dow Jones International News*, December 4, 2001. See also "Sony faces PS One dilemma in Europe," *Consumer Electronics*, December 10, 2001. For more information on Sony's work on corporate social responsibility, see Sony, *CSR 2004*, http://www.sony.net/SonyInfo/Environment/environment/communication/report/2004/qfhh7c000000lv99-att/CSR2004_E.pdf.

[6] For details, see "Directive 2000/53/EC Of The European Parliament And Of The Council of 18 September 2000 on end-of life vehicles," *Official Journal of the European Communities*, L 269/34, October 21, 2000.

[7] See Duncan Austin, Niki Rosinski, Amanda Sauer, Colin Le Duc, *Changing Drivers: The impact of climate change on competitiveness and value creation in the automotive industry* (Washington, D.C.: World Resources Institute (WRI) and Sustainable Asset Management (SAM), 2003). Estimates are based on WRI's methodology of assessing the risks and opportunities of carbon constraints due to increased regulation needed to achieve emissions reductions and fuel efficiency demands. The 'value exposure' assessment measures risk due to increased costs from improving fuel efficiency of vehicles already sold.

forecasted to need only an extra US$24 per vehicle to meet new standards,[8] partly due to more fuel-efficient vehicles.

The global chemicals industry is also a ripe target for environmental regulations. If enacted as expected, by 2006 a new piece of EU legislation – Registration, Evaluation and Authorization of Chemicals (REACH) – would force producers to track up to 30,000 of an estimated 100,000 chemicals.[9] Companies would have to register these substances *and* prove they are safe. Moreover, the plan is to extend regulatory requirements to customers downstream in the supply chain, thereby affecting nearly every manufacturing industry.

The impact of this kind of legislation is daunting. While analysts say the United States government is lobbying hard to weaken or stop the new laws,[10] they predict such efforts will only be a stopgap measure. As health issues continue to be uncovered, companies in the leading industrial economies should expect to see more such legislation in the near future.

The challenge of responding to these kinds of new national, regional or global compliance demands is magnified by the fact that few companies are able to take a holistic look at their business and end up responding in a suboptimal way. Segregated management of functional, business unit, and geographic divisions of most companies means that opportunities for significant improvement in areas such as global supply chain redesign or tax-efficient global intellectual property management are rarely pursued. This "silo" mentality is sometimes furthered by the often short-term considerations of capital markets to which companies respond. Designing and optimizing a network of operations takes time, and while many short-term results can be achieved, most benefits accrue over the life of an investment.

---

[8] See Duncan Austin, Niki Rosinski, Amanda Sauer, Colin Le Duc, *Changing Drivers: The impact of climate change on competitiveness and value creation in the automotive industry* (Washington, D.C.: World Resources Institute and Sustainable Asset Management (SAM), 2003).

[9] Registration, evaluation and authorisation of chemicals (REACH) was outlined in a February 2001 white paper and subsequently adopted by the European Commission. See The Commission of the European Communities, White Paper, "Strategy for a Future Chemicals Policy," COM(2001) 88 Final, February 27, 2001. For further information, see http://europa.eu.int/comm/environment/chemicals/reach.htm. The REACH proposal will replace more than 40 existing directives and regulations with a single, integrated system in which 30,000 chemicals would need to be registered. Under this system, companies that produce and import chemicals will need to assess the risks to the human health and the environment and take steps to manage any risks identified, thereby shifting the burden of proof for ensuring the safety of chemicals on the market from public authorities to industry. See Karen Wontner, "Far-reaching proposals," *Supply Management*, January 8, 2004.

[10] See Demetri Sevastopulo, "Concern at US efforts on chemicals law," *The Financial Times*, April 6, 2004.

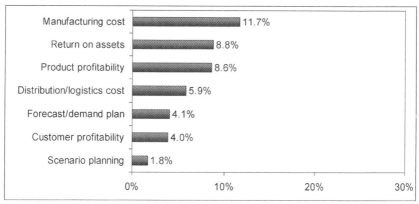

**Percentage of Respondents that are Highly
Satisfied with Information Availability in Area**

Figure 2.4. "Flying Blind: The Challenge of Visibility in Complex Global Networks.

Rationalizing and managing operations in customer-facing, product innovation, and supply chain areas from a global, holistic view is a major undertaking. The internal resistance to shutting down operations, changing processes and reporting relationships, and serving markets in new ways can be immense. The risk of doing the optimization wrong can also be significant.

That fact is that most companies lack fundamental capabilities necessary for monitoring, designing, and effectively restructuring their networks on an ongoing basis. For example, fewer than 12 percent of the companies in our study say they are "highly satisfied" with their information on critical metrics such as manufacturing cost, customer service levels, and product profitability (Figure 2.4). Without this information, it is no wonder that most manufacturers improve operations on a "local" basis – i.e., creating efficiencies one link at a time as our research indicates they do.

Given the complexities, one might ask: Is it worth it? Should global manufacturers even consider such wrenching change? The answer is that they indeed should, for several reasons. Global expansion is inevitable. Vast new markets await most manufacturers in areas such as China, India, Eastern Europe, and South America. This will pressure companies to move their supply lines and demand-generation activities quickly. With the rapid acceleration in new product introductions and the need to leverage R&D expenditures on a global scale, companies will face mounting pressure to boost the efficiency of their global networks – not just every five or 10 years, but on an ongoing basis. And they must ensure that new initiatives are always implemented in alignment with current and future optimal global network structures. This will help them minimize or avoid costly future network changes and gradual loss of competitive position due to poorly structured operations. This does not mean, of course, that companies should consider constantly moving pieces of the global supply chain, or restructuring flows at every little turn of event. The cost of constantly changing locations or the directions of physical or information flows simply would be too high. Rather it means that companies should ensure that the global network structure is optimal today while positioned appropriately for future changes and major new investments.[11]

So how can large, global manufacturers overcome the barriers and generate the extraordinary benefits from optimizing their networks?

2.3.   Profiting from Continuous Network Optimization

Optimizing global networks is not a trivial task.[12] However, it is becoming a key capability of some of the world's leading manufacturers. In our research, we have identified a small group of global manufacturers that has significantly outdistanced the competition through superior capabilities for managing complex global networks. We call these companies "complexity masters" (Figure 2.5). These companies have very complex, global operations–measured as the spread of the four main parts of the value chain (R&D/engineering, sourcing, manufacturing, and marketing/sales) across 13 countries/regions around the world. But they also have better capabilities for managing their global value chains–measured by an index of 10 capabilities. (For more details on the methodology and classification of complexity masters and other companies in the global database, see appendix "Defining Complexity Masters.")

Comprising just 7 percent of all companies in the analysis – and less than 15 percent of the most global companies (quadrant 3 and complexity masters combined) – complexity masters are a select group. With higher asset returns, faster growth, and profit levels up to 73 percent higher than their competition, the complexity masters clearly outperform their competitors. Compared with the peer group of the most globalized and complex companies in quadrant 3, complexity masters are nearly 50 percent more profitable. Indeed, because they have better capabilities for managing their global value chains, complexity masters can design their operations more holistically when determining where

---

[11] See also, Deloitte Research, *Performance Amid Uncertainty in Global Manufacturing: Competing Today While Positioning for Tomorrow* (New York, 2002).

[12] On network optimization, see also an extensive treatment by David Simchi-Levi, Philip Kaminsky, and Edith Simchi-Levi, *Managing the Supply Chain: The Definitive Guide for the Business Professional* (New York: McGraw-Hill, 2003).

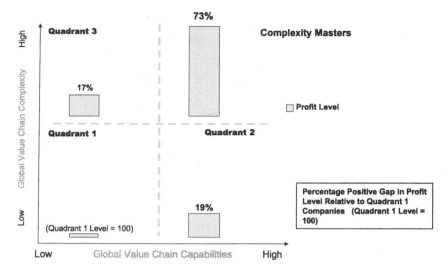

Figure 2.5. Global Value Chain Capabilities Matter to Performance in Complex, Global Networks.

to place and how to manage manufacturing, distribution, R&D, sales, marketing, and other activities. The result is a more optimized business with a better balance of growth goals, cost reductions, and risk.

By analyzing the performance of complexity masters and the leading practices of companies around the world, we are able to take a closer look at how companies can manage investments, capabilities, and practices to optimize their global operations. (See appendix B for more details on the methodology and profile of these companies.)

2.3.1　Taking a Holistic View

The most successful manufacturers take a holistic approach to managing their global networks. While they still have a long way to go, complexity masters are ahead of their competitors in building the capabilities for continuously optimizing global network investments (Figure 2.6). Indeed, industry leaders such as Procter & Gamble, Toyota, and Dell are more deliberate about including a broad set of relevant factors (customer service levels, lead time, flexibility, cost, risk, tax, regulatory issues, and environmental issues, etc.) into major

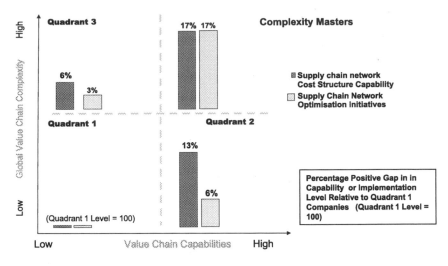

Figure 2.6. Supply Chain Network Structure Matters to Competitive Differentiation.

decisions on sourcing, manufacturing, new product introductions, or entry into new markets.[13]

Crucially, they understand that building these factors into the design of their networks is key to optimization. They know that excluding any of them (such as R&D, tax, and other regulatory issues) can expose the firm to higher costs or additional risks down the road. This is equivalent to the process of designing of new products. If sourcing and supply chain design are not taken into account early in product development, a manufacturer can lose significant (70 percent to 80 percent by some estimates) of its future cost reduction or revenue enhancement opportunities over the lifecycle of its product. The reason: Certain product features and designs can make it harder to reduce cost and incorporate new features and functions quickly and inexpensively later. This principle can be applied to most other investments in the value chain as well.

As companies restructure their networks to introduce new products, bring in new suppliers, or enter new markets, the optimal design of the network is bound to change as well. And as each restructuring moves beyond the design and planning stage, the ability to optimize the entire network efficiently becomes limited. Continuously optimizing the network to ensure that each new initiative is designed and implemented with the overall network structure in mind has become critical to remaining competitive (Figure 2.7).

---

[13] See e.g. Gary Rivlin, "Dell bucks the outsourcing trend," *The New York Times*, December 20, 2004;

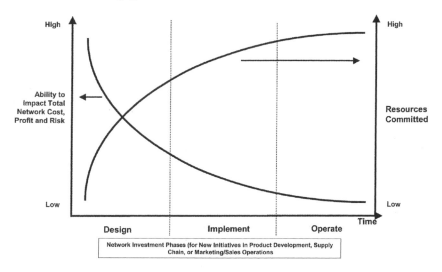

Figure 2.7. The Case for Continuous Network Optimization.

Doing this requires significant "visibility" into all parts of the value chain – a picture that is aided by improved technologies and information processes. Without the organizational, process, and technology infrastructure to support global optimization, most initiatives will fail. It requires not only better management processes for decision-making and execution but it also that top management oversees and supports continuous network optimization. It is not surprising, for example, that complexity masters are up to 50 percent more likely than other groups studied to have one executive in charge of the overall supply chain.

From our research, it is clear that successful companies are making significant investments to ensure continuous optimization of their global business by taking a number of actions (Figure 2.8).

- **What**? They incorporate all relevant competitive (markets, product development, supply chain) and compliance (regulation, taxation, etc.) drivers into optimizing major network investment initiatives and restructuring efforts.
- **When**? They ensure early and constant identification of current and future opportunities to redesign and optimize the global network.
- **How**? They build a network optimization infrastructure to align people, organizational incentives, and processes to make it easier to continuously redesign and restructure the network.

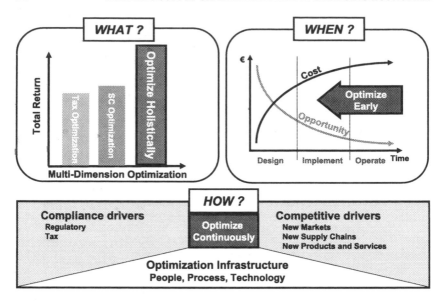

Figure 2.8. The Continuous Optimization Model.

   – They establish an integrated, flexible technology infrastructure to gain visibility and dynamically support changes in the network structure.

   – They enforce ongoing communication about global optimization goals and opportunities across the organization to boost awareness and influence local or functional initiatives on new market entry, low-cost sourcing, or product development. They realize that each initiative represents a low-cost opportunity to optimize the global network.

At one of the world's largest industrial product manufacturing companies, a global network optimization initiative boosted performance dramatically. Pressured by competition in an oversupplied market, the company benchmarked its value chain against those of other companies around the globe. The findings showed that the company lagged behind in a number of key areas. The manufacturer then traced the roots of its deficiencies and discovered that one downside factor was its decentralized operations. Because of misaligned incentive structures, from country to country, executives rarely worked together to reduce costs or improve operations. Each plant operated largely independently and focused on reducing its own costs. There was little standardization from business unit to business unit in business processes, procedures, performance metrics, and software applications. Production allocations to plants were also suboptimal. For example, the company was manufacturing high-margin products in locations with the highest tax rates.

Even after building a leading-edge plant in one of its largest markets with the aim of reducing costs, the company could not deliver the expected savings because it had inadvertently increased lead times to markets in other countries. With high variability in demand, the company quickly realized that increased inventory costs consumed the unit cost savings it had achieved in manufacturing. So it was back to the drawing board. After assessing key value chain functions, the company launched a global initiative to restructure and continuously optimize its business. It included the following actions:

- Streamlining organizational, decision-making, and incentive structures to align managers' objectives and metrics with company goals.
- Realigning and standardizing supply chain processes and systems on an enterprisewide basis.
- Consolidating and reorganizing factories – including shift of production from higher-cost/higher-tax-rate plants to larger, more strategically located facilities with lower-costs and tax rates – and the adoption of lean manufacturing processes to increase productivity and reduce inventory costs.
- Creating a global sourcing center to pool purchases by business units, gain volume discounts, and improve efficiency; upgrading interfaces and information technology infrastructures with trading partners; and identifying and implementing outsourcing opportunities.
- Adopting results of global optimization analysis, including establishment of a global freight management center.
- Improving customer management to enable the sales force to focus on marketing a more profitable mix of products to customers around the world.
- Creating global shared services capabilities for finance and accounting, order management, HR, and other headquarters activities.

The result of the efforts so far has been a stunning 75 percent increase in profits. How so? Optimization is enabling the company to deliver the same product volume at a significantly lower overall cost base, while ensuring the sale of a more profitable product mix.

Despite some success stories, most companies have a long way to go. In the following section we discuss the key design elements of optimization, the challenges and opportunities, and how a number of companies are optimizing their networks.

2.3.2   Designing the Global Network

Companies have struggled for decades to optimally design and restructure their value chain networks.[14] Yet, the need for network redesign and restructuring has increased for two primary reasons.

1. **Competitive Drivers.** Not only are the networks themselves more complex than ever, they are also changing at an unprecedented rate.[15] The reasons are many. Product and technology cycles are becoming shorter, time to market is shrinking, new locations for low-cost sourcing are emerging, and customer demand grows more fickle by the day. This forces companies to constantly rethink the optimal location and configuration of their facilities.
2. **Compliance Drivers.** Factors such as the increasing complexity and changes in national regulations, taxation, and international trade and investment regimes (e.g., new WTO rules and admission of new WTO members such as China) can wreak havoc with existing network designs.

Indeed, manufacturers are not standing still. Most have major initiatives planned or already under way to change sources of supply, overhaul manufacturing operations, and enter new markets over the next three years. In such an environment, it is highly unlikely that a network designed five or 10 years ago is optimal today. Yet, as we have shown, few companies are taking advantage of the opportunity to optimize their global operations fully.

Leading companies, however, are pushing forward with a more comprehensive view of optimization, including both competitive and compliance drivers. Consider the efforts drug giant GlaxoSmithKline has made to restructure a global production and distribution network that was hampered by the proliferation of product variations and production complexity. The company moved from a country-based manufacturing approach to a global approach with fewer plants that are dedicated globally to specific parts of a drug's lifecycle, such as the ramp-up phase. In total, the annual savings resulting from the restructured global network is about US$500 million a year – approximately the bottom-line impact of a blockbuster drug.[16]

---

[14] See Virginia Postrel, "Operation everything: it stocks your grocery store, schedules your favorite team's games, and helps plan your vacation. A primer on the most influential academic discipline you've never heard of," *The Boston Globe*, June 27, 2004.

[15] For evidence, see e.g. Deloitte Research, *Mastering Complexity in Global Manufacturing: Powering Profits and Growth through Value Chain Synchronization* (New York and London, 2003); and Deloitte Research, *The Challenge of Complexity in Global Manufacturing: Critical Trends in Supply Chain Management* (New York and London, 2003).

[16] See Deloitte Research, *Mastering Complexity in Global Manufacturing: Powering Profits and Growth through Value Chain Synchronization* (New York and London, 2003).

2.3.3   The Role of Competitive Drivers

For simplicity, we discuss "competitive drivers" as they affect the three broad business areas that make up a manufacturer's value chain:

- Demand chain (marketing, sales, and service)
- Supply chain (sourcing, manufacturing, and logistics)
- Innovation and product lifecycle management (R&D, design, engineering, development, and launch).

Demand Chain: Optimizing Marketing, Sales, and Service from a Global Perspective.   Companies are constantly on the hunt for new growth opportunities – launching new products to expand existing markets or entering new markets with existing or new products. These efforts, however, put significant strains on the organization's global network. For example, bringing supply chains into a new market not only increases complexity and costs, it also can create challenges for marketing and sales strategies in other markets. While the initiative may have been local, the repercussions on the network are global.

While new, potentially large markets can appear to be irresistible, the increased complexity creates significant challenges for even the best companies.[17] Preparing for the long haul is almost always a necessity when entering new markets such as China and India. For example, with more than a dozen major regions and over 2,000 languages and dialects, India is not the unified market one might think. Similarly, China, the new epicenter of the globalizing economy, consists of more than 60 distinct regions with varying regulatory, legal, infrastructural, cultural, and language differences.

As global mobile-phone makers Nokia, Motorola, and Samsung have learned, in order to drive growth and market share in emerging markets, distribution must be extended to ever smaller markets. For example, in China, Nokia is developing relationships with specialty retailers, consumer electronics chains, and small, regional or city-level distributors rather than working simply with a limited set of national distributors. Colin Giles, Nokia's senior vice president for customer and market operations in the region, says: "China is so big and diverse that it's not possible to classify it as a single market. That

---

[17] Before the end of the second quarter of 2004, automakers will have recalled more than 14 million units in North America, exceeding the total of 2003 by 2.5 million, with warranty expenses exceeding manufacturers' annual profits. Source: AMR Research and National Highway Traffic Safety Administration (NHTSA). See Kevin Mixer, Joe Souza, and Fenella Scott, "Early warning solutions: A transformation roadmap," AMR Research, June 21, 2004). Narishiko Shirouzu and Sebastian Moffet, "As Toyota closes in on GM, quality concerns also grow," *Wall Street Journal*, August 4, 2004.

is what we did three years ago. Today, we look at every market as different and we look for the best distribution or business model to suit that market."[18]

To capture these markets profitably, companies will find that planning, persistence, and flexibility are key. From our research, however, it is clear that few companies effectively build and leverage their global network when entering new markets. Not surprisingly, for most the major challenge to supply chain flexibility is forecast error, followed by long lead times, product proliferation, and supply chain visibility. These are all fundamental issues in building a profitable, sustainable business; without strong capabilities in these areas, global companies can rapidly lose the edge to smaller, national or regional competitors.

Carlsberg, one of the world's largest beer producers, acquired Poland's Okocim to gain access to a valuable brand and expand into new markets in Central and Eastern Europe. To guarantee the future viability of the expansion, Carlsberg quickly realized that it had to restructure Okocim's operations. Through an extensive assessment of production, packaging, distribution, and sales and marketing operations, the new network design includes a reduction in production sites from four to three, packaging sites from 12 to seven, and warehouses from 12 to six. Overall, the network optimization is expected to reduce total supply chain costs 15 percent while positioning the company for sales growth.

Supply Chain: The Global Pursuit of Lower Manufacturing and Supply Costs.   Sourcing from low-cost countries is the obsession of the day at multinationals around the world. Pushed by maturing markets and price competition from competitors sourcing in those low-cost locations, companies in all industries are aggressively assessing new locations for sourcing components and manufacturing goods.

Our research suggests that China is at the top of everyone's list for sourcing. Over the next three years, 55 percent of North American manufacturers and 39 percent of Western European manufacturers plan to enter or expand their sourcing in China. While the promises are great, the obstacles for leveraging the opportunities are vast. Consider the supply chain challenges alone. By some estimates, in China, 30 to 45 percent of the cost of goods sold is logistics cost – double the level in Europe and the United States – and there are few national third-party logistics providers. Road, air, and rail transportation systems have trouble keeping up with the requirements for a 21st-century supply

---

[18] See Phil Tinari, "Hung up: Multinational cell phone makers figured they knew best how to sell their products. Then they got to China," *Wall Street Journal*, September 27, 2004.

chain.[19] In addition, quality risks are plentiful. One company that redirected the sourcing of critical pumps to a Chinese supplier experienced a 70 percent defect rate – seven out of 10 pumps failed – with a severe impact on ongoing operations.

Nevertheless, the opportunities for low-cost sourcing and selling to vast and fast-growing markets are too great to ignore. Companies in just about all industries need to find a way to incorporate emerging economies like China and India into their global networks.

Indeed, the importance of taking a holistic view of new sourcing and supply chain opportunities cannot be overstated – as Ingersoll-Rand has found. The global, diversified manufacturer identified US$200 million in potential savings if it increased purchasing from low-cost countries to 30 percent of its total sourcing expenditures. The company initially assumed that transportation, duties, and taxes would add about 13 percent to the cost of imported materials. But a detailed assessment showed average ranges of 13 to 24 percent. Tony Bozzuto, Ingersoll-Rand's director of global logistics, pointed out: "We found cases where it was 200 percent" if antidumping penalties or other special fees were included in the calculation.[20]

Innovation and Product Lifecycle Management.   Despite having product innovation at the top of their growth agendas, few manufacturers are organizing R&D and product lifecycle management operations from a global perspective.[21]

For example, many companies fail to take advantage of regulatory and tax issues when deciding where to locate R&D facilities, or to consider intellectual property rights in making such decisions. The result: suboptimal investments and possible leakage of critical product or process technologies, a major concern, especially in developing markets. In India, it is estimated that counterfeit products make up 20 to 30 percent of the automotive parts market.[22] One automotive manufacturer calculates that 50 to 60 percent of counterfeit products with its trademark are made in China.[23] An inspection of a Chinese auto parts

---

[19] See also Deloitte Research, *The World's Factory: China Enters the 21st Century* (New York: Deloitte, 2003).

[20] See Merrill Douglas, "The total cost of global sourcing: Ingersoll-Rand scopes out the full cost of sourcing from different overseas suppliers," *LIT Toolkit*, February 2003.

[21] For more evidence of this "innovation paradox," see Deloitte Research, *Mastering Innovation: Exploiting Ideas for Profitable Growth* (New York, 2004).

[22] See "Spurious automobile parts industry turning 'organised'," *India Business Insight*, December 5, 2003.

[23] See Tim Trainer, "Counterfeiting and theft of tangible intellectual property: challenges and solutions," International Anti Counterfeiting Coalition Inc., Washington DC, March 2004.

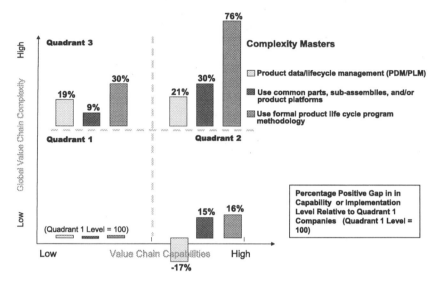

Figure 2.9. Managing the Product Lifecycle: Processes Key in Complex Global Networks.

factory found 7,000 sets of counterfeit brake pads destined for exports.[24] Not only does counterfeiting result in revenue loss, it can also jeopardize product and consumer safety and the integrity of the brand. Nokia reported there was a possibility that counterfeit batteries used with its mobile phones could explode.[25]

Some companies are better than others at guarding and exploiting their product innovation and process techniques. They build closer links between R&D, supply chain, and marketing and sales to improve the global design of their networks and maximize profitability over the lifecycle of products and services. Key action points include:

- Building supply chain considerations into product development processes early to make it easy and less expensive to upgrade products. Not surprisingly, complexity masters are much further ahead in using product data and lifecycle management technologies and processes to design more flexible product structures (Figure 2.9).[26]

---

[24] See Joann Muller, "Stolen cars," *Forbes*, February 16, 2004.

[25] See "Bogus batteries pose safety threat," *The Vancouver Sun*, December 12, 2003.

[26] For more details, see Deloitte Research, *Mastering Innovation: Exploiting Ideas for Profitable Growth* (New York and London, 2004).

- Incorporating intellectual property matters into supply chain design initiatives early on to protect patents and manage regulatory and tax issues. After experiencing copyright infringements by competitors from China, Invacare, a U.S.-based medical product manufacturer and distributor with US$1.5 billion dollar in sales, is now including the extensive involvement of lawyers in the product design process to ensure protection of intellectual assets.[27]

- Maximizing the value of R&D and intellectual property through better buying, selling, and licensing of technology. With nearly 76 percent of its value in intellectual property and intangible assets, by one estimate, 3M is one of the most innovative companies in the United States.[28] As one example of leveraging innovation, the company has built a technology transfer website to license and find new applications for more than 20,000 of its patented technologies.[29]

2.3.4   Factoring Compliance Drivers into the Network Design

Compliance drivers are external factors such as regulatory and taxation issues, which a manufacturer must consider before designing or restructuring a global value chain. Unfortunately, most organizations treat these matters as an afterthought. Typically, this happens when manufacturers perceive such drivers to be static and assumes that they will affect all competitors in similar ways, thereby leveling the playing field.

But this perception is far from the truth. While it might be easier to design a network by ignoring complex current regulatory and tax regimes, and potential future changes, these factors have profound impacts on the efficiency and effectiveness of the global value chain.

Regulatory Issues in Global Optimization.   For any global company, complying with regulatory issues is no small burden; such compliance requires dealing with local, national, and international regulatory matters on an ongoing basis. Ensuring that the global network is managed and optimized appropriately from a regulatory perspective renders the task even more challenging.

The list of regulatory issues impacting the design of a global supply network is long. It includes product traceability, end-of-life recycling and disposal, labeling, product liability, food safety, subsidies, labor laws, and corporate governance and financial reporting issues such as those surrounding the

---

[27] See also Eric Sherman, "Taking intellectual property seriously: there's more at stake in protecting and selling your IP than you may think," *Chief Executive*, Vol. 203, November 2004.

[28] See Gordon V. Smith, and Russell L. Parr, *Valuation of Intellectual Property and Intangible Assets*, 3rd Edition (New York: John Wiley & Sons, 2000).

[29] Source: 3M Annual Report 2003.

Sarbanes-Oxley regulations. Consumer perceptions of regulatory issues also count. Creating the wrong image in the eyes of the consumer can be extremely costly, as many consumer product makers have found over the years.

Failing to account for new regional and national regulatory issues in the network design can greatly worsen business performance. For example, consumer product manufacturers have recently run up against barriers to entering or expanding into new product markets in Europe due to specific national product regulations. The European Union is also stepping up enforcement of its antidumping rules and has increased by more than fivefold its staff dedicated to pursuing antidumping cases. Multinational companies are thus finding it increasingly risky to source from low-cost countries that violate the rules.

Furthermore, local content requirements that reduce taxation and import duties can force companies to rearrange their supply lines. In established markets of Western Europe and North America and emerging markets like China, Russia, and India, governments continue to put demands on multinationals that want to enter their markets. As Harley-Davidson's Chairman, Jeffrey Bleustein, said recently: "We cannot sell a motorcycle in China unless we are willing to manufacture it there and, frankly, I don't think Harley-Davidson with its Americana image and the kinds of quality and features we put into the motorcycle would have the same cache even in China if it were built in China."[30]

Calculating local content requirements, however, is no easy matter. And designing a network that balances these requirements with possible future changes can challenge even the best analysts. As products and industries evolve, product classifications can change as well, with significant impact on business design. Take the example of LCD and plasma monitors for computers. In the past, European Union customs authorities assigned them a zero percent duty rate. But given the fact that such screens are now used increasingly in high-definition-television sets, customs authorities recently decided that such monitors with video connectors should be classified as TVs, which are levied 14 percent import duties.[31]

Leading companies are more proactive when factoring in regulatory issues in the design of their global networks. They not only consider regulatory constraints as a cost of doing business, they leverage them for competitive ad-

---

[30] See "Let's have a level playing field: Not even Harley-Davidson can penetrate China's market," *Manufacturing and Technology News*, April 2, 2004.

[31] For a discussion, see "EU customs distinction between plasma monitors and televisions under discussion," *TDC Trade*, March 19, 2004. http://www.tdctrade.com/alert/eu0405.htm. See also Jennifer L. Schenker, "EU may redefine the computer screen," *International Herald Tribune*, May 20, 2004.

vantage.[32] For example, Toyota, one of the market leaders with its hybrid gas-electric engine technology, may boost earnings before interest and taxes (EBIT) 10 percent by 2015, according to some estimates, due to current and future regulatory demands for low-pollution vehicles.[33] In vehicle-part recycling, Toyota and Denso, one of the world's largest automotive suppliers, are collaborating with DuPont on a new process – Composite Recycle Technology – to produce parts from 100 percent recycled material. The goal is a recycling rate of 95 percent by 2015.[34]

Managing Taxation from a Global Optimization Perspective.   Business taxation can be a touchy subject. Minimize taxes and risk the ire of governments and public opinion; ignore it and wait for competitors with lower tax bases to exploit it to their advantage.

Simply *complying* with tax regulations around the world – be it direct corporate taxes or indirect taxes (including sales tax, value added tax, customs duties, and environmental taxes) – is a big challenge. Not surprisingly, as customs authorities around the globe move away from paper-based systems toward electronic processing, technology is playing an increasing role in managing compliance. With authorities imposing stiffer penalties for noncompliance, companies will need to integrate customs software into their enterprise resource planning (ERP) systems and key supply chain modules such as warehouse and transportation management. Being at the forefront of responding to demands like these, complexity masters are ahead of the game in adopting global technology platforms such as advanced planning and scheduling, and transportation management systems (see Figure 2.11 further below).

Corporate tax treatment and rates can vary significantly from country to country. For multinationals it matters where profits are accrued. It is thus critical for managers to understand the differences – in compliance and reporting,

---

[32] See also "HP, Dell, IBM and Leading Suppliers Release Electronics Industry Code of Conduct; Global Supply Chain Standards Promote Socially Responsible Business Practices," *Business Wire*, October 21, 2004.

[33] See Duncan Austin, Niki Rosinski, Amanda Sauer, Colin Le Duc, *Changing Drivers: The Impact Of Climate Change On Competitiveness And Value Creation In The Automotive Industry* (Washington, D.C.: World Resources Institute (WRI) and Sustainable Asset Management (SAM), 2003). See also "The Drivers: How to Navigate the Auto Industry," Deutsche Bank AG, July 31, 2002, for a discussion of the strategic implications of Toyota's leadership in environmental technology.

[34] See Duncan Austin, Niki Rosinski, Amanda Sauer, Colin Le Duc, *Changing Drivers: The Impact Of Climate Change On Competitiveness And Value Creation In The Automotive Industry* (Washington, D.C.: World Resources Institute (WRI) and Sustainable Asset Management (SAM), 2003).

tax rates, treatment, and tax credits – when considering expansion, R&D investments, restructuring of existing networks, or mergers and acquisitions.

Transfer pricing issues become a critical component in those decisions. With intangible assets such as patents, know-how, and brand value becoming critical components of the value of products and services, identifying the "right" transfer price between entities under common control is no small task. However, understanding the issues involved is critical to designing an optimal network. Over the last two decades, the value of intangible assets as a share of total market capitalization of U.S. companies has nearly doubled to more than 70 percent.[35]

To boost growth and attract high value-added activities, dozens of countries now offer companies tax credits on R&D activities, not only for laboratory research but also for other types of innovation (including process improvements in manufacturing, production trials, and software integration). For example, new investments in radio-frequency identification (RFID) technologies for improving the supply chain may be candidates for R&D tax credits.[36] Incorporating these tax credits into global supply chain development and restructuring decisions can have a significant impact on the bottom line.[37]

This also includes intellectual property (patents, trademarks, trade secrets, trade names, copyrights, etc.), intellectual assets (such as codified knowledge, contracts, permits, licenses, and non-competes), and intellectual capital (human capital, organizational capital, customer capital, distributor relations, and supplier relations). Different countries treat the valuation and returns on those assets very differently for regulatory, tax, and other purposes.

While most media discussion is about corporate – or direct – taxation, indirect taxation is often more important. Indirect taxes are not levied directly on income but rather on ongoing business activity. They can take many forms – import duties, antidumping levies, safeguard measures,[38] value-added taxes (VAT), goods and services taxes (GST), excise/consumption taxes, agricultural

---

[35] Kevin G. Rivette and David Kline, "Discovering new value in intellectual property," *Harvard Business Review*, January-February 2000.

[36] See e.g. Larry Shutzberg, *RFID in the Consumer Goods Supply Chain: Mandated Compliance or Remarkable Innovation?*, Industry White Paper, October 2004. http://www.packagingdigest. com/newsite/Online/RFID_IWP.pdf

[37] See e.g. "Dupont Canada Develops Breakthrough Product at Half the Cost Thanks to Canada's R&D Tax Credits," Advanced Materials Draft Paper, Canadian Chemical Producers Association, September 10, 2002, Calgary, Alberta, Canada.

[38] A World Trade Organization (WTO) member may take a "safeguard" action (i.e., restrict imports of a product temporarily) to protect a specific domestic industry from an increase in imports of any product which is causing, or which is threatening to cause, serious injury to the industry. For further details, see http://www.wto.org/english/tratop_e/safeg_e/safeg_e.htm.

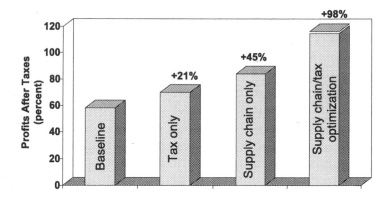

Figure 2.10. Taking a Holistic View: Building Tax Issues into Optimization.

levies, and stamp taxes. Having to deal with such complexities on a global scale, managers can easily become overwhelmed with the job of managing indirect taxes in a holistic, optimal manner.

Despite the increased complexity, however, there are significant benefits from taking a comprehensive approach to taxation in network optimization. From studying the design and restructuring of dozens of global supply chains, we have found that incorporating global tax issues into restructuring efforts typically can double bottom-line results as compared to restructuring without tax considerations (Figure 2.10).[39]

Consider the case of a U.S.-based global consumer goods manufacturer. To streamline operations and reduce costs, the company wanted to centralize purchasing across Europe into a pan-European procurement center. While the benefits from optimizing the supply chain alone would be worth several hundred millions of dollars, when it factored in both direct and indirect taxation, the company found the bottom-line benefits were twice as high.

---

[39] Though restructuring of the network considering both operational and taxation drivers can actually increase a company's total direct taxes, other factors will offset this result: Total profits before taxes increase faster and the net profits after taxes are higher, thus reducing the effective tax rate for the company. Note: The effective tax rate is calculated as total income taxes paid as a percentage of total taxable income before taxes.

2.3.5   Optimization Infrastructure – the Role of People, Process, and Technology

As we have seen, optimizing a global network can generate significant benefits – particularly if all relevant factors are taken into account early on in the design phase. However, most companies do not have the global infrastructure to support the design, implementation, and management of an optimized network. This includes capable organizational structures, business processes, and information systems.

Why is this the case? Most multinationals have grown organically and through mergers and acquisitions. Entering new markets and accessing new sources of supply around the world often creates new silos of operation. These silos seldom adopt the same systems and processes of the rest of the company, because during their creation they were focused on seizing new opportunities. International mergers and acquisitions only exacerbate the problem because they create even larger constellations of dissimilar global infrastructures. The result is often a spaghetti-like web of different processes and systems around the world.

Consider the example of General Motors. In 1996, before undertaking a major initiative to streamline supply chain, product development, and order management systems, the giant automaker found it had more than 7,000 discrete information systems. Most of them were legacy, silos of information that did not integrate data flow across the enterprise. For example, design engineering used 22 different engineering systems, hindering collaboration among the global product development staffs. Since then, GM has dramatically overhauled its global infrastructure in an attempt to reduce or eliminate such problems and is building a platform for continuing optimization of its global network.[40] To optimize its global logistics operations, GM and global logistics company CNF have established a joint venture, Vector SCM.[41] Launched 5 years ago, Vector SCM has grown rapidly and plans to manage US$4.8 billion of GM's US$5.5 billion total logistics spending worldwide in 2005. Vector SCM has built strong capabilities for continuously analyzing, designing, implementing, managing, and monitoring a global supply chain. This includes a

---

[40] From P. Koudal, H. Lee, B. Peleg, P. Rajwat, S. Whang, and R. Tully, "General Motors: Building a Digital Loyalty Network Through Demand and Supply Chain Integration," Stanford Case Study, January 1, 2003. See also Deloitte Research and Stanford Global Supply Chain Management Forum, *Integrating Demand and Supply Chains in the Global Automotive Industry: Building a Digital Loyalty Network at General Motors* (New York, 2003). For further information, see Laurie Sullivan, "Business integration–car maker takes global approach," *Computing*, September 30, 2004.

[41] Based on research by Stanford Global Supply Chain Management Forum and Deloitte Research.

global staff with deep technical skills, and a technology infrastructure that includes global network optimization applications and a multi-terabyte-size data warehouse for recording and monitoring information on global network operations and performance.

Vector SCM is not only helping GM optimize its existing investments, it is also helping the automaker to assess, support, and implement new global sourcing initiatives, logistics networks, and new product introductions. In addition, Vector SCM is helping GM expand into new markets such as China. With improvement percentages in spending efficiency ranging from the single to double digits, the total impact on the network is dramatic. It is expected to increase further as Vector SCM continues to help GM optimize and restructure its supply chain.

The organizational and people issues also matter. Major network restructuring efforts often require expensive staff relocation, which can be especially counterproductive if the most qualified staff members – those with years of experience and specialized skills – must be replaced. Our research suggests that the constraints imposed by current organizational structures, such as the location of regional headquarters, can significantly reduce opportunities for improvement. A European consumer products business that reorganized its supply chain estimates that due to the constraints involved in relocating managers it had to forgo about 50 percent of the benefits it might otherwise have

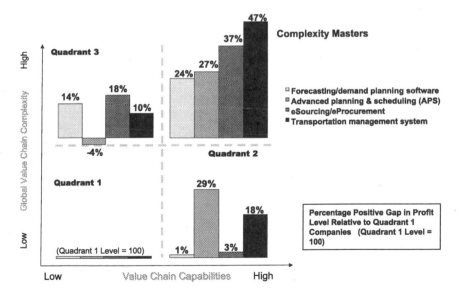

Figure 2.11. Staying Ahead on the Technology Curve.

realized from optimization. It is clear that organizational structure and change management issues need to be built into ongoing network optimization efforts up front to avoid wrong choices and allow for flexibility down the road. While complex, global and continuous optimization can actually make life *easier* for managers. When they make business decisions in light of what is good for the entire global network, they can help their companies avoid major mistakes and wasted efforts.

While the cost of instituting a global infrastructure to support network optimization is large, the experience of leading companies shows the benefits are worth the cost – and the effort. Perhaps not surprisingly, our research shows that complexity masters are far ahead in creating the systems, methodologies, and processes for the effective management of global networks (Figure 2.11).

2.4.   Conclusion

Unlocking the value of globalization through continuous network optimization is important today – and will grow in importance tomorrow. To effectively optimize their networks, companies must factor in both competitive and compliance drivers. Without deep insight into operations, including current and future customers, distributors, and suppliers and in-depth knowledge of the effects of regulatory, tax, and other issues, it will be difficult for organizations to consistently make the right decisions. To get continuous network optimization off the ground, companies must carefully consider their investment in the needed people, organizational, process, and technology capabilities.

Optimization of the global network can no longer be done just every three, five or 10 years. With dramatic changes continuing to impact global networks, leading companies are building the capabilities to look holistically at their operations on an ongoing basis. This form of continuous network optimization is the *Holy Grail* – a key competitive advantage that can help to substantially differentiate global business networks. As our research shows, companies making the greatest strides in global optimization are reaping the benefits through higher growth, profitability, and shareholder value.

Appendix: Defining Complexity Masters

To determine the practices of manufacturers that manage complexity well, we focused on a subset of the total survey population – the more than 300 sur-

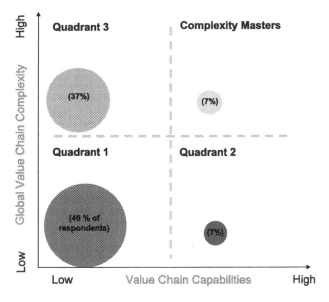

Figure 2.A. The Quadrants of Mastering Complexity.

vey respondents with annual revenues ranging from at least US$200 million.[42] We then divided the survey population by two dimensions (Figure 2.A):

- Degree of global value chain complexity. This was based on measuring the geographic diffusion (low or high) of four value chain functions (sourcing, manufacturing, engineering, and marketing/sales operations) across 13 geographic regions. We created a global value chain complexity index, scoring companies on a scale from 1 to 52. A manufacturer's score depended on the extent to which it scattered the four value chain activities across the 13 geographies.[43]
- Level of value chain capabilities. This axis gauges the relative competitiveness of each company on 10 value chain capabilities. We created a universal measure by taking a composite score of each respondent's ratings in product innovation, time to market, sourcing effective-

---

[42] For more on the complexity masters, see Deloitte Research, *Mastering Complexity in Global Manufacturing: Powering Profits and Growth through Value Chain Synchronization* (New York and London, 2003). See also Deloitte Research, *Mastering Innovation: Exploiting Ideas for Profitable Growth* (New York, 2004).

[43] The 13 countries/regions are: Australia/New Zealand, China, India, Japan, Korea, Other Southeast Asia, Western Europe, Central Europe, Eastern Europe, Africa, United States/Canada, Mexico/Central America, and South America.

ness, product quality, manufacturing flexibility, manufacturing productivity and cost-effectiveness, manufacturing lead time, logistics effectiveness, customer service, and supply chain cost structure. Manufacturers scored themselves against primary competitors on a 5-point scale (1 equals "significant disadvantage," 3 is "equal capability," and 5 is "strong advantage." Based on the 10 metrics and the 5-point scale, we then created a value chain capability index in which a company could score between 10 and 50.

By grouping survey respondents along the two axes, four groups result:

- Quadrant 1. Companies with low global value chain complexity (scoring below 20 on the global complexity index) and low-to-medium value chain capabilities (scoring below 40 on the value chain capability index). These manufacturers make up nearly half (49 percent) of the base.
- Quadrant 2. Companies with low global value chain complexity (scoring below 20 on the complexity index) but high value chain capabilities (scoring 40 and above on the value chain capability index – on average, exceeding the capabilities of their primary competitors across our 10 metrics). Only 7 percent of the respondents fell into this category.
- Quadrant 3. Companies with high complexity (scoring 20 and above on the complexity index) but low-to-medium capabilities (scoring below 40 on the value chain capability index). This group accounted for about 37 percent of all companies.
- Quadrant 4. Companies with high complexity (scoring 20 and above on the complexity index) and high capabilities (scoring 40 and above on the value chain capability index). We refer to this group as the "complexity masters." Just 7 percent of the sample fell into this category.

Hau L. Lee and Chung-Yee Lee (Eds.)
Building Supply Chain Excellence
in Emerging Economies
©2007 Springer Science + Business Media, LLC

Chapter 3

SHANGHAI OR CHARLOTTE?
The Decision to Outsource to China and Other Low Cost Countries[*]

David F. Pyke
The Tuck School of Business
Dartmouth College, USA

Abstract: The decisions of whether and how to outsource to low cost countries (LCCs) are often very troubling and complex. The popular business press is full of success stories and sometimes shrill stories of failure and of jobs lost to other countries. Managers need to cut through the noise and understand how to approach these issues from both strategic and tactical perspectives. By tracing the experiences of four manufacturing companies, introduce a framework for the outsourcing decision process – examining corporate strategy, operations strategy, total landed cost, and risk. Throughout, we illustrate the framework with examples. We conclude with some brief comments on the distinction between low cost country (LCC) and domestic sourcing.

3.1.   Introduction

One can hardly pick up a newspaper or business publication today without seeing an article or editorial about outsourcing to China or other developing countries.[1] Some opinion pieces and blogs predict dire consequences if the outsourcing trend continues, while others highlight the increased standard of living that has followed similar trends in the past.[2] These articles reflect the

---

[*] The author gratefully acknowledges helpful comments from Andrew Grimson, Chung-Yee Lee, Hau Lee, Susan Pyke, and David Robb.

[1] See Dawson (2005) for a very interesting example. Also, see Venkatraman (2004).

[2] Friedman (2005) is a recent best seller that addresses these issues in depth.

challenges faced by workers, policy-makers and managers. Many employees fear losing their jobs oversees, and yet flock to Wal-Mart to buy cheap imported products. Politicians worry about constituents' jobs, trade imbalance and global competitiveness, while recognizing the stabilizing effect of strong trade relations. And managers try to balance the needs of their employees and communities with the demands of an often brutally competitive marketplace.

It is no wonder that managers are hungry for information about operations in China, India and other developing countries. They are eager to understand how to approach the complex issues of whether to outsource, where to outsource, and how to outsource to achieve the maximum benefits for their companies. They wonder if they should purchase components or finished goods, or if they should do their own manufacturing by acquiring, merging, or even starting new operations. Unfortunately, many managers make these decisions based on limited information and without a strategic context. They often do not understand the full cost of doing business overseas, choosing instead to focus only on astonishingly low unit costs.

This chapter is intended to help managers with these issues by tracing the experiences of four manufacturing companies. We begin by briefly describing the four companies and their specific concerns. We then introduce a framework for the outsourcing decision process – examining corporate strategy, operations strategy, total landed cost, and risk.[3] Throughout, we illustrate the framework with examples. We conclude with some brief comments on the distinction between low cost country (LCC) and domestic sourcing.

3.2.   The Four Companies

Firm A is a large multinational diversified industrial company with annual revenues of about $10 billion (Table 3.1). In spite of having a manufacturing presence in over 25 countries and sales in more than 120 countries, managers, prior to 2003, were remarkably U.S. centric. Procurement from low cost countries lagged behind competitors. In early 2003, the CEO issued a mandate to increase the volume of purchases from LCCs by a factor of seven by 2008, while reducing the total cost of purchased components by 20%. After this initial decision to outsource from LCCs, and in particular from China, managers had significant concerns about the effect on quality and delivery performance. Furthermore, they were apprehensive about the effect of LCC sourcing on their ability to manage the global supply chain. It had only been a few years since

---

[3] Total landed cost captures the entire cost of producing and transporting products from the origin to the destination.

Table 3.1. The Four Firms.

	Firm A	Firm B	Firm C	Firm D
Firm Size	$10 billion	$80 million	$60 million	$100 million
Products	Diversified industrial: Auto parts Electrical parts	Production equipment; Consumables	Molds; Plastic parts	Plasma metal cutting tools
Current Manufacturing Base	25 countries Many factories	3 countries 4 factories	2 countries 5 factories	1 country 1 factory
Firm Strategy	Technological leadership; Consistent quality; JIT delivery; Low cost	High quality; New technology; Reliable delivery; Premium price	Consistent quality; Reliable delivery; Complex parts; Low cost	Consistent quality; Technological leadership; Premium price
Markets	120 countries	U.S. and Europe	U.S.	U.S., Europe, Asia
Impetus for LCC sourcing	CEO mandate; Cost pressure	VP Operations' gut feel that they should "be in China"	Cost pressure; Future markets	Shortage of capacity; Huge sales in Asia; Develop "Asia strategy"
Concerns	Quality of LCC parts; Delivery of LCC parts; Managing global supply chain; Intellectual property	Effect on the domestic workforce and community; Quality of LCC parts; Delivery of LCC parts; New product introduction	Quality of LCC parts; Delivery of LCC parts; Managing the global supply chain; Management time and expertise	Impact on sales growth and customer/market concentration

Firm A had created the position of vice president of supply chain and aggressively pursued supply chain integration, and they were worried that their recent advances would be at risk.

Firm B is a manufacturer of equipment for high speed production lines for the pharmaceutical and food industries. This $80 million company also produces highly profitable consumable products used with their equipment. They operate two factories in the U.S., one in the U.K., and one in continental Europe. The U.K. factory is mostly an assembly operation, with all components purchased from the surrounding area. The other factories manufacture and purchase components and perform final assembly and testing. Firm B has no sales in China or elsewhere in Asia because their products have exceptionally high quality, and hence are priced at a premium. In late 2003, senior managers began wondering if a manufacturing presence in China would open up the Chinese market, at least to some extent. The primary driver of the LCC discussion, however, was the vice president of operations who felt strongly that they needed to "be in China" to manufacture existing products or components. He and other senior managers, including the CEO, continually debated whether this was really necessary, and if so, whether they should purchase components from China, merge with or purchase a Chinese manufacturer, or build a new plant in China.

Firm C is a manufacturer of injection molded products for the automotive and health care industries. This $60 million firm makes molds in their own tool shop for internal use, and also buys additional molds from small job shops in the U.S. and China. However, their primary business – over 90% of revenues – is complex molded parts. Firm C operates four factories across the U.S., in addition to a new plant just opening in a new industrial park in Mexico. The auto industry, in general, exhibits constant price pressure and active outsourcing to LCCs, while the health care industry is a bit less cost competitive. Quality requirements in both industries are very high, and health care customers sometimes even require clean room capability. Cost pressure was the initial motivation for this firm to look to China. Like Firm B, they are known for high quality and reliable delivery, but unlike Firm B, the likelihood of selling molded parts in China is low, at least in the near term. In the longer term, they anticipate that their primary customers will open or expand assembly operations in China and will require suppliers to operate nearby.

Firm D is a $100 million manufacturer of plasma metal cutting tools, in both manual and large mechanized versions, as well as the consumable parts used with the tools. They operate one factory in the U.S., but their business is expanding rapidly and there is an immediate need for additional capacity. Approximately 50% of sales are outside of the U.S., with the majority of foreign sales in Asia to transplants and Asian customers. Managers therefore want to develop "an Asia strategy" that will help them understand whether to expand domestically or to open operations in Asia, and what the risks and benefits of each location will be. They also want to be certain that they consider all the relevant factors in their decision.

3.2.1   Options

These four companies ranged in the outsourcing decision process from Firm A, which had already decided to aggressively pursue LCC procurement, to Firm D, which was just beginning an investigation. The rationale for examining LCCs also varied widely, from Firm D, which had significant sales in Asia, to Firm B, where the VP of Operations simply had a feeling that they should be in China, to Firm C, which was facing serious cost pressure. They all shared one key characteristic, however. None of the four firms had extensive experience in Asia, so the management challenges were significant, and the risks seemed high.

These firms considered a range of options, including

- Purchasing components
- Purchasing finished products
- Merging with an existing firm
- Acquiring an existing firm
- Opening a new factory

The required risk, investment and commitment all increase from the top to the bottom of this list. Furthermore, the management expertise required to purchase components from China is trivial compared to that required when opening and operating a new plant in a different culture, halfway across the world. Yet, all options were on the table, to varying degrees, for each firm.

3.3.   A Framework for the Decision Process

Many firms jump on the LCC bandwagon without putting the decision in a strategic context. They see reports of hundreds of firms going to China, and they assume that they have to do the same. Or they have managers who, ignoring quality and delivery performance, feel that China is the sole answer to cost pressure. Our four firms pursued the answers to their LCC questions in a variety of ways, and yet their experience suggests a framework that is generally applicable.

The framework (Figure 3.1) begins with a review of corporate strategy and examines the LCC questions in that light. If the LCC initiative does not support corporate strategy, there is no need to delve into deeper questions and

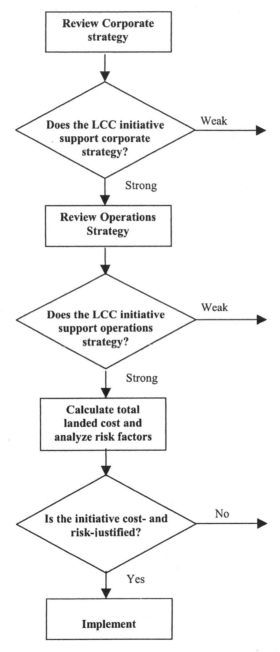

Figure 3.1. Framework for the LCC Decision.

implementation issues. The framework then reviews the operations strategy and again investigates how the LCC initiative supports it. Finally, it analyzes the total landed cost and risks of the different options. Once the decision has been made, a team is selected to handle implementation. In this chapter, we are focused on the decision process, leaving detailed implementation issues aside for now.

3.4. Corporate Strategy Review

The LCC decision can be a major one indeed. Firm B, for instance, had been a significant presence in the same small town for almost 100 years. Any talk of China created anxious hallway discussions, not to mention concern in the local community. In any company, one can be certain that the decision will be scrutinized closely by all parties affected. They will ask whether this move is really necessary, whether alternatives have been considered, and what it will do to help the company and workers thrive. To answer these questions, and of course to make the best decision for all stakeholders, managers should set the context by reviewing, and perhaps updating, the corporate or business unit strategy.

See Figure 3.2 for a schematic of the strategy review process. [4] The process is designed to

- Review and update the corporate mission, values and principles
- Carefully consider global trends that could impact the company's future
- Understand the competitive landscape
- Come to a shared commitment to a set of performance targets
- Engage discussion of key strategic initiatives, including the LCC outsourcing issue
- Come to a shared commitment to these initiatives
- Develop a set of performance targets for each functional area that support the company's overall targets
- Develop tactical initiatives for each function that will enable them to achieve their targets.

The first task is always to review or update the corporate mission and statement of values and principles. Any strategy review must be built on this foundation. Then, in light of competition and global trends, managers should agree on a set of performance targets for the corporation. These targets might be very specific – to grow the company by 50% in the next five years, for instance, or

---

[4] See Gupta & Govindarajan (2003), Grant (2005) and Collis et al. (2001) for more on corporate strategy.

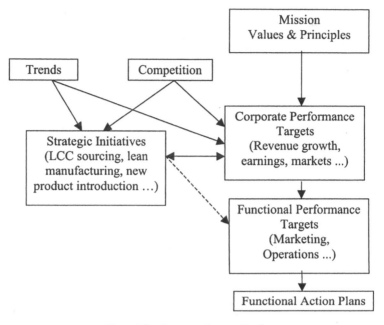

Figure 3.2. Corporate Strategy Review.

quite general – to increase the pace of new product introduction, say, in the automotive business. These targets, in turn, set the context for a discussion of key strategic initiatives, which might include lean manufacturing, LCC outsourcing, opening new markets, and so on. Once the team has committed to a set of initiatives, they should set or refine performance targets for each functional area. The operations group, for instance, might have a goal to reduce unit cost by 10% or improve yield to 99.5%, while the marketing group might target a certain percentage increase in market share. Likewise, an initiative to speed new product introduction will have direct implications for the performance targets of the engineering team. Finally, these targets drive a set of initiatives that set the agenda for each functional area for the next several years.

This strategy review process is broadly applicable and can be particularly helpful to companies facing a dynamic environment. Of course, strategy research has proliferated over the past several decades, providing managers with a wealth of analytical tools, from SWOT (strengths, weaknesses, opportunities, threats) analysis and five forces to hypercompetition.[5]

---

[5] See for instance Porter (1983) and D'Aveni (1994).

3.4.1   Firm C's Corporate Strategy Review Process

Firm C directly follow the strategy review process outlined here. They held a retreat with the goal of creating a new five year strategic plan – a far broader mandate than simply addressing the LCC sourcing issue. Participants included the Chairman of the Board and senior managers from all functional areas including operations, sales and marketing, human resources, and finance. The CEO chaired the meeting, which was facilitated by an academic (this author). After several preparatory meetings among the CEO, CFO and facilitator, the retreat was convened.

The retreat began with a review of the mission, values and principles – a discussion that took a surprising amount of time considering the fact that most participants had been with the company for many years. The most vigorous debate pertained to several possible additions to the values and principles, although the group settled on a largely unchanged version. The participants then engaged in an active and wide-ranging discussion around trends and the competitive landscape. Any idea was entertained, but in depth discussion was reserved for the most pertinent issues for the firm. Important trends included, for instance, the possible shift of major automotive customers' operations to Asia. Managers also discussed whether health care customers would shift assembly operations there as well. If so, would it be a competitive advantage to have operations in-country? If not, is such a shift on the horizon? Sales and marketing personnel, who were in close contact with these customers, weighed in heavily on these discussions.

After agreeing on a core set of trends and competitive issues that were directly relevant to the firm, the discussion moved to key corporate performance targets. These included five-year targets regarding global presence, revenue and profitability. Customer composition (percent automotive, for instance) and concentration (percent to any one customer) were prominent as well. Clearly, the LCC decision must be founded on such long term corporate goals. A goal to double revenues in the next five years could lead to very different outsourcing decisions than a goal to grow revenues by 10%. Because these major targets must be shared by all senior managers, the agenda allowed for significant discussion and airing of views. After coming to agreement on a set of about five fundamental corporate goals, the discussion proceeded to strategic initiatives.

The CEO, CFO and facilitator had previously developed a short list of major strategic initiatives that would be introduced, but the participants were asked to brainstorm any initiatives that they felt were worth discussion. Following common brainstorming techniques, each participant was given a total of three votes that they could cast on one, two or three of the initiatives. The LCC decision was one on the short list, and it received a large number of votes. Other initiatives discussed included facilities decisions (open new plants, close existing plants), new markets, six sigma, lean manufacturing, and new capabilities, among others. The outcome of the discussion of each initiative was an action plan, if appropriate, that identified the manager responsible for carrying it forward.

The group debated the LCC issue at great length, focusing on China in particular. It was an expansive discussion that touched on issues of management time and expertise, product quality, customer expectations, labor relations, and many others. Ultimately, the group decided that CEO and CFO would pursue discussions with an investment bank in Beijing, which had already been identified, in a search for a mold-making firm. The goal was to buy a controlling interest in a high quality, technically advanced firm that would make molds for sale in the U.S. and China, and that could eventually expand into injection molded parts for sale in China. The group felt that achieving their aggressive growth goals required a deliberate, but expeditious, approach to this search.

As we concluded the conversations of strategic initiatives, we ran a simple exercise designed to highlight the issue of management time requirements. For each initiative that was moving forward, we asked each person to raise a hand, and keep it up, if the initiative would significantly affect their job or require considerable time from them. The lean initiative, for instance, would have only a minor effect on sales and marketing managers, but it would require a major effort from operations managers. We moved through each initiative, keeping hands up, and then asked participants to stand if both hands were already raised. At the end of the exercise, a few people were sitting with no hands raised, most had one hand raised, and two people were standing up with both hands in the air. Senior managers readily saw the potential pitfall of limited management time and energy.

The retreat continued by focusing on each functional area, with a particular emphasis on operations, marketing, engineering and human resources. Senior managers from each function listed the performance targets that they currently tracked, and then the entire group identified a few additional measures that should be included. Finally, functional managers went to breakout sessions to discuss their functional area strategies, i.e. to set measurable targets that would help achieve the corporate goals identified earlier, and to create action plans designed to meet those targets. These strategies and action plans were shared with the entire group at the close of the meeting. Because this retreat was focused on a new five year strategic plan, these discussions covered all aspects of the business, with the China initiative playing a moderate, but significant role.

3.4.2 Corporate Strategy Review: Summary and LCC Concerns

Because the impetus for Firm C's examination of LCC sourcing was cost pressure and possible future customer requirements, the strategic discussion focused on operations and marketing issues. For other firms, the strategic LCC discussion will center on different topics. For instance, Firm A has a technological lead in many of its markets, and as a result, is often the sole source supplier of high margin products. When managers engage the conversation about LCC sourcing, they inevitably raise concerns about protecting their intellectual property and patents. Even sourcing a simple, time-insensitive component may hasten the creation of stronger competition.[6] They also know, however, that their technological advantage will not last forever, and that their competitors are aggressively pursuing Asia strategies. Therefore, to prepare for the future, they need to maintain R&D spending, build engineering talent, and consider cost reduction initiatives that include purchasing components from LCCs. Firm B has a variety of issues that should receive attention at this stage of the analysis, but their presence in the same community for nearly 100 years means that human resource issues should be underscored. Firm D, on the other hand, is shipping so many products to Asia that they should focus on, among other things, possible marketing effects, sales growth in different regions of the world, effects of local content, and the impact on customer and market concentration.

Senior managers should provide ample opportunity for functional managers to raise concerns and questions as they pertain to LCC sourcing (Table 3.2). R&D and engineering managers will likely raise issues involving quality, intellectual property, difficulties with designing components that will be made in multiple countries, and design software integration. Operations managers should address capacity, total landed cost, delivery performance, quality, inventory requirements, and other supply chain issues. They should focus on current and future performance because today's low cost developing country may be tomorrow's high cost developed country. Likewise, today's low quality source may become tomorrow's high quality source – or competitor. Marketing managers, on the other hand, should highlight market opportunities in the foreign country, brand and image issues in the home country, pricing developments, and competitive threats and responses. And, human resources managers will certainly question the effect on the current workforce, as well as communications strategies moving forward.

---

[6] See Amaral et al. (2005) who warn that companies should carefully consider "the underlying means, motive and opportunities" of potential sources to take control from the domestic firm.

Table 3.2.   Concerns of Different Functions.

Function	Potential Issues and Concerns
R&D and Engineering	Intellectual property Patents Product design across different countries and cultures Quality of product designs Design software integration Creation of a future competitor

Operations and Supply Chain Management	Capacity
	Total landed cost
	Delivery performance
	Quality
	  Process capability
	  Six sigma or other quality programs
	Inventory requirements
	  Longer and more variable lead times
	Managing the supply chain
	  Collaboration
	  Forecasting
	  Production scheduling

Marketing	Market share Effect on the brand Effect on pricing Competitive threats and responses Creation of a future competitor
Human Resources	Effect on morale of the current workforce Communications with the workforce and community

In summary, the corporate strategy review is designed to identify major performance targets and strategic initiatives in the context of industry and competitive trends. As noted above, if outsourcing to a low cost country somehow violates the corporate strategy, or is not otherwise dictated by competition or other strategic initiatives, there is no need to further the conversation. For many firms, however, the opposite is true, and they should proceed to in-depth reviews of their functional strategies.

3.5.   Operations Strategy Review

Functional strategies are derived from the corporate strategy and initiatives. Our focus will be on the operations strategy, although as should be clear from the previous discussion, engineering, marketing and human resources strate-

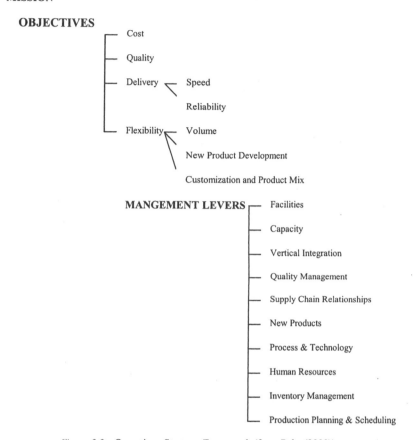

Figure 3.3. Operations Strategy Framework (from Pyke (2000)).

gies may require similar attention. The operations strategy is composed of three levels – a mission, operations objectives, and management levers (See Figure 3.3, Pyke (2000), and Slack & Lewis (2002)). The *operations mission* defines the direction for the operations function, as opposed to the corporate mission, which defines the direction for the company as a whole. The operations mission states a purpose for the operations and sets priorities among the objectives. It specifies the primary task that must be achieved for operations to succeed. The *operations objectives* – cost, quality, delivery (speed or reliability) and flexibility (volume, new product development, customization/product mix) – are measurable targets that should be well defined and rank ordered. We cannot emphasize enough how important it is to rank order these objectives so that all operations employees know how to make trade-offs when they inevitably arise. And the objectives must be measurable so managers know whether they have achieved them. The *management levers* are the many and varied tools that managers use to achieve these objectives. We define ten levers, although these can change over time as new developments arise. (For instance, the supply chain lever is a relatively recent addition to the list.) Managers should specify current policies for each lever, update the policies as appropriate, and rigorously analyze whether there are inconsistencies among the policies or between the policies and the operations objectives. Our experience suggests that this process will likely identify opportunities for improvement as performance measures are created and refined, and as policies associated with the levers are adjusted to support those objectives.

An operations strategy review can include an examination of global trends, competition and key strategic initiatives, especially if these have a direct impact on the firm's operations or supply chain.[7] In the case of Firm C, the operations strategy review followed immediately from the corporate strategy review, so there was no need to have a separate discussion. Firm B, on the other hand, engaged an operations strategy review independently, and therefore, it was appropriate to discuss trends, competition and initiatives, such as LCC sourcing, in that context.

3.5.1   Firm B's Operations Strategy Review Process

Firm B had recently updated their corporate strategy, with a new emphasis on innovation and new product development, but they had not reviewed their operations strategy for several years. Therefore, the vice president of operations called a 2½ day retreat for all his plant managers, supply chain managers and senior engineering managers. The vice president, a plant manager, and the facilitator (again, this author) spent many hours interviewing senior managers from all functional areas, as well as the CEO, COO and president of the firm. They also sent out a questionnaire to each person who would be attending the retreat. (See Exhibit 1, which asks a series of questions designed to ascertain the current operations strategy and to provide a brief definition and summary of the three levels.)

3.5.1.1   Operations Mission and Objectives.   The group spent very little time working on the operations mission because it so closely aligned with the new corporate mission. However, they took several hours to fully work

---

[7] In fact, some authors (e.g. Miltenburg (1995)) include "initiatives" or "projects" as a fourth level of operations strategy. These projects typically apply to more than one management lever and through them to the objectives.

through the operations objectives. Following the questionnaire in Exhibit 1, the group was asked to carefully define what they meant by quality, how they measured delivery performance, their current industry standing with regard to new products, and so on. They also highlighted differences between new equipment and consumables for each objective. After a vigorous debate, they rank ordered the objectives as follows: 1) quality, 2) delivery, 3) flexibility, and 4) cost. In spite of the prominent role of new product development in the corporate mission, it became clear from the discussion and the mission itself that quality and delivery still took precedence.

Firm B used the term "agility" instead of flexibility, so their ranking gave rise to an acronym, QDAC, which was used throughout the retreat. Every time a new initiative was discussed, several people would emphasize that the initiative should *not* violate QDAC. This is precisely how the operations objectives should influence the conversation. In fact, after the retreat, the firm ordered polo shirts with letters "QDAC" prominently displayed.

Much like Firm C's corporate strategy retreat, Firm B engaged a discussion of global trends, although in this case, the focus was a bit more operational. Nevertheless, they did raise the questions of future markets and competitive moves in Asia. These trends and the operations objectives set the context for a discussion of major initiatives, including radio frequency identification (RFID), lean manufacturing, new product development, and of course, entry into China.

As noted above, for Firm B quality was absolutely the most important objective. Reliable, on time delivery was a close second. They supply equipment for other manufacturers, and they *will not* delay the opening of a high speed production line because their equipment is late. And they will not shut down a customer's line because of a machine failure. Their new corporate mission had highlighted new product introduction as being critical to continued success. Therefore, while cost was an important consideration, it was not the principal driver of performance. They could charge a 10 – 20% premium because of their technology, quality and delivery performance. This context was critical to the discussion of a China strategy. If the primary reason for "being in China" was to reduce cost, it should be clear from QDAC that such a move must not damage quality, delivery performance, or their ability to efficiently introduce new products. Therefore, the group focused intensely on the effect on product quality if components were to be manufactured in China. They also noted that delivery times would increase from a week to about eight weeks, which would require careful planning and more inventory. Engineering managers, furthermore, carefully considered the challenges of designing and developing new products if some components were to be sourced from half way around the world. Of course, these factors do not preclude a China strategy, but they do suggest that thorough planning is necessary.

It would be appropriate to note at this point that Firm B decided not to open manufacturing operations in China. Nor did they immediately pursue a joint venture or acquisition. Rather, they decided to purchase a fairly complex, but non-critical component as a way to begin understanding the process of doing business in China. They knew that manufacturing in China was not an immediate need, but they also knew that their competitors were moving fairly aggressively into low cost countries. Therefore, they felt strongly that they needed to begin the process in a low risk way so they could be positioned for the future. This fruitful discussion followed directly from a close examination of the ranking – QDAC – and the measurements of the operations objectives.

3.5.1.2   Ten Management Levers.   The next level of operations strategy contains the ten management levers (Figure 3.3 and Exhibit 1). During the operations strategy retreat, managers from Firm B worked through each of the ten levers, citing current policies and looking for inconsistencies, opportunities and points of leverage. This was a lengthy discussion that we will not review in detail here. We will, however, provide one example. The plant manager in the major U.S. plant had been working on lean manufacturing for several years, and progress was remarkable. Flow times had decreased from three weeks to 26 hours, quality had improved, inventories had decreased, and customer satisfaction was higher. It had been a stunning success. One of the principles of lean manufacturing is to reduce work-in-process inventories in conjunction with a pull, or Kanban, system, and it had served Firm B well. One concern that surfaced during the retreat, however, was that their renewed emphasis on innovation and new products might be at odds with their lean initiatives. To effectively operate a lean system, a firm should have limited variety and a fairly predictable and stable demand pattern.[8] A rash of new products, unfortunately, will by definition be less predictable and stable, at least until they have been on the market for some time. Therefore, the plant manager was tasked with preparing their lean manufacturing system for an increased number of new products, probably by increasing inventory levels and process task times. Again, this is exactly the right outcome – adjusting production planning and inventory policies to be consistent with the operations objectives.

After a review of the ten levers in general, it is often desirable to use them as a tool for examining major initiatives. For instance, when considering LCC sourcing, managers can use the ten levers to help identify all the relevant operational issues. The inventory lever, for example, will highlight the required

---

[8] See Silver et al. (1998), Chapter 16.

adjustments to inventory policies if lead times increase or become more variable. The supply chain lever should raise the issue of collaboration with suppliers and the possibilities or difficulties of doing so with a new partner. If collaboration decreases, how will that impact quality, forecasting, inventory management or production scheduling? Using the quality management lever, managers can ask whether the new supplier has six sigma or other quality programs in place. Are their systems compatible with ours? Will our customers care? The new products lever should focus on the processes used to introduce new products. What software is used? How do they handle the handoff from design to manufacturing? Are their processes compatible with ours? As a final example, the process and technology lever raises the question of production layout and equipment. Is their equipment capable of consistently meeting the required tolerances? These discussions can take a significant amount of time, so during the retreat with Firm B, we raised the issues, engaged brief discussions, and then assigned relevant managers to follow up with more detailed information.

Thus far, we have considered corporate and functional strategies, with a particular emphasis on operations strategy. The goal of these extensive discussions is to be certain that any proposed LCC outsourcing decision complements and advances competitive success. Furthermore, they are designed to illuminate potential pitfalls, inconsistencies, and requirements for management attention. If the proposal passes muster at this point, it is time to examine total landed cost and risk in complete detail.

3.6.   Total Landed Cost

Most firms will have a fairly good sense of the cost of outsourcing to an LCC before engaging the corporate and functional strategic discussions. Firm D, for instance, was just beginning the LCC discussion, but they had a clear sense of the current cost of transporting products from their U.S. plant to their customers in Asia. They also were developing numbers on the cost of manufacturing in Asia, but these were still fairly rough. As they began discussions of an "Asia strategy" retreat, the total landed cost was clearly part of the agenda, although they recognized the need to set the context with the larger strategic considerations. Supply chain managers at Firms A and B had already created total landed cost models that captured most of the relevant costs, although they did not include some of the soft costs discussed below. A total landed cost model should include:

- *Inbound materials.* The cost of buying raw materials and components for the LCC factory. Many companies are able to capture significantly lower raw materials and component costs by buying from other LCC factories. However, if those sources are not available, or do not have the required quality, the firm may not be able to take advantage of these sources.

- *Inbound logistics.* The cost of moving materials and components to the LCC factory. Factories that are close to raw materials sources will benefit here, but again, if the current domestic source is the best choice, the firm may find itself moving materials across the ocean, and then shipping finished product back.

- *Manufacturing at the LCC site.* Consider labor, assembly and equipment costs (such as molds or other asset specific investments). Yield rates, setup times and costs, and quality costs may be taken into account as well. It may not be necessary to break out all these costs if the per unit cost is sufficient for the analysis.

- *Overhead at both the LCC site and domestically.* Consider information technology infrastructure, communications, duplicate functions, legal personnel, and so on. Be careful not to double count the costs if some of these components appear in other categories.

- *Customs, duties and taxes.* These figures clearly change over time as nations modify their trade relations. Plant location can make a difference if there are special short or long term tax advantages to certain regions. Finally, some countries give incentives for not repatriating profits, so it is important to involve accountants, lawyers and tax experts.

- *Inventory at the LCC site.* Raw materials, work in process and finished goods inventory. Consider who owns the inventory, how much is required to meet the throughput needs, and the cost. Note that regulations in the LCC may influence inventory ownership.

- *Outbound logistics.* Consider the costs and lead times for the following:
  - LCC plant to the port
  - LCC port to domestic port
  - Domestic port to distribution centers
  - Pick and pack operations at the distribution centers (and plants, if appropriate)
  - Distribution centers to customers

- *Domestic inventory.* The cost of inventories at the domestic site will increase if lead times increase or are more variable. Many companies use inventory formulas that are based on average demand, forecast errors, service level targets, and lead times. If these formulas are already employed, it is straightforward to calculate the increased inventory cost due to an LCC supplier.[9]

---

[9] See, for example, Chapter 7 of Silver et al. (1998).

- *Travel.* The travel costs for managers to visit suppliers. One manufacturer did not include these costs in its decision, yet they sent two managers to Taiwan several times each year to meet with their supplier. As it happens, this cost did not overwhelm the unit cost savings, but it clearly should be included in an outsourcing decision.
- *Translation.* Firm C hired a person fluent in Chinese to help insure that all documents were accurate in both languages. These costs can be fairly minor, but Firm C was also considering a full time person for the first year or two of operation – although this person would serve other functions as well. These costs clearly should be included in the analysis.
- *Relationship management.* Often these costs are captured in overhead or the procurement function. However, the time and cost associated with managing a relationship with a company in a different time zone, culture and language can be significantly higher than managing a domestic supplier. This category should focus on the incremental cost of managing the LCC relationship, to the extent that it has not been accounted for in overhead, translation, travel or other costs.
- *Soft costs.* These include, for instance, the cost of managers getting up at 3:00 a.m. to place phone calls to their supplier. Management time and stress fit here, if they have not been captured in the relationship management costs. These costs can be exceptionally difficult to specify, and yet when we talk to managers, they rise to the surface immediately.

The LCC outsourcing decision should account for as many of these costs as can feasibly be specified. Our experience suggests that for many firms, even after accounting for all these costs, LCC sourcing is significantly cheaper than their current domestic source. Nevertheless, it would be shortsighted to ignore any of these components. Typical models of total landed cost will not include soft costs or some of the other components, such as translation costs. We include them here not to suggest that they should be used in every model, but that managers should incorporate them, quantitatively if possible, in their decision process.

3.7. Risk

If the LCC initiative is consistent with the corporate and operations strategy, and it appears that the total landed cost is sufficiently lower than the current domestic source, one would expect managers to move ahead aggressively. Before doing so, however, it is critical to carefully examine multiple risk factors that will likely arise with LCC sourcing.

Of course, *currency risk* increases with any offshore source, as firms that lived through the Asia financial crisis of the late 1990s will attest. Related to currency risk is *political risk*, a broad topic that is worth careful thought.[10] In the post–9/11 world, for instance, imports from Muslim countries may be treated much more stringently than those from other countries, even if there is no evidence associating that country with terrorist activities. In the case of China, analysts wonder if its corrupt political system can manage its massive economic growth over the long term. Will changes in the banking system, or in agricultural policies, spur layoffs and widespread strikes? How will China deal with the Taiwan issue, and what effect will that have on stability in the region? Similar questions can be asked of any low cost country, although the answers will vary widely and by definition will be highly uncertain. Nevertheless, we would strongly encourage managers to assess political risk and the potential fallout from large scale disruptions to their supply chains. At the very least, they should have backup plans that can be quickly adopted in the event of a disruption.

We have discussed *quality risk* in the context of operations strategy and Firm B's QDAC priorities. As another example, Firm A had an experience that animated their discussion of LCC sourcing. They changed steel suppliers after many years with a single source whose quality and delivery performance were excellent, but whose price had increased well above the competition. When the new supplier, chosen after a careful review and qualification process, delivered its first shipment, operators at Firm A found that acceptable quality had decreased from well above 90% to less than 50%. And that was a new *domestic* supplier! The procurement manager responsible for shifting millions of dollars of business to LCC sources was understandably concerned with quality risk, and he made sure the group heard about it. Firm A's response was to identify experienced partners to help find and qualify sources so that quality risk would be minimized.

Lead time risk increases as well. Longer lead times tend to amplify the well known "bullwhip effect," which can cause difficulties with forecasting and inventory management.[11] Furthermore, if a firm has been buying domestically for years, its procurement managers may not have the tools to analyze the effects of long and variable lead times. Assuming airfreight is too expensive, they will have to make adjustments to inventory and production planning policies. Even with correct inventory policies, however, the risk of shortages can increase dramatically because of potential disruptions to a long supply chain. And managing that risk can be difficult when the supplier resides in another country, in a vastly different time zone, and speaks another language. Several

---

[10] See Bremmer (2005) for an in-depth discussion of political risk, and www.aon.com/politicalrisk for a political and economic risk map.

[11] See Lee et al. (1997).

years ago, a hiking boot company, which bought almost its entire product line from Asia, spent $200,000 in airfreight charges because a key supplier was going to deliver late. A year later, they ordered early to avoid the risk of more airfreight costs. As it happens, this time the supplier delivered right on time (i.e. much earlier than the true need), and the firm ran out of storage space. Boxes of boots were stored in the back of 18 wheel trucks in the parking lot. For weeks, workers had to dig through piles of boxes to ship orders to customers.

One other risk that should be highlighted is *intellectual property risk*, which, as noted above, was an important concern for Firm A as a technological leader in its industry. Certain countries are known for piracy of intellectual property, with little recourse in the courts.[12] Many companies who outsource in these countries therefore choose products and components that are not on the leading edge of technology. Costs may increase in the near term, but intellectual property risk is reduced. Alternatively, they open and manage their own facilities to better maintain control.

3.7.1 Risk and Uncertainty

Ideally, a firm would adjust all their costs by the relevant risk factors. So, for instance, they would modify inventory calculations to account for risk and variability in lead times. Unfortunately, some risk factors create such high levels of uncertainty that such adjustments are infeasible.[13] What is the probability of China attacking Taiwan? If they do attack, what is the probability of a complete collapse of trade between the U.S. and China, at least for a time? How long would that collapse last? These questions are extraordinarily difficult to answer. Therefore, we suggest dividing the risk factors into two categories – a) short term or quantifiable and b) long term or difficult to quantify.

The first category includes currency, lead time and quality risks. Rigorous due diligence in choosing suppliers can certainly mitigate quality and lead time risk. Residual uncertainty associated with these risks, and with currency risk, is often quantifiable in the form of probability distributions. The currency exchange problem, for instance, is well studied, and firms typically assign their treasury departments to develop hedging strategies. Moreover, based on the dynamics of currency exchange rates, sophisticated firms are able to quantify the option value of excess plant capacity. Likewise, inventory policies easily can be adjusted to account for variability in lead times.[14] These methods generally

---

[12] See Dietz et al. (2005) for comments on protecting intellectual property, and Massey (2006) for a history of intellectual property rights protection in China.

[13] The Analytical Hierarchy Process may be useful in weighing tangible and intangible factors. See Saaty (1990).

[14] See Silver et al. (1998), Section 7.10 for instance.

rely on expected value calculations or the use of constraints in an optimization model. As long as the underlying uncertainty is represented by a fairly well understood probability distribution, these tools can be quite effective.

If a dramatic event occurs, however, by definition it will not be captured by commonly used probability distributions. The Asian currency crisis of the late 1990s, for instance, caught many managers by surprise. The best hedging strategies were useless in the face of a crisis of this magnitude. This leads to the second category of risk factors – those that are long term or difficult to quantify. In this category, we would consider political and intellectual property risks, as well as crisis events that dramatically affect currency exchange rates, lead times and quality. A political crisis in China could totally interrupt supply lines, resulting in unacceptably long lead times. Whereas inventory policies can easily be adjusted for moderately increased lead time variability, such adjustments would be woefully inadequate during a crisis. But accurately adjusting inventory policies for extremely rare events is very difficult. Therefore, we recommend that, in the normal course of business, managers use tools appropriate for the risks in the first category, while at the same time explicitly preparing for risks in the second category. Tools for this second category often focus on careful backup plans and alternate sources of supply, all developed in advance.[15] For this reason, we regularly advocate that managers considering China or another LCC source should maintain a domestic backup. Other tools include insurance coverage and retaining key knowledge and skills in house. Short term costs may increase, but long term value, even the likelihood of survival, may be enhanced.

3.8.   Domestic Sourcing

One may wonder what the fundamental difference is between domestic sourcing and LCC sourcing as described here. Why not use the same framework for any decision to outsource, or to change from one domestic supplier to another? We would suggest that the framework does indeed apply to any sourcing decision. Procurement and supply chain managers need to account for total landed cost, regardless of the source, and place the decision in the context of the firm's operations and corporate strategy. Nevertheless, there are some essential differences.

Of course, as noted above, LCC sourcing may dramatically increase risk. Logistics and supply chain problems can be magnified, as can issues with

---

[15] See Lee (2004) and Chopra & Sodhi (2004) for excellent discussions on managing supply chain risk.

taxes, customs and duties.[16] Furthermore, LCC sourcing intensifies management stress because of differences in culture, business climate, negotiating styles, time zones and language. These differences are quite distinct from domestic sourcing and have been discussed at length in this chapter. Managers should not underestimate them.

LCC sourcing, on the other hand, may create opportunities that are not available from domestic sources. For instance, sourcing in a country can open the local market due to local content regulations. Furthermore, product configurations developed for the local market can help balance capacity and possibly even reduce customs. Finally, tax incentives are often available in developing countries as they try to create jobs and transfer technology.

In our experience, LCC sourcing just *feels* different for companies that do not have extensive global experience. The firms discussed in this chapter took this decision very seriously and were willing to take ample time to set it in a strategic context and to understand the implications for their management teams. The result was carefully weighed decisions with paced and deliberate plans for moving forward. For Firm A, this meant shifting millions of procurement dollars offshore, whereas for Firm B it meant buying just one component from China. Nevertheless, based on total cost and strategic analyses, Firm A maintained domestic sources for many parts, which highlights the fact that LCC sourcing is not an all-or-nothing decision.

3.9. Summary and Conclusions

Too often, firms take a purely tactical approach to low cost country sourcing. They focus entirely on unit cost and justify the decision on this factor alone. In our experience, this is by far the most common LCC sourcing pitfall, although implementation problems are frequent as well. In particular, due to weak preparation and analysis, managers are caught off guard by late deliveries, poor quality, insufficient capacity, culture or negotiation conflicts, and so on. We submit that the four stage decision process described in this chapter captures a best practices approach.

First, excellent firms approach the LCC decision in the context of a corporate strategy review. Is the decision consistent with the mission, values and principles of the firm? Does it respond adequately to competition, global trends and specific corporate performance targets? It may not be necessary to undertake a full corporate strategy review for each LCC decision, but managers should be very clear that it is consistent with existing strategy.

---

[16] It is worth noting that in some industries domestic lead times are *more* variable than those from international sources. The same inventory formulas referenced above could be used to analyze either case.

Second, these firms conduct a careful operations (and marketing, if appropriate) strategy review. An analysis of the four operations objectives – cost, quality, delivery, and flexibility – will force managers to highlight important operational considerations other than cost. As tempting as LCC unit costs may be, firms should be meticulous in examining the effect of the decision on quality, delivery performance and new product introduction. Furthermore, the ten management levers serve to elicit a broad array of potential issues, problems and opportunities. Best practice firms will ensure that the ten levers are adjusted for the new LCC source and that they are consistent with the other levers and with the operations objectives.

Third, best practice firms develop a comprehensive total landed cost model that includes easily quantifiable costs, such as customs, duties, inventory, and inbound and outbound logistics, as well as soft costs, such as relationship management and management stress. If a cost is difficult to quantify, it is still wise to include it in the list, even without a specific number attached. It can thus serve as caution for decision makers.

Finally, these firms carefully examine the multiple risk factors that arise with an LCC decision. For the quantifiable ones, such as increased lead time variability, they adjust the relevant cost accordingly. For the difficult to quantify ones, they employ a number of tools including explicit backup plans and alternate sources of supply.

The LCC sourcing decision can be very difficult and emotionally charged. The four stage decision process described in this chapter will not eliminate the intense anxiety that workers and managers often feel, but it may well diminish it. And it will certainly ensure that the decision is grounded in careful analysis.

Exhibit 1: Operations Strategy Questionnaire

1. How would you define *cost* for Firm B? Is your goal to be the low cost provider, competitive with the industry, or do you charge a premium?
   a. What measurements do you use for *cost*?
2. How would you define *quality* for Firm B?
   a. What measurements do you use?
3. How would you define *delivery* for Firm B? Is it important to have rapid delivery of products? Services? Service parts? Or is reliable delivery more important?
   a. What measurements do you use for delivery?   ·
4. How would you define *flexibility or agility* for Firm B? How often do you introduce new products? Is this important? Is your product line broad enough? Should it be reduced? Do you have issues with seasonality?
   a. What measurements do you use for each of these?
5. Assuming that sometimes it is necessary to make tradeoffs among these four objectives, how would you rank order them in terms of importance to Firm B?

        Cost _____
        Quality _____

Delivery ____
Flexibility ____

6. For the *facilities* lever
   a. Where are your facilities located?
   b. Should you consider a different location?
   c. How are they focused? In other words, what activities are performed at each location?
   d. Are these policies consistent with the objectives you defined above?
7. For the *capacity* lever
   a. When do you expand or contract capacity? Before or after demand swings?
   b. Do you need to make expansion or contraction decisions now?
   c. Are these policies consistent with the objectives you defined above?
8. For the *vertical integration* lever
   a. How do you make decisions to outsource components or production?
   b. Are you considering new outsourcing decisions now?
   c. What is outsourced currently?
   d. Are these policies consistent with the objectives you defined above?
9. For the *quality management* lever
   a. How do you pursue quality now?
   b. Do you have a TQM program, or another similar initiative?
   c. Do you use teams or other decentralized quality initiatives?
   d. Do you use Statistical Process Control?
   e. Are these policies consistent with the objectives you defined above?
10. For the *supply chain relationships* lever
    a. What supply chain initiatives are you currently using?
    b. What are you considering?
    c. Do you employ e-procurement? Strategic alliances? Vendor managed inventory? Coordinated forecasting, planning or replenishment?
    d. Are these policies consistent with the objectives you defined above?
11. For the *new products* lever
    a. Do you use cross-functional teams for new product development?
    b. Do you involve suppliers?
    c. Do you have a formal system of milestones?
    d. Are these policies consistent with the objectives you defined above?
12. For the *human resources* lever
    a. Do you use a bonus system in manufacturing?
    b. Do you cross train you people?
    c. What other HR policies do you employ?
    d. Are these policies consistent with the objectives you defined above?
13. For the *inventory* lever
    a. When do you trigger inventory orders?
    b. How many do you order at a time?
    c. How often do you review inventory status?
    d. What service targets do you set?
    e. Do you use inventory management software?
    f. How are forecasting decisions made?
    g. Are these policies consistent with the objectives you defined above?
14. For the *production planning and scheduling* lever
    a. When do you trigger production orders?
    b. How many units are ordered at a time?
    c. Do you employ MRP? Kanban? Other lean manufacturing tools?
    d. Are these policies consistent with the objectives you defined above?
15. Are the policies in each of these levers consistent with those in other levers?

References

Amaral, J., Billington, C.A., and Tsay, A.A. (2005). Safeguarding the Promise of Production Out-sourcing. *Interfaces*, forthcoming. http://www.aon.com/politicalrisk.

Bremmer, I. (2005). Managing Risk in an Unstable World. *Harvard Business Review* (June), 51–60.

Chopra, S., and Sodhi, M.S. (2004). Managing Risk to Avoid Supply-Chain Breakdown. *Sloan Management Review* (Fall), 53–61.

Collis, D.J., Pisano, G.P., and Rivkin, J.W. (2001). *Strategy and the Business Landscape: Core Concepts.* Upper Saddle River, NJ: Prentice-Hall.

D'Aveni, R. (1994). *Hypercompetition.* New York: The Free Press.

Dawson, C. (2005, February 21). A 'China Price' for Toyota. *Business Week,* 50–51.

Dietz, M.C., Lin, S.S.-T., and Yang, L. (2005). Protecting Intellectual Property in China. *The McKinsey Quarterly*(3), http://www.mckinseyquarterly.com/article_page.aspx?ar=1643&L1642=1621&L1643=1635&srid=1617&gp=1640.

Friedman, T.L. (2005). *The World Is Flat: A Brief History of the Twenty-First Century.* New York: Farrar, Straus and Giroux.

Grant, R.M. (2005). *Contemporary Strategy Analysis: Concepts, Techniques, Applications* (5th ed.). Malden, MA: Blackwell Publishers.

Gupta, A.K., & Govindarajan, V. (2003). *Global Strategy and Organization.* New York: John Wiley & Sons, Inc.

Lee, H.L., Padmanabhan, P., & Whang, S. (1997). The Bullwhip Effect in Supply Chains. *Sloan Management Review, 38*(3), 93–102.

Lee, H.L. (2004). The Triple-a Supply Chain. *Harvard Business Review, 82*(10), 102–.

Massey, J.A. (2006). *The Emperor Is Far Away: China's Enforcement of Intellectual Property Rights Protection, 1986–2005.* Forthcoming, Summer, *Chicago Journal of International Law.*

Miltenburg, J. (1995). *Manufacturing Strategy: How to Formulate and Implement a Winning Plan.* Portland, Oregon: Productivity Press.

Porter, M.E. (1983). Industrial Organization and the Evolution of Concepts for Strategic Planning: The New Learning. *Managerial and Decision Economics, 4*(3), 172–180.

Pyke, D.F. (2000). *A Framework for Operations Strategy.* Unpublished Note, Dartmouth College, Hanover, NH.

Saaty, T.L. (1990). *Multicriteria Decision Making: The Analytic Hierarchy Process.* Pittsburgh: RWS Publishers.

Silver, E.A., Pyke, D.F., & Peterson, R. (1998). *Inventory Management and Production Planning and Scheduling* (3 ed.). New York: John Wiley & Sons.

Slack, N., & Lewis, M. (2002). *Operations Strategy.* New York: Prentice Hall.

Venkatraman, N.V. (2004). Offshoring without Guilt. *Sloan Management Review, Spring,* 14–16.

Hau L. Lee and Chung-Yee Lee (Eds.)
Building Supply Chain Excellence
in Emerging Economies
©2007 Springer Science + Business Media, LLC

Chapter 4

LIFE-SAVING SUPPLY CHAINS
Challenges and the Path Forward

Anisya Thomas and Laura Rock Kopczak
Fritz Institute, USA

Abstract: There is a great need to improve the world's supply chains for disaster relief. This paper provides background on the current state of logistics in the humanitarian environment and the factors that have limited the evolution of knowledge and the performance of humanitarian supply chains. We consider the external pressures that aid agencies (AA) are feeling from donors, local aid agencies, governments and corporations, as well as the internal limitations that have impeded progress in logistics. We then recommend five strategies that together define a path forward for aid agencies, building on the experience of supply chain evolution in the private sector. An appendix provides information on resources for academics that would like to get involved in this exciting field.

4.1.   Introduction

On December 26, 2004, the earthquake in Sumatra and the destructive Tsunami that it unleashed riveted the attention of the entire world. Each day for the first two weeks, the toll of the dead and missing crept higher and heart-wrenching stories of families devastated beamed into the living rooms of people in every country. The response was immediate, and unprecedented amounts of relief money were collected.

While the drama unfolded on television, the disaster relief infrastructure at the local, national and international levels sprang into action. Volunteers from the affected communities began to clear the dead, and relief agencies began to provide food and shelter to those made vulnerable. International relief agencies activated their assessment teams. Supplies to provide the basic necessities to large numbers of people across a broad geographic span were ordered and transported from locations around the world.

Unfortunately, disaster relief is and will continue to be a growth market. Both natural and man-made disasters are expected to increase another five-fold over the next fifty years due to environmental degradation, rapid urbanization and the spread of HIV/AIDS in the developing world. According to the Munich Reinsurance group, the real annual economic losses have been growing steadily, averaging US$75.5 billion in the 1960's, US$138.4 billion in the 1970's, US$213.9 billion in the 1980s and US$659.9 billion in the 1990's.

The world's poor and least developed bear the brunt of the burden. Of the world's disasters during 1990–98, 94 percent occurred in developing countries which also endure two-thirds of the economic losses. Adverse consequences include destruction of physical infrastructure, loss of lives and livelihoods, forced migration and disruption of economic activity. The estimated loss as a result of disasters was $65 billion each year over the period 1992–2001.

One of the notable aspects of the relief efforts following the 2004 Asian Tsunami was the public acknowledgement of the role of logistics in effective relief. In the immediate aftermath of the Tsunami, as relief goods flooded the airports and warehouses in the affected regions, aid agencies struggled to sort through, store and distribute the piles of supplies while disposing of those that were inappropriate. In Sri Lanka, the sheer number of humanitarian flights with supplies paralyzed the capacity to handle goods at the airport. Downstream, relief agencies struggled to locate warehouses to store excess inventory. In India, transportation pipelines were bottlenecked. In Indonesia, the damaged infrastructure combined with the flood of assistance from the military representatives from several countries and large numbers of foreign aid agencies, created a coordination and logistical nightmare. As a European Ambassador at a post-Tsunami donor conference said, "We don't need a donors' conference, we need a logistics conference".[1] Similarly, a spokesman for Doctors Without Borders, in announcing their decision not to accept any more money for the relief operations said "What is needed are supply managers without borders: people to sort goods, identify priorities, track deliveries and direct the traffic of a relief effort in full gear".[2] Humanitarian Logistics, the function that is charged with ensuring the efficient and cost effective flow and storage of goods and materials for the purpose of alleviating the suffering of vulnerable people, came of age during this Tsunami relief effort.

This paper provides background on the current state of logistics in the humanitarian environment and the factors that have limited the evolution of

---

[1] New York Times, January 6, 2005.

[2] Economist.com Global Agenda, January 5, 2005.

knowledge and the performance of supply chains for humanitarian relief. We consider the external pressures that aid agencies (AAs) are feeling from donors, local aid agencies, governments and corporations, as well as the internal limitations that have impeded progress in logistics. We then recommend five strategies that together define a path forward for AAs with regards to logistics.[3]

4.2.   The Scope, Importance and Challenge of Humanitarian Logistics

Humanitarian Logistics is defined as the process of planning, implementing and controlling the efficient, cost effective flow and storage of goods and materials as well as related information from the point of origin to the point of consumption for the purpose of alleviating the suffering of vulnerable people. The function encompasses a range of activities, including preparedness, planning, procurement, transport, warehousing, tracking and tracing and customs clearance.[4]

Humanitarian Logistics is central to disaster relief for several reasons. First, it is crucial to the effectiveness and speed of response for major humanitarian programs, such as health, food, shelter, water and sanitation. Second, with procurement and transportation included in the function, it can be one of the most expensive parts of a relief effort. Third, since the logistics department handles tracking of goods through the supply chain, it is often the repository of data that can be analyzed to provide post-event learning. Logistics data reflects all aspects of execution, from the effectiveness of suppliers and transportation providers, to the cost and timeliness of response, to the appropriateness of donated goods and the management of information. Thus, it is critical to the performance of both current and future operations and programs.

The importance of logistics to humanitarian relief can be illustrated by the efforts of the International Federation of the Red Cross and Red Crescent Societies (IFRC) to provide assistance to the Indian Red Cross after the Gujarat earthquake. On January 26th, 2001, a 7.9 Richter earthquake struck Gujarat, where 41 million people were preparing to celebrate Republic Day in

---

[3] The views presented here are based on extensive research conducted by Fritz Institute over the past three years, including case studies, interviews and conferences with the leading aid agencies, technology and process development with select AA's and surveys of the response to the 2004 Tsunami disaster that affected south Asia.

[4] The definition and tasks of logistics was based on a 2004 sector wide survey of humanitarian logisticians from headquarters and the field that worked with a broad range of aid agencies.

India. Thousands of people were killed, the local airport was destroyed, the infrastructure severely damaged, and very little information was available in the early stages of the disaster. Nonetheless, within the first 30 days of the earthquake, along with the assistance of 35 partner organizations, the International Federation of the Red Cross's Logistics Emergency Unit arranged the delivery of 255,000 blankets, 34,000 tents, 120,000 plastic sheets, and large quantities of other items such as kitchen sets that were distributed to beneficiaries by the Indian Red Cross. More than 300 other global, regional, national and local NGOs and UN agencies similarly mobilized their staffs and resources.[5]

Thus, the supply chain for relief is the ultimate sense-and-respond supply chain. Once a disaster occurs, an AA sends in a team of experts to complete an initial assessment within 24 hours and an appeal for donations within 36 hours. The appeal lists specific items and quantities needed to provide immediate relief to the affected populations. Emergency stocks of standard relief items are sent in from the nearest relief warehouses. Calls are made to traditional government donors and the public and commitments for cash and/or in-kind donations secured. Suppliers and logistics providers are lined up, and mobilization of goods from across the globe begins. When supplies arrive, local transportation, warehousing and distribution have to be organized.

This is a tremendous feat to accomplish, given the remote places in which disasters tend to occur, the uniqueness of the requirements for each disaster in terms of both expertise and goods, and the fact that the disaster site is often in a state of chaos. Physical infrastructure such as roads, bridges and airports is often destroyed. National and local government, through which aid agencies must often coordinate their activities, may be severely impacted, or even uprooted in the case of a conflict situation. Transport capacity may be extremely limited, or non-existent.

4.3.   The Humanitarian Sector Ecosystem

Aid agencies are the primary vehicle through which governments channel as much as $6 billion in annual aid targeted at alleviating suffering caused by natural and manmade disasters.[6] The period from 1990 to 2000 saw total humanitarian aid from governments double in real terms from approximately $2.1 billion to $5.9 billion.[7] In the aftermath of the Tsunami, it is estimated that

---

[5] Choreographer of Disaster Management: The Gujarat Earthquake, Kuldeep Kumar, Ramina Samii & Luk Van Wassenhove (2002).

[6] *Financing International Humanitarian Action: A Review of Key Trends,* Humanitarian Policy Group briefing paper, November 2002.

[7] *Financing International Humanitarian Action: A Review of Key Trends,* Humanitarian Policy Group briefing paper, November 2002.

the aid budget might actually have grown to $12 billion. While the largest AAs are global in scale, there are also many smaller regional and country-specific AAs.

Most global aid agencies engage in a mix of development and relief activities.[8] Relief refers to the emergency food, shelter and services provided in the immediate aftermath of a natural or man-made disaster. An example of relief would be the initial 90–120 days of services provided by the various aid agencies to assist the people affected by the Tsunami in December 2004. By contrast, development refers to the longer-term aid aimed at creating self-sufficiency and sustainability of a community. An example of a development program would be the Area Development Programs executed by World Vision India to feed and school children and teach women basic business skills in the slum areas outside Chennai in southern India.

The aid agencies receiving donations from this global community can be classified into three broad categories: entities operating under the United Nations umbrella such as the World Health Organization (WHO) and the United Nations High Commission for Refugees (UNHCR), international organizations such as the International Federation of Red Cross and Red Crescent Societies (IFRC), who operate as a federation with country offices that are auxiliary to country governments, and global non-governmental organizations (NGOs) like CARE and World Vision. NGOs also maintain country offices, but their offices are not affiliated with the country governments.

International aid donations from each country are then channeled through the international AA's to local partners in the affected countries. In most cases, it is these partners, closest to the affected population and of the same culture that provide the relief services to the affected populations. For example, appropriations from the US Congress are channeled to USAID which in turn funds the U.S. based aid agencies or UN organizations. This money is then used to implement aid programs in countries affected by disasters. In most cases, the relief activities are then carried out by local organizations from the affected areas with local staff who can speak the language and communicate with those who are receiving the help.

Donors, both government and the public are becoming increasingly demanding with respect to performance and impact. With an increasing number of aid agencies, the competition for donor funding is getting more intense, and data demonstrating impact is likely to be the differentiator. Further, donors are

---

[8] For purposes of this report, the focus will be on disaster relief in the aftermath of a natural disaster or humanitarian emergency. Although we acknowledge the critical importance of development, the dynamism and velocity of the relief scenario present specific exigencies which merit investigation.

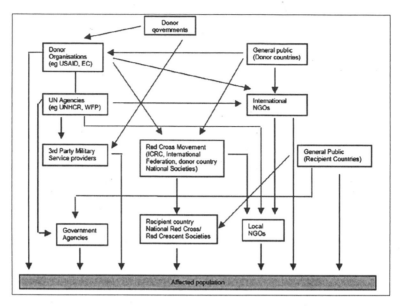

Figure 4.1. Humanitarian Sector Funding Flows.[9]

becoming less tolerant of obvious and expensive duplication of effort, and are strongly encouraging aid agencies to collaborate around the creation of common services. As a consequence, aid agencies have become more aware of the need to strategically use their resources and to be efficient in addition to being effective.

4.3.1 Humanitarian Logistics: Pain Points

Our research over the last four years suggests that certain common challenges face the field of humanitarian logistics and have impeded its evolution. These challenges are in many ways similar to those faced by corporations two decades ago.

Lack of Recognition of the Importance of Logistics.   Most aid agencies have two broad categories of activities, programs and support services. Programs refer to the front-line activities in relief and development, the provision of services such as food, water, shelter, sanitation etc. Support services refer to the activities of the back room which support the front-line: logistics, technology, finance, communication, human-resources etc. Funds are usually

---

[9] Adapted from a 2001 DAC report.

allocated by donors to programs with a certain percentage allowed for administration, which includes support. Thus, the focus is on direct relief rather than investment in systems and processes that will reduce expenses or make relief more effective over the long-term. As a consequence, logistics and other support services may not have adequate funding for strategic disaster preparedness, and infrastructure such as information systems.

A related challenge has to do with the fact that most decisions during an operation are made by the program staff who control the budgets. Our research suggests that the assessment team which is sent to determine the needs of the population that has been affected by a disaster or humanitarian crisis often does not include a logistician. Based on the assessment, the program staff determines the supplies that need to be procured in order to provide relief services, and then inform logistics that they are responsible for the immediate procurement and transport to the field. Since they are not often consulted in the decision process, difficulties that may ensue delay the relief and put the logistics departments on the defensive.

Lack of Professional Staff. In general, aid agencies are defined by their personnel, who share a common value system relating to the alleviation of suffering of victims of disasters and humanitarian emergencies. People who choose a career in this world come from diverse and varied backgrounds. Our sample of head logisticians included an actor, an osteopath, an extreme sports enthusiast, a nurse and a country manager. These people are driven by a desire to resolve crises and do good in the world. They have achieved their positions by trial and error and experience in multiple disaster theaters over several decades.

However, as the operations of international aid agencies expand to simultaneously include multiple geographies, organizations are struggling to find adequate numbers of people who can manage the complex supply chains of relief. For example, in order to effectively respond to the Tsunami, 88% of organizations we surveyed had to pull their most qualified logistics staff from other crisis areas such as Darfur.[10]

In 2004, Fritz Institute in conjunction with Erasmus University and APICs, a widely recognized training and certification body for commercial logistics, conducted a survey of approximately 300 humanitarian logisticians at the field, regional and headquarters levels of major aid agencies. The purpose was to identify existing training and certification programs and the range of logistics functions that they encompassed. Respondents to the survey[11] represented a

---

[10] *Logistics and the Effective Delivery of Humanitarian Relief*, Fritz Institute (2005)

[11] 30% response rate/92 respondents

wide variety of organizations including the UN, the Red Cross movement and international and regional NGO's from headquarters as well as the field.

Over 90% of the respondents indicated that they felt training was directly linked to performance on the job and that standardized training would be useful in the field. However, only 73% had access to any logistics training. For those with access, training was most often provided by coworkers on the job, or by in house training staff. However, respondents noted that job training within organizations tended to be non-standardized, the content largely dependent on the trainer. The respondents indicated frustration with lack of consistency in training, lack of ways to measure the effectiveness of training, lack of funding for training and lack of specific training in humanitarian logistics.

Ineffective Leveraging of Technology.   Our survey of logisticians from international aid agencies that participated in the Tsunami relief operations showed that only 26% of the respondents had access to any tracking and tracing software. The remainder used Excel spreadsheets or manual processes for updates and tracking of the goods arriving in the field. Despite this, 58% stated that they received accurate and timely information of what was in the pipeline!

In the humanitarian sector, particularly at the field level, logistics and supply chain management is still largely manual. Lack of existing software that can handle the dynamism of humanitarian supply chains, limited internet access, computer shortages and lack of trained staff as well as the inability of IT staff at headquarters to understand the imperatives of the field, and the need to keep networks secure are the main reasons that humanitarian logisticians provide as driving the slow evolution of IT. However as a consequence there are systemic deficiencies in humanitarian supply chains. These include:

- Data must be written out onto multiple forms and keyed into multiple spreadsheets
- Budget control is inadequate; funds may be misspent as a result
- Usage of funds is not tracked to the extent that donors have requested
- Procurement procedures are difficult to enforce; integrity is lacking
- Tracking and tracing of shipments is done manually using spreadsheets
- There is no central database of history on prices paid, transit times, or quantities received/purchased
- Reports are done manually; therefore little reporting and performance analysis is performed, other than reporting to donors on quantities of relief items delivered for a given operation

In the private sector, supply chain technology has enabled the transformation of the function from a peripheral to a strategic one. By accumulating data about the supply chain, decision makers have new ways to create efficiencies. Historical data also allows greater effectiveness through the tracking of supplier performance, cycle times, inventory levels and turns etc. More and more aid agencies are now exploring ways in which this type of technology can be leveraged in their organizations.

Lack of Institutional Learning. The organizational culture, lack of technology, insufficient funds for infrastructure and training and high employee turnover create an environment in which there is a lack of institutional learning. Once a crisis is dealt with, humanitarian heroes are immediately assigned to the next mission, rather than taking the time needed to reflect and improve. Or they leave. Input from the organizations we interviewed suggested that turnover of field logistics personnel was as high as 80% annually. Thus, while logisticians have a remarkable track record for getting the job done under the most adverse and extreme circumstances, the lessons learned from one disaster to the next are often lost. The experience of the occasional veteran logistician is largely tacit and difficult to communicate to the next generation, nor is it transferred from one field context to another.

This situation compounds the human resource challenges that face organizations. Often crises are in parts of the world where the organization may not have operated before, and finding experienced people in the location of the disaster is difficult. Thus, organizations rely on core staff or international rosters, as the few experienced people are in high demand. In every relief situation scores of innovative solutions and ways of working are quickly improvised and devised. However, since the staff move on, these are rarely recorded or preserved for the next event.

Limited Collaboration. With the emerging competition for funding among major relief organizations, the heads of logistics tend to each fight their own battles with little collaboration. Although many of them face the same challenges and know each other, or of each other, they do not often meet or talk to one another except in an actual disaster. While they actively collaborate and help each other during a disaster, few inter-organizational links exist for the collaboration to continue after an operation, in preparation for the next event. For example, we found that several of them were thinking of deploying a regional warehouse structure for faster response. Coincidentally, three were actually talking with warehouse providers in the same city. Similarly, two others had commissioned expensive analyses to select a fleet management system and three were wrestling with the idea of a training program for field logisticians. None knew that their counterparts had the same objectives and, therefore, there was little collaboration or resource sharing.

4.4.   The Path Forward

Today's underdeveloped state of logistics in the humanitarian sector is much like corporate logistics was 20 years ago. At that time, corporate logistics suffered from underinvestment, a lack of recognition, and the absence of a fulfilling, professional career path for people performing the logistics function. Over the last 20 years, corporate logistics has found its voice with top management. Under the rubric of supply chain management, it has established itself as a core discipline whose best practices are taught and researched at top business schools and promulgated by leading consulting firms.

In our conversations and convenings we ask logisticians from global, national and regional aid agencies about their aspirations for themselves and their function. It is not surprising that their most significant priority is a knowledge based field with a clear career track, collaboration with peers across organizations and the ability to demonstrate the value of logistics with unambiguous measures and metrics that tie with organizational strategy. The way for logistics to strengthen its power and be recognized is by showing results and systemic improvement, by clearly demonstrating over time how it is contributing to the effectiveness of the overall organization and facilitating its response to external pressures.

4.4.1   The Five Strategies

This section details five strategies we recommend for moving forward to improve humanitarian logistics. Figure 4.2 and Table 4.1 show the relationship

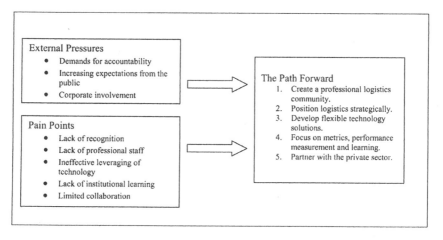

Figure 4.2.   Development of a Path Forward.

Table 4.1. How the Five Strategies Address the Pain Points.

	Lack of Recognition	Lack of Professional Staff	Ineffective Leveraging of Technology	Lack of Institutional Learning	Limited Collaboration
Create a professional logistics community.	✓	✓	✓	✓	✓
Focus on metrics, performance measurement and learning.	✓			✓	
Position logistics strategically.	✓		✓		
Develop flexible technology solutions.			✓	✓	
Partner with the private sector.	✓	✓	✓		✓

between the pain points and the strategies, and how each strategy addresses particular pain points. The five strategies each contribute as follows:

- Leveraging the **community of practice** will enable humanitarian logisticians to exploit shared knowledge and experience on common issues and to create a consistent, powerful voice with all members of the ecosystem. It will also create an environment where collaborative solutions can be discussed, if not implemented.
- Focusing on **metrics, performance measurement and learning** capabilities will empower logisticians to continuously improve the effectiveness of relief operations. It will also help them communicate the strategic value of logistics to senior managers and donors.
- **Positioning logistics strategically** will enable logisticians to create awareness of the contribution that logistics makes and to obtain needed funding and resources.
- Developing **flexible technology solutions** will improve responsiveness by creating visibility of the materials pipeline and increasing the effectiveness of people and processes. Furthermore, advanced information systems will create the infrastructure for knowledge management, performance measurement and learning.
- **Partnering with the private sector** will enable aid agencies to draw on resources and knowledge from a sector with a different but related set

of experiences. It will also allow aid agencies to leverage specific best practices that are already well developed in the private sector and tailor them to the humanitarian context.

The remainder of this section discusses each of the five strategies in more detail.

4.4.2   Creating a Professional Logistics Community

Corporate supply chains have evolved rapidly in recent years, largely due to the fact that all major universities around the world now train people in supply chain management. Professors and academics devote time to this subject and professional associations abound, making continuous improvement in skills available to practicing supply chain managers.

Creating a professional logistics community among aid agencies and encouraging greater cross-fertilization between humanitarian logistics and corporate logistics can encourage the sharing of experience and the transfer of knowledge. Over the last three years, several cases in supply chain management have been written at leading business schools, and are being used in classrooms to expose supply chain managers in training to the field of humanitarian logistics (see Appendix 1 for a list of cases).

In the aid community, formalizing knowledge by creating standardized training programs, a professional association and events and forums with peers from the private sector can all be helpful in addressing the lack of accumulated and institutionalized knowledge about humanitarian logistics. Common training and the use of tools like e-learning would open up the possibility of creating a cadre of logisticians in the field and promote standardized logistics practices and in-country logistics capacity.

4.4.3   Metrics, Performance Measurement and Learning

In general, aid agencies have focused on "getting the job done," and have put little effort into performance measurement, other than reporting to donors on the amount of relief and usage of funds for a given relief operation. Furthermore, learning has been approached on an ad-hoc basis, through post-operation evaluations, rather than through a structured process of continuous improvement, as has been used successfully in the corporate sector since the 1980's.

The potential for improving logistics effectiveness by categorizing and strategizing about disasters and responses can be understood by considering the following example. Hurricanes are cyclical disasters in some parts of the world. Roughly speaking, 14 storms can be expected to hit somewhere in 10 vulnerable Central American and Caribbean countries between June and November each year.[12] While the location, nature and extent of damage for a particular hurricane cannot be known in advance, by reviewing information from past operations and implementing regional strategy based on flexible, readily deployable mechanisms, logistics can be transformed from an activity that is almost exclusively reactive to one based on preparedness and experience-based action. The IFRC recently established a regional approach for hurricanes. Based on analysis of past events, the IFRC defined standardized relief item descriptions, guidelines for acceptance of in-kind donations, and frame agreements with international and local suppliers for key relief items. The frame agreements were blanket purchasing agreements that set terms and required suppliers to stock a certain level of inventory at their own premises. The IFRC also improved preparedness by implementing regional warehouses and kitting operations that could kit items to create general kitchen sets or water/sanitation sets on quick notice that would be appropriate to a range of expected situations.

Lessons from the corporate sector tell us that the key to transforming logistics from a peripheral function to a strategic one is demonstrating its importance to organizational outcomes. Thus, logisticians in aid agencies must learn to build a business case for their function by leveraging the power of measurement. Having solid data on the performance of a supply chain, and tracking improvements and cost savings is a powerful way to change donor and management perceptions of the criticality of logistics.

4.4.4   Positioning Logistics Strategically

Having created an organizational environment in which performance is measured, knowledge is built and kept, systemic improvement occurs rapidly, and communities share and leverage their capabilities, logistics managers must take their story to all stakeholders – donors, their own organizations, corporate partners and the media – regularly and with pride. This will allow the logistics organization to showcase its contribution and garner continued support and resources to further develop and improve.

The logistics organization must do three things well to demonstrate its strategic value to the organization. First, it must develop and execute plans to steadily improve the efficiency and effectiveness of operations. Showing this improvement through key performance measures will let upper management

---

[12] IFRC: Choreographer of Disaster Management, INSEAD teaching case, Fontainebleau, France, 2002.

and donors know that logistics is creating value through better use of resources and through improved service to beneficiaries.

Second, logistics must develop, execute and publicize long-term initiatives that create systemic change that supports the evolution of the organization's mission. Examples of strategic long-term initiatives are:

- Decentralization of supply chain tasks and decision-making from headquarters to the regions
- Development of track and trace capability from source to beneficiary
- Development of Internet-based real-time reporting systems accessible to donors and partners
- Disaster preparedness efforts that leverage learning across operations
- Identification and management of supply and pricing of key relief commodities
- Programs to improve relief cycle times

Third, logistics must position its initiatives within the context of what other AA's are doing. This will demonstrate to top management that logistics is taking advantage of all available ideas and technology and is setting the organization up well to compete against other AA's based on competitive advantage in logistics.

4.4.5   Developing Flexible Technology Solutions

Aid agencies have a common need for integrated information technology (IT) solutions that support procurement, distribution through a pipeline, tracking and tracing of goods and funds, flexible and robust reporting and connectivity in the field. AA's have large numbers of items that might be needed in short-order at the outset of a disaster. For example, The IFRC/ICRC catalog of relief items includes 6,000 items from cranial drills for surgery at disaster sites to plastic sheeting for shelter. The UNICEF warehouse in Copenhagen includes an inventory of $22 million in relief supplies at any given time; these supplies are procured from over 1,000 vendors worldwide. Procurement involves global sourcing, drop shipment, using commercial transportation, as well as third-party logistics firms, chartering aircraft or procuring local transportation such as mules and donkeys, tracking shipments and monitoring prices for commodities around the world.

Developing flexible technology solutions will improve responsiveness by creating visibility of the materials pipeline and increasing the effectiveness of people and processes. Furthermore, advanced information systems will create the infrastructure for knowledge management, performance measurement and learning. Information systems can be used to provide online catalogs, for correspondence and communication with suppliers and other partners, for performance measurement based on transaction history, and as a basis for collaboration.

The Humanitarian Logistics Software (HLS) developed by Fritz Institute in partnership with the IFRC automated its supply chain creating automated standardized processes for mobilization of supplies, procurement and tracking. Using best practices from the private sector and adapted to the humanitarian sector, the software has significantly improved the speed, accuracy and capacity of the IFRC supply chain. These improvements were particularly evident during the Tsunami when the software was used to mobilize, procure and track supplies for the massive 7 country relief operation that was coordinated by the IFRC.

4.4.6 Partnering with the Private Sector

Recently there has been increased interest in partnering with the private sector on humanitarian problems. These partnerships seek to leverage private sector resources and knowledge to create systemic change within the humanitarian world.[13] The benefits companies receive from these partnerships include opportunities for employees and management to focus their skills on humanitarian problems and the goodwill generated with customers by doing good in the world. They also benefit from the exchange of ideas with HRO logistics managers.

Aid agencies are now benefiting from increased interest in disaster relief from the corporate sector. Partnerships between aid agencies and the corporate sector are allowing the importation of tried and true methods and technologies to be adapted and applied to disaster logistics. For example, the relationship between the global logistics provider, TNT, and the UN's World Food Programme, is creating an awareness of the benefits of supply chain technology not only at WFP, but throughout the entire humanitarian sector.[14]

Partnering has its risks, however. One ongoing issue is the donation of unneeded goods, or "push-based giving." AA's need to create and enforce rules, processes and incentives that drive companies toward need-based giving and eliminate the donation of inappropriate and unsolicited items. Another issue is brand identity. The AA must be wary that its brand name is not co-opted or weakened through association with or use by the private sector. In many

---

[13] Corporations and Disaster Relief: How to Partner with Heroes, Working paper, Anisya Thomas and Lynn Fritz, 2005.

[14] Moving the World: The TPG-WFP Partnership by Rolando Tomasinsi and Luk Wassenhove; The TPG-WFP Partnership by Ramina Samii and Luk Wassenhove.

cases, there is greater public goodwill towards the AA brand than towards the company brand.

4.5. Conclusion: Learning as the Basis for Strategic Contribution

The large-scale relief operations after the Tsunami highlighted the need for effective logistics. In India, Indonesia, Sri Lanka, Thailand and all the other developing countries affected by the Tsunami, the need to get relief supplies from within the country and from the rest of the world was urgent. Collapsed infrastructure and the chaotic environments complicated the challenge for relief logistics. Nonetheless, the relief operations continued and the innovation and resilience of humanitarian logisticians accomplished the task.

The challenges experienced in the delivery of supplies during the Tsunami prompted a systematic evaluation of the challenges faced by people who manage the supply chains for relief. This paper outlines five ways in which those challenges can be addressed. The challenges and solutions are equally applicable at the global or local level. While moving relief items to disaster sites will continue to be an important role for logistics, providing timely information, analyzing that information to garner insight as to how to improve operations, and learning internally and with others must be the strategic focus. Establishing a community that shares and invests jointly in advancing the field can leverage each logistician's efforts many-fold. It is through these two mechanisms of information and community that humanitarian logistics can find its voice and create its future, rather than limit itself to responding to the present.

4.6. Appendix: Resources and Direction for the Academic Community

The role of academics in humanitarian logistics is to help raise awareness, to develop pedagogical materials, to build the science of humanitarian logistics and to support the logistician.

Table 4.2 provides information on resources and research direction for academics interested in working in this field. Teaching cases and articles about humanitarian logistics are listed in the References section.

Table 4.2. Resources and Research Direction for Academics.

WEB SITES	ACADEMIC RESEARCHERS
www.fritzinstitute.org	Jo van Nunen, Erasmus University
www.fmreview.org	Luk van Wassenhove, INSEAD
	Beatriz Ayala-Öström, Cranfield University
	Rolando Tomasini, INSEAD
	Ramina Samii, INSEAD

JOURNALS
Forced Migration Review 18, *Rethinking Humanitarian Logistics* (September 2003)

RESEARCH TOPIC INPUT FROM AID AGENCIES
• Performance measures for humanitarian supply chain efficiencies
• Cost benefit analysis of information technology systems
• Study of determinants of success or failure in systems implementation
• The impact of transport on development programs
• The impact of donor directives on fund allocation to the creation of new programs within aid agencies.
• Impact of logistics on ability to meet disaster relief program goals

References

Managing Information in Humanitarian Crises: The UNJLC Website, Case Study, Rolando M. Tomasini & Luk Van Wassenhove (April 2005).

Can Heroes be Efficient?, Case Study, Laura R. Kopczak & M. Eric Johnson (October 2004).

The TPG-WFP Partnership II – Learning How To Dance, Case Study, Ramina Samii & Luk Van Wassenhove (April 2004).

Genetically Modified Food Donations and the Cost of Neutrality: Logistics Response to the 2002 Food Crisis in Southern Africa, Case Study, Rolando M. Tomasini & Luk Van Wassenhove (March 2004).

Moving the World: The TPG-WFP Partnership I – Looking for a Partner, Case Study, Rolando M. Tomasini & Luk Van Wassenhove (February 2004).

Coordinating Disaster Logistics after El Salvador's Earthquakes, Case Study, Rolando M. Tomasini & Luk Van Wassenhove *(October 2003).*

Logistics: Moving the Seeds of a Brighter Future (UNJLC's Second Year in Afghanistan), Case Study, Ramina Samii & Luk Van Wassenhove (September 2003).

The United Nations Joint Logistical Center: The Afghanistan Crisis, Case Study, Ramina Samii & Luk Van Wassenhove (May 2003).

The United Nations Joint Logistical Center: The Genesis of a Humanitarian Relief Coordination Platform, Case Study, Ramina Samii & Luk Van Wassenhove (April 2003).

Choreographer of Disaster Management: Preparing for Tomorrow's Disaster (Hurricane Mitch), Case Study, Ramina Samii & Luk Van Wassenhove (2002).

Choreographer of Disaster Management: The Gujarat Earthquake, Case Study, Ramina Samii & Luk Van Wassenhove (2002).

Forced Migration Review 18, *Why Logistics?,* Anisya Thomas (September 2003).

Forced Migration Review 18, *The Academic Side of Commercial Logistics and the Importance of this Special Issue,* Ricardo Ernst (September 2003).

Forced Migration Review 18, *Humanitarian Logistics: Context and Challenges,* Lars Gustavsson (September 2003).

Forced Migration Review 18, *A Logistician's Plea,* John Rickard (September 2003).

Forced Migration Review 18, *Towards Improved Logistics: Challenges and Questions for Logisticians and Managers,* Donald Chaikin (September 2003).

Forced Migration Review 18, *UN Joint Logistics Centre: A Coordinated Response to Common Humanitarian Logistics Concerns,* David B Kaatrud, Ramina Samii and Luk N Van Wassenhove (September 2003).

Forced Migration Review 18, *The Central Role of Supply Chain Management at IFRC,* Bernard Chomilier, Ramina Samii and Luk N Van Wassenhove (September 2003).

Forced Migration Review 18, *The World Food Programme: Augmenting Logistics,* Peter Scott Bowden (September 2003).

Forced Migration Review 18, *Logistics Under Pressure: UNICEF's Back to School Programme in Afghanistan, Paul* Molinaro and Sandie Blanchet (September 2003).

Forced Migration Review 18, *Coordination in the Great Lakes,* George Fenton (September 2003).

Forced Migration Review 18, *Lean Logistics: Delivering Food to Northern Ugandan IDPs,* Margaret Vikki and Erling Bratheim

Forced Migration Review 18, *Food Aid Logistics and the Southern Africa Emergency,* Jon Bennett, (September 2003).

Forced Migration Review 18, *The Humanitarian Use of the Military,* Rupert Wieloch (September 2003).

Forced Migration Review 18, *Marrying Logistics and Technology for Effective Relief*, H Wally Lee and Marc Zbinden (September 2003).

Forced Migration Review 18, *Humanitarian Mapping*, Rupert Douglas-Bate (September 2003).

Forced Migration Review 18, *Complex Emergency- Complex Finance*, Guy Hovey and Diana Landsman (September 2003).

Forced Migration Review 22, *Logistics Training: Necessity or Luxury*, Anisya Thomas and Mitsuko Mizushima (January 2005).

Asia-Pacific Development Review, Humanitarian Logistics: Matching Recognition with Responsibility, Anisya Thomas (June 2005).

International Aid and Trade Review, Elevating Humanitarian Logistics, Anisya Thomas (January 2004).

Forced Migration Review 21, *Leveraging Private Expertise for Humanitarian Supply Chains*, Anisya Thomas, (September 2004).

Hau L. Lee and Chung-Yee Lee (Eds.)
Building Supply Chain Excellence
in Emerging Economies
©2007 Springer Science + Business Media, LLC

Chapter 5

DUAL SOURCING STRATEGIES
Operational Hedging and Outsourcing to Reducing Risk
in Low-Cost Countries

M. Eric Johnson[1]
Center for Digital Strategies
Tuck School of Business
Dartmouth College, USA

Abstract: Sourcing strategies that employ operational hedging can reduce the risk of operating in low-cost countries. This article examines the sourcing strategy of toy-
maker Mattel. Like the high technology industry, toys suffer from many supply
chain ailments including short product life, rapid product turnover, and seasonal
demand. Coupled with long supply lines and potential political and economic
turmoil in Asia, toymakers face an unusually complex set of risks. Managers
in many businesses can learn valuable lessons in managing uncertainty from
toymakers. Set during the Asian financial crisis, the case describes a facility
location decision for Hot Wheels and Matchbox cars. Besides the international
location decision, the case illustrates: 1) How toy makers manage demand uncertainty; 2) Mattel's outsoucing strategy in Asia; 3) How Mattel integrates its
marketing and supply chain strategy.

5.1.   Introduction

Sourcing in South East Asia offers the possibility of radical cost reductions
for many products. However, exploiting the promise of low-cost sourcing requires rethinking your supply chain strategy. With the benefits are risks and
hidden costs that some firms only discovered after making significant investments. Firms in any industry would be wise to learn from organizations with

---

[1] This article was written with research assistance from Tom Clock.

deep history and experience in the region. When I think of an industry with such experience, I think of toys. Toys are one of the world's oldest consumer products. Over the past five decades the toy industry has steadily matured from a cottage industry into a global market of over $50 billion. Yet investors know that the industry is far from tranquil. Key features that have long characterized the toy business are its rapid change and uncertainty. Demand for fad-driven products can balloon overnight and then suddenly pop as the next hot product sweeps the market. Constant product innovation, short life cycles, and high cannibalization rates are typical. Supply chains that span the globe and include many emerging countries add currency and political risk that can disrupt supply and change cost structures with little notice.

Take a tour of any industrial park in China, Malaysia, Indonesia, and Thailand and you will find factories building Hot Wheels cars next door to ones producing flash drives, printers next to Barbie dolls, Furbys next to cell phones – all experiencing the benefits and risks of operating in low-wage countries. How should firms manage these risks? In this article, we examine a case study of Mattel and its decision process to add production capacity to a network of both outsourced and Mattel-operated facilities. Set during the Asian financial crisis, the case illustrates: 1) How toy makers manage demand and supply uncertainty; 2) Mattel's outsourcing strategy in Asia; 3) How Mattel integrates its marketing and supply chain strategy.

5.2.   Company Background

Based in California, Mattel, Inc designs, manufactures, and markets a broad variety of toy products. The company's product lines include Barbie fashion dolls, Hot Wheels die-cast toy vehicles, and Fisher-Price preschool toys. Mattel produces all of these toys overseas, primarily in Southeast Asia, with many wholly owned manufacturing facilities in these locations including China, Malaysia, Indonesia, Mexico, and Italy.

Mattel was founded in 1944 by Elliot and Ruth Handler. By 1955, annual sales reached $5 million and the Handlers decided to take a gamble that would forever change the toy business. In what seemed at the time a risky investment, the Handlers signed a 52 week contract with ABC Television to sponsor a 15-minute segment of Walt Disney's Mickey Mouse Club at a cost of $500,000 – a sum equal to Mattel's net worth at the time. Up until this move, most toy manufacturers relied on retailers to promote their products. Prior advertising occurred only around the holiday season. The popular daily kids show made the Mattel brand well known among the viewing audience, translating quickly into sales. The success of the Handlers' pact with kids TV started a marketing revolution in the toy industry.

Mattel made toy industry history again in 1959 with the introduction of Barbie. With the success of Barbie, Mattel made its first public stock offering and, by 1963, was listed on the New York Stock Exchange. In the next two years Mattel's sales skyrocketed from $26 to $100 million. The introduction of Hot Wheels miniature model cars in 1968 was another spectacular success making Mattel the world's largest toy company by the end of the decade. In 1987, CEO John Amerman charted a new strategy for Mattel, closing many of the company's US manufacturing capacity, focusing the company on its core brands such as Barbie and Hot Wheels, and by making selective investments in the development of new toys – particularly within core products like Barbie. The Barbie make-over was so effective that from 1987 to 1992 sales shot up from $430 million to nearly $1 billion, accounting for more than half of the company's $1.85 billion in sales. At that time, Mattel estimated that 95% of all girls in the United States aged 3 to 11 owned Barbie dolls. Finally, in deals lauded by Wall Street analysts, Mattel acquired Fisher-Price in 1993 and Tyco in 1997, boosting Mattel's revenue to $4.8 billion.

Over the years, the ability to create new products and quickly meet demand remained nonnegotiable requirements for success in the toy industry. Manufacturers had to live with the reality that inventory in times of hot sales could reap large rewards, but often became worthless overnight. Mattel introduced hundreds of new toy products. Many of the new toys reflected increased demand among core product lines – for example, the market's renewed interest in collectible Barbie and Hot Wheels products. Beyond core products, there remained a large, lucrative segment of non-core toys whose market life was typically less than one year. Many of these products were related to popular movie characters. More and more, filmmakers and toy manufacturers combined their efforts to market their products to the public. These were high turnover products where time to market was critical. Mattel typically produced core product lines in-house and outsourced the production of non-core lines to a network of vendors. Outside vendors gave Mattel the needed flexibility to handle hot products and the seasonal changes in toy sales. In the US, toy sales historically followed strong seasonal trends with nearly half of all sales coming in November and December.

Ron Montalto, who had lived and worked in Hong Kong for fifteen years, was Senior Vice President responsible for company's Vendor Operations Asia division (VOA), which managed Mattel's outsourced production. Mattel began the vendor program in 1988 hoping to add flexibility to the company's traditional in-house manufacturing. Montalto spent ten years developing VOA into one of Mattel's most valuable strategic assets. By 1997, it was responsible for manufacturing products that generated nearly 25% of the toy company's total revenue.

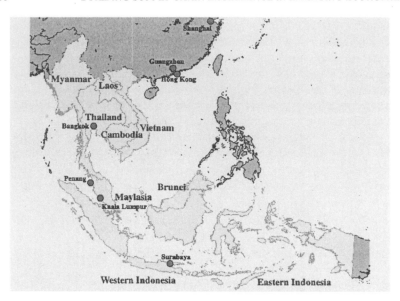

Exhibit 5.1. Current and Potential Die-Cast Plant Locations.

The Tyco merger resulted in VOA manufacturing products that generated an additional $350 million in revenues for the Mattel organization. The majority of those revenues came from a combination of Tyco's Matchbox die-cast cars, its line of radio-controlled (RC) cars, its View Master® series, and products from its Sesame Street license. As part of reorganization after the merger, Montalto picked up the responsibility of all die-cast operations. With demand for Matchbox cars at 64 million units in 1997 and growing, die-cast capacity was a concern. Tyco manufactured the cars through joint-venture arrangements in Shanghai and Bangkok. Both of the joint ventures were minority share partnerships which raised questions for Mattel in the future. What's more, the quality of Matchbox products had been eroding for years and was at an all-time low. The production equipment and steel molds used in the manufacturing plants were becoming obsolete. Though it might be possible to upgrade the existing Tyco operation in Bangkok, Montalto saw little hope of expanding the Shanghai operation.

Mattel owned a state-of-the-art die-cast facility that was operating at full capacity in Penang, Malaysia (see Exhibit 5.1). Expanding that facility significantly beyond its volume of 120M cars would be expensive and complicated. There was no room for further building on the site and no available land adjacent to the plant. After performing a significant analysis over the summer of 1997, Montalto championed a proposal to solve the capacity problem by building a new China facility. However, before the plant was approved, a financial storm began sweeping across Asia. Throughout the fall and winter, the plant decision was debated. Some executives inside Mattel argued that they should reconsider building a new plant in Malaysia to concentrate die-cast production in a single country. Others felt that they should consider Indonesia as a way to take advantage of low labor costs and very attractive exchange rates. Mattel already operated a plant in Indonesia that produced Barbie® dolls. Montalto had to decide whether Mattel should go forward with the new China plant, build a plant in Malaysia or Indonesia, expand one of the existing facilities, or outsource the surplus die-cast volume through VOA.

5.3. Miniature Car Market

Die-cast 1:64 scale miniature cars have been a long-standing favorite among children and adults. Matchbox cars were introduced by a small company founded in 1947 by two unrelated school friends, Leslie Smith and Rodney Smith. Few would have imagined that the company, Lesney Products, had created a term that would later become the generic name for any small toy replica of a car or truck. In 1982, the company met with financial difficulties and the Matchbox brand was sold to a Hong Kong based holding company, Universal International which later became a subsidiary of Tyco Toys.

Mattel introduced Hot Wheels in 1968 and quickly became the market leader, often gaining market share while other companies lost market share – or worse – went bankrupt. By 1997 there were few major competitors in the 1:64 category other than Racing Champions® and Hasbro's Winner's Circle® which both focused primarily on replicas of racing cars including NASCAR. In Europe, both MIRA and Bburago competed with wider size offerings, producing cars at 1:43, 1:25, and 1:18 scale. Larger cars were often purchased by collectors and there were also several other small Japanese and English companies that marketed these high-end replicas.

While both 1:64 scale miniature car replicas, Hot Wheels and Matchbox competed in very different market segments (see Exhibit 5.2). Matchbox cars emphasized realism in both scale and detail. For years they had been manufactured entirely of metal, making them heavier and more durable. These elements made the car more appealing to younger children, typically 2–4 years old. Moreover, much of the Matchbox sales were outside of the US while Hot Wheels were an American phenomena. Hot Wheels cars featured more fantasy designs both in form and decoration. With a larger creative element, they appealed to older children who participated in more imaginative play patterns.

Prior to 1994, sales of die-cast cars, including Hot Wheels, were relatively flat. However, over the course of the next three years, demand for the Hot

Exhibit 5.2. Hot Wheels and Matchbox Products.

Wheels skyrocketed to 155 million units in 1997, while Matchbox saw much slower growth. Mattel attributed much of the growth to a new rolling mix marketing strategy. In the past, Mattel relied heavily on retailer's POS data to help forecast future demand and make replenishments throughout the supply chain. Starting in 1994, Mattel incorporated a new marketing strategy to sell diecast cars. Mattel determined that variety was the key driver of sales. If customers saw new products every time they went in the store, they were more likely to buy. The company implemented a rolling mix strategy by shipping retailers a 72-car assortment mix with SKU contents that changed 7–8% every two weeks. Stock keepers at various retail outlets shelved the individual Hot Wheels blister packs directly out of the 72-car master carton. Over the course of a year the product line changed over two times entirely. This strategy developed an organized, non-reactionary method of new product introduction and old product obsolescence. New products varied from brand new 'First Edition' cars, to redecorated models of cars already produced. By rolling the mix, Mattel was able to market a much broader range of SKUs without requiring any additional retail shelf space.

Mattel also found that it could educate the consumer and encourage buying patterns based on product introduction. Marketing began introducing 'Series Cars', a set of four cars sold individually and released every month. Each series would stay on the retailers' shelves for five months and then be permanently discontinued. The strategy created urgency among consumers to buy the products while they were available. Series cars also helped promote the existing collector market. In addition, Mattel played to the collector market by introducing 'Treasure Hunt' cars. These cars were only manufactured in lots of 20,000 and were extremely rare. One new Treasure Hunt car was made each month. They were randomly inserted into a retailer's assortment pack. These cars made it into the hands of a lucky few and were highly prized as collectible items. In 1996, a limited number of Treasure Hunt assortment packs (all 12 cars) retailed at FAO Schwartz for $150. A year later, the same assortment sold for over $1,000 between collectors.

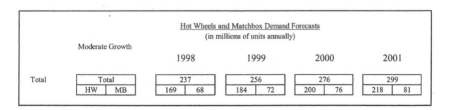

		Hot Wheels and Matchbox Demand Forecasts (in millions of units annually)			
	Moderate Growth	1998	1999	2000	2001
Total	Total	237	256	276	299
	HW | MB	169 | 68	184 | 72	200 | 76	218 | 81

Exhibit 5.3. Market Projections.

Through its rolling mix strategy Mattel no longer had to rely on POS data to forecast market demand for specific SKUs, but rather to plan the changes to the mix. Since Mattel guaranteed its retailers that the mix would sell, the retailers stocking problems were simplified to merely purchasing assortment packs and stocking the store shelves. Mattel believed it could incorporate the same strategy into the newly acquired Matchbox line and experience similar results (see Exhibit 5.3 for market forecasts of both Hot Wheels and Matchbox cars). No other manufacturer had the capability to offer consumers Mattel's level of variety.

5.4.　Die-Cast Manufacturing

The manufacture of die-cast cars involved well-defined production steps that could be performed either in-house or by third parties. Among die-cast manufacturers, there was a continuum in terms of the degree to which the processes and manufacturing steps were conducted in-house, as opposed to being subcontracted to other firms. While most firms had in-house die-casting, plastic injection molding, and basic painting and decorating processes, there was wide variation for other processes, including electroplating, vacuum metalizing, and package printing.

In the first step, a press injected molten zinc into a mold to create the body of the vehicle and/or the chassis (unless one or both of those parts were plastic). Mattel made most of its own die-casting molds at a facility in Malaysia, but also outsourced them to firms in Hong Kong. Presses could be outfitted with two different types of molds – conventional or unit die. Conventional molds usually had one car body cavity or two chassis molds. Unit dies were smaller than conventional molds traditionally used in the die-casting process and they offered quick changeover. Most importantly, two dies (or molds) could be fit into each machine. For every machine "shot"[2] two car bodies, four chassis, or some combination could be produced. Die-cast molds had a useful life of about 1.5 million shots, after which time the seams of the mold often began to leak creating excessive wasted zinc called "flash" and eroding the quality of the car.

The delivery of molten zinc could be machine specific (individual machines equipped with their own melting pots) or a more complicated central furnace and feeder system. The furnace and feeder system reduced energy costs associated with changing temperature settings on individual machine furnaces and

---

[2] Shots refer to each time molten zinc is pressed into a mold cavity, allowed to cool, and released into a waiting bin. Shot times for 35T die cast machines were 9-10 seconds each.

maintained the zinc at a more uniform temperature, thus improving the cast quality.

The bodies and chassis were then removed from the press by the operator. Bodies and chassis would be separated from the excess metal that flowed through the mold ducts into the cavities. This excess metal would be removed and recycled. The bodies and chassis would then be deflashed, deburred, and polished by vibrating the parts with smooth ceramic stones in a large bowl for 30 minutes. This process removed all the unwanted metal while smoothing sharp edges and seams.

The decoration of the car involved an electrostatic application of base and top coat to the car body via a painting system. A common system was supplied by Ransburg and could be used to paint any metallic surface.[3] Die-cast cars were attached by hand to a "tree" that hung from a conveyor line which carried the cars through the painting and drying processes. Each tree carried up to 72 cars. The trees themselves were spaced 16 inches apart and run at the conveyor speed of 7 feet per minute. On the other hand, chassis were electroplated to prevent corrosion and to maintain a shiny appearance.[4] The electroplating process involved dipping the metal chassis in a series of chemical baths to deposit a thin layer of shiny metal.

After applying the base color, additional decorations were applied to the car body and other parts using a "tampo" machine. Aside from the zinc weight of a die-cast vehicle, the major source of variance in the cost[5] of a car was the number of tampo operations the car under-went. Each "hit" by a tampo machine added one color to one surface of the car. Highly decorated cars with dozens of colors, like NASCAR replicas or highly detailed collectibles, tended to cost more than vehicles with fewer colors and decorations. The determination of how much decoration to apply to a product was purely a marketing decision.[6] Standard Hot Wheels and Matchbox cars typically sold for under $1.00 in US retail stores, while NASCAR and other collector edition cars were usually priced at $3.00 or more.

In addition to die-cast parts, most mini-vehicles included plastic injection-molded parts, notably the interior, the windows, the wheels and sometimes the

---

[3] Ransburg and other electrostatic painting systems are used in many industries including the automobile industry, to paint metal products.

[4] Many mini-vehicles, including many Hot Wheels cars, had plastic chassis in order to reduce zinc cost, and thus did not use electroplating.

[5] The number of moving parts, i.e., moving doors and hoods, can also affect cost significantly. Most of the basic vehicles produced by Mattel did not have moving parts.

[6] As a marketing ploy, Matchbox enclosed an unpainted, untrimmed "first shot" car in the same box with the corresponding, finished collectible to illustrate the "before and after" effect of decorating the car.

chassis. These parts were produced on conventional plastic injection molding machines that were commonly used to produce other small plastic toys as well as thousands of other products. As with die-cast machines, there were many types and sizes of plastic injection molding machines. Plastic injection molds typically had 2 cavities per mold and a useful life of about two million shots. [7] 70 ton injection mold machines would be required to produce plastic chassis, windshields, interiors, engines, etc. 110 ton machines were needed to produce the wheel components. Each car required one wheel mold and an average of 2.5 molds for other plastic parts[8]. Wheels were typically produced on a 32-cavity mold. Cycle time for the 70 and 110 ton injection mold machines was typically 16 and 20 seconds respectively.

Plastic parts were sometimes finished using vacuum metalizing (VUM) to impart a silvery metallic sheen to the parts. The plastic parts were first painted with a base coat of lacquer. Next, a thin film of metal was applied to the plastic parts by ionizing lengths of tungsten metal in a vacuum chamber. One system would typically satisfy all volume demand up to 100 million units of production and cost approximately $1.2 million. While some Hong Kong vendors had electroplating systems, most would choose not to purchase VUM systems, but rather outsource that process for the relatively few vehicles having VUM parts. After VUM, the plastic bodies would be given a top coat of clear lacquer to preserve the finish. If a colored metallic was desired, the clear coat could be dyed (for example red or gold).

After molding, wheels were decorated in a hot stamping process used to apply the metallic appearance to the hub cap area of the plastic wheels. The assembly of the wheels and axles, called the "barbell" assembly, was traditionally performed by hand. Because Mattel's Malaysia factory was located in a relatively high labor cost area, Mattel had developed machines to automatically insert the pins into the wheels to form the barbell assembly. This process was unique to Mattel.

The assembly of the various pieces of the vehicle into a final product was performed manually by unskilled labor. This operation often involved conveyor belt systems, or small 2–6 person manufacturing cells, where the main piece of equipment employed was a device that fastened the body and chassis of the car together (a process called "staking") after it was manually assembled.

---

[7] Most plant processes were planned to run one 8-hour shift per day, however, both the injection molding and die casting processes would run three 8-hour shifts. Production calculations for the three shift processes used a 22 hour day, or 7.3 hour shift, to account for downtime and breaks.

[8] This figure varies from car to car. The engineering standard for Hot Wheels averaged 2.5 molds per car.

Packaging the product, usually in blister packs, was often carried out at the manufacturing facility. Most vendors had heat sealant machines which sealed plastic blisters to pre-printed "blister cards," and used those devices to package a variety of other toys and products in addition to mini-vehicles. The printing of the blister cards or other packaging, and the vacuum forming of the blister was often outsourced, but could be performed in-house, depending on a vendor's preference.[9]

The process of manufacturing a mini-vehicle was labor intensive and involved machine production processes that were, for the most part, modular in nature. Operating in low labor cost countries like China or Malaysia, labor cost typically represented 10–20% of the product cost. With the possible exception of the Ransburg painting system (and the more rarely used electroplating and VUM systems) most segments of the production process could be expanded incrementally as needed, without creating significant excess capacity at any step in the process or requiring significant capital expenditures. In fact, whether a vehicle was all plastic or part die-cast metal and part plastic, the production process was generally not susceptible to large economies of scale – aside from the usual economies associated with spreading facility and plant management costs over a large number of products. Mattel's own experience as well as that of the vendors Mattel had engaged, demonstrated that multi-product production was sufficient to obtain much of the possible production economies. Aside from facility and management overhead costs, most of the mini-vehicle production process could be described as proportional to the incremental machinery that was added to the plant as production needs increased. Transportation costs from Asia to Los Angeles varied between $3,000-$4,000 for a shipping container that could hold up to 300,000 cars.

5.5. Outsourcing Strategy – Vendor Operations Asia

Vendor Operations Asia (VOA) was the outsourcing arm of Mattel, Inc. Montalto and his personal assistant started operations in 1988 with very little capital and a lot of faith. The vendor concept was initiated following an extensive competitive study by McKinsey and Company. The study recommended that Mattel differentiate between core and non-core products, manufacturing its core products in-house and outsourcing all non-core products. Mattel originally decided that its Barbie and Hot Wheels products were core. In the following years, the company added selective Disney and Fisher-Price lines to the list. Non-core products tended to be promotional items, or toys with short

---

[9] A new vacuum forming machine cost approximately $105,000.

life cycles that were often introduced together with a children's television series (examples include The Mighty Ducks, and Street Sharks). Non-core toys experienced the fashion-like demand typical in the toy industry.

By 1997, VOA employed over 400 staff and generated sales revenues in excess of $1.4 billion. The group operated through a network of approximately 35 vendors that were contracted to manufacture Mattel products. Vendors were typically registered Hong Kong companies with manufacturing facilities and political expertise in mainland China. VOA selected vendors to produce new toys based on expected time to market, a vendor's manufacturing competence, unique process capabilities, and price.

VOA enabled Mattel to produce a large number of short life-cycle toys without the capital commitments required in wholly owned manufacturing. Moreover, it enabled Mattel to push certain risks onto its suppliers. These risks included demand variability and product diversity. Supplier metrics were based on the ability to produce high quality goods at a competitive price, and to deliver them to end-users on-time. Toy sales were directly related to the number of new product introductions and speed to market. In recent years, Mattel had introduced roughly 300 new, non-core toys each year.

The strength of VOA rested on its vendor relationships. Mattel was a marketing driven company that demanded high product quality and precise design conformance. Montalto's organization had been challenged for almost a decade to help individual vendors develop the internal capabilities necessary to satisfy Mattel's standards. It was an ongoing process that spanned multiple types of manufacturing, from the assembly of plush toys (like Winnie-the-Pooh) to the fabrication of technology goods such as children's tape recorders and cameras (sold under the Fisher-Price brand).

The new toy development process began at Mattel's corporate headquarters in California. Design teams created a *Bid Package* that contained the new product's blue print, engineering specifications and often a physical model. The *Bid Package* was sent to VOA for vendor quotation and selection. After a vendor had been selected *Tool Start/Debug* began. Each new toy required a set of tools for manufacture. The most common tools were hardened steel molds used in plastic injection and die casting. Shortly after *Tool Start* came *Tool Let*. This was a scheduling milestone and was considered day one of the production process. *Tool Let* was the point at which Mattel assumed liability for the tooling costs. Tooling costs varied considerably based on the complexity of the toy – tool sets for past toys ranged from $50,000 to $2,000,000. After the tools were completed the production process began. Step one or *First Shots* (FS) was typically a run of 50 units to determine what mold/process modifications were required. This was also the point at which a commitment date by the vendor was established. Step two, or *Engineering Pilot* (EP), was for touch-up. There could be a second or third EP if necessary, depending on the toy's complexity. Step three was the *Final Engineering Pilot* (FEP) that established complete test durability. Step four was *Production Pilot* (PP); typically 1,000 units were run at this stage and the manufacturer used the entire assembly line to run the product. When the new toy met design compliance, step five, *Production Start* (PS) began.

5.6. Production Options

5.6.1 Guangzhou

By the summer of 1997, Mattel was close to a decision to build a new plant in Southern China to handle the increased demand for Hot Wheels and to consolidate Matchbox production. Labor in the Guangzhou region was cheap and plentiful. Including benefits such as dormitories and educational programs, the fully loaded rate was less than $0.50/hour (see Exhibit 5.4). To avoid mainland China's 21% import duty on capital equipment, Mattel planned to locate the facility in one of the special Industrial Zones. The most promising site under consideration was located in the Guangzhou Baiyun Industrial Zone. The Baiyun zone was in Luogang township, east of Guangzhou. It was 12 miles from Baiyun International Airport and 3 miles from Huangpu New Harbor. A medium-sized cargo railway station was located in the zone.

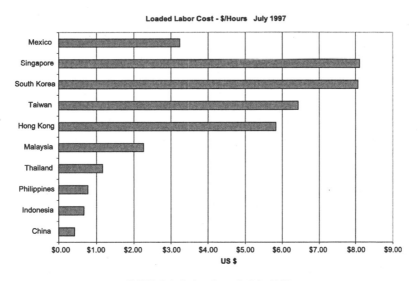

Exhibit 5.4. Labor Rates in July 1997.

Based on estimates from MMSB, the single story facility required about 325,000 square feet to accommodate 100 million units of production per year. Contractor quotes for building the factory shell were $10 per square foot. Bringing the shell to usability in terms of water pipes, telephone lines, electrical wiring, etc. was conservatively estimated at 50% of the shell's cost. Mattel would also be responsible for building dormitories to house the factory workers. Dormitories would each have six floors (maximum height without elevators) and approximately 2500 square feet per floor. Based on its other manufacturing sites in South East Asia, Mattel was committed to providing a minimum of 40 square feet of living space per direct labor employee. Staff labor would require a minimum of 100 square feet per employee.

The idea of building the China plant had been analyzed for nearly a year. By July, Montalto's team had developed a capital expenditure request that was circulating at the corporate headquarters in California. The plan included three options for the initial size of the plant (50, 100, 150M cars). It appeared that one of the options would certainly be approved and that construction would commence in the beginning of 1998, with first production in 1999. Then overnight the environment changed. Starting with South Korea and spreading quickly throughout the region, plunging currencies and stockmarkets turned the fast growing Asian economies on their ears. It happened so quickly that companies like Mattel were caught by surprise. Reflecting on the rapid changes, the *Economist* lamented,

> "If anybody had predicted a year ago that Indonesia, South Korea and Thailand would have to go cap in hand to the IMF, they would have been thought mad. This was, after all, the East Asia whose economic policies the international financial community was forever applauding: a world away from Latin America or Africa, where trouble was always on the cards."[10]

By January, many of the East Asian currencies had been sharply devalued (see Exhibit 5.5). Yet China, whose currency was not fully convertible and thus fixed by the central government, held steadfast. Thus, in relationship to other countries in the region, China no longer looked as inexpensive and the plant decision was back out on the table at Mattel.

5.6.2   Indonesia

With the rapid devaluation of Indonesia's currency, some inside Mattel felt it should be considered again as a possible site for a new plant. Indonesia had very low labor rates and was thus suitable for high labor products. Because of this, Mattel had already built a doll factory in Jakarta in 1996. The reduction in

---

[10] "Frozen Miracle," Economist, March 7 1998.

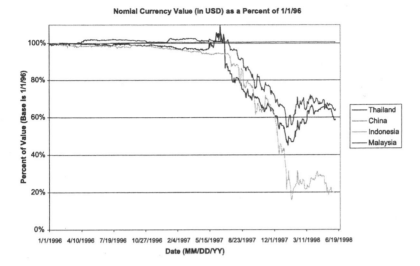

Exhibit 5.5. Exchange Rates.

currency value had made the labor even cheaper. However, labor productivity was low and managers at Mattel felt it was unlikely that productivity levels could ever be improved to Malaysian levels. Earlier investigations had identified Surabaya as a possible plant site where the costs of building a plant were similar to those in China. In addition to standard return on investment criteria, Mattel was also trying to diversify risk. There was inherent volatility in dealing with third world countries, due to both internal changes in regulations and external pressures. Adding Indonesia gave Mattel a diversification advantage its competitors didn't have, while at the same time allowing the company sufficient economic leverage to maintain some influence with local governments. In principle, these same advantages would apply to a new die-cast facility. In addition, Mattel's experience in running an operation in Indonesia would be a significant advantage when starting up a new facility. However, Indonesia's government was under intense public reproach and it was not clear if the long-time president could survive the crisis.

5.6.3 Penang

Located in Penang, Mattel Malaysia Sdn Bdh (MMSB) was the only Mattel facility that manufactured Hot Wheels vehicles. Mattel acquired the plant from GEC (of the UK) in September 1980. At the time of its acquisition, the plant was an 80,000 square foot facility used to manufacture TV sets. Mattel began production at MMSB in January 1981. Total start-up costs amounted to approximately $5 million (in 1980 dollars), and production volume at MMSB for the first two to three years averaged 30 to 35 million mini-vehicles per year. In 1984, Mattel added 180,000 square feet to the plant and began manufacturing male action figures. The plant was again expanded in 1994 by an additional 5000 square feet.

In 1996 the plant was dedicated to mini-vehicle production providing a significant capacity expansion. The 1996 expansion effectively used up the available space for die-cast car production at MMSB, resulting in Mattel's determination in June of 1996 to begin outsourcing incremental mini-vehicles requirements (11 million vehicles in 1996) from vendors in China. China vendors provided nearly 35 million vehicles to Mattel in 1997 and were expected to provide between 40 and 50 million vehicles in 1998. Throughout 1997 Arun Kochar, VP and plant director, worked to increase MMSB capacity by improving the production process. By the end of the year, MMSB was producing over 10 million units per month, based on two shifts per day, six days a week. Kochar felt that another 10–20% improvement might be possible in 1998, but doubted further sustainable increases could be achieved.

Labor at the Malaysian plant was very productive with high quality output. As compared with other poorer countries in East Asia, labor in Penang was more skilled and expensive. The higher skill translated into a high quality product and allowed Mattel the flexibility needed to support the rolling product mix that changed weekly. Unfortunately, the labor market was getting tight. To keep a steady flow of labor, Kochar had to regularly recruit workers from the small towns in the countryside. Workers were predominantly young women, many of whom stayed in Mattel furnished housing. Mattel was very sensitive to labor conditions and often over compensated both in age requirements and working conditions. For example, the plant had recently installed air conditioning to increase worker comfort, yet very few workers had air conditioning in their own homes.

5.6.4   Kuala Lumpur

Another possible site for a new plant was in Kuala Lumpur (KL), Malaysia. Mattel already had a doll factory in KL and the existing die-cast plant in Penang. Adding another die-cast facility in KL would offer the company single country manufacturing and greater managerial control. Economies of scale would come in the form of internal tool production and inter-plant exchange, management staff, material input costs, and distribution. In addition, the labor population in Malaysia was, on average, more productive than anywhere else in Southeast Asia. There were two downsides to making KL a future plant site – labor availability problems and higher labor costs.

5.6.5 Bangkok

Under Tyco, the manufacturing of Matchbox toys was divided between two factories, one in Bangkok and one in Shanghai (Shanghai Universal Toy Company or SUTC). Excess demand beyond the capacity of these two plants was outsourced to a pool of south China vendors. Over recent years, Tyco management led by Rug Burad (VP of Tyco Manufacturing) had been gradually phasing out much of the Bangkok plant's production due to management costs and poor quality. Many of the conventional molds used to produce Matchbox cars had been moved to Shanghai. When Mattel took over the partnership position in Bangkok, the factory was producing only 21 million units in a building that could accommodate equipment and workers for production of 50 million units. The Matchbox plant was brought under the management of Kochar. Much of the remaining equipment was old and the presses were equipped to handle only conventional molds. Retrofitting the machines to accept unit dies would be expensive. Since Hot Wheels were made almost exclusively with unit dies, the plant could not effectively take on Hot Wheels volume without further investment. Labor costs in Thailand were half of Malaysia but labor productivity was significantly lower.

5.6.6 Shanghai

SUTC carried the bulk of Tyco's die-cast car production, producing 33 million Matchbox units in 1997 with about 1000 workers. The die-cast presses were operating at full capacity and further expansion would require significant equipment investment. The plant not only offered Mattel a production facility but also a domestic distribution license. This non-transferable license enabled Mattel to sell die-cast cars in China as long as it continued operating SUTC at its original location. In 1997, total vehicle sales in China was about three million units. Since the cars were inexpensive and durable, many inside Mattel felt that the market could grow significantly as Chinese parents increased their toy purchases. Closing or relocating the plant would jeopardize the distribution agreement. Moreover, if Mattel closed the plant, it would be forced to pay the Chinese government $5000/employee in severance. Nevertheless, Montalto was concerned with SUTC's fit with Mattel's future manufacturing strategy. One of the main problems was the minority share partnership position Mattel inherited from Tyco. In addition, the quality standards at SUTC were far below any Hot Wheels producing facility. Strategically within China, Shanghai made a poor location choice for a toy manufacturer because of the city's emphasis on developing technology-based industries and its relatively high labor cost (over $1.00/hour). Labor productivity was about one half of that in Penang. As with Bangkok, the plant employed conventional molds, which would require retrofitting the machines to accept unit dies.

5.6.7   VOA

Ideally, Mattel could outsource die-cast production until its own facilities were established. However, the one area where VOA had not developed extensive vendor capabilities was in die casting. There were very few South China vendors in the die-cast business and fewer still that could produce high quality products. Die-casting was a cruel business that required large capital investments and offered meager returns. For a vendor to be able to produce Mattel quality cars, a large capital investment (between $10 and $30 million) was required. Montalto found it exceedingly difficult to persuade his vendors to take on this new business and the risk associated with it. One notable firm was Zindart – a Hong Kong company that had been recently listed on the NASDAQ exchange. Zindart produced a wide range of die-cast cars for many different toy firms as well other non-toy die-cast products. Nevertheless, Montalto worried that there just wasn't enough high-quality, die-cast capacity in the vendor base to meet the Matchbox demand.

5.6.8   Making a Decision

Montalto was confident that the Marketing Department's demand forecasts were accurate, especially under the moderate growth scenario. The increased demand for mini-vehicles was expected to come in significant part from Europe where Mattel was re-launching Hot Wheels products. Mattel desperately needed additional die-cast capacity and it was Montalto's job to recommend a way to find it. The fastest way to increase production would be to expand capacity in the existing Mattel facilities. Since Mattel produced Matchbox cars in Bangkok and Shanghai, either one of these factories could be expanded to accommodate more production. The other expansion option concerned VOA itself and the amount of core business Mattel wanted to outsource. A longer-term solution would be to build new capacity, but the question remained where? Malaysia, Indonesia and China were all viable alternatives for a new die-cast factory.

5.7.   Lessons from Mattel

With the currency crisis raging, Mattel decided to put its decision to build a new plant in Guangzhou on hold so that it could reanalyze the options and watch the Asian economies cope with the changes. While some executives felt

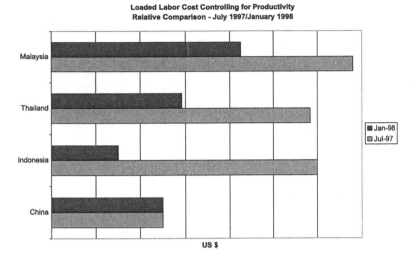

Exhibit 5.6. Impact of Currency Devaluation on Labor Cost.

that the crisis could have lasting impact, Mattel's economists argued that the economic forces of purchasing price parity would, over time, bring the real labor costs back towards pre-crisis levels. Indeed, after a few months, inflation within Indonesia began driving real labor costs back up. Moreover, by January the exchange rate depreciation bottomed out and many Asian currencies began to slowly rise against the US dollar. Productivity and quality also had a significant impact on the decision. Even with the very low wage rate in Indonesia, factoring labor productivity into the analysis made the total cost difference between China and Indonesia much smaller (see Exhibit 5.6). As exchange rates began to stabilize in January, the total labor cost (controlling for productivity and quality) in both Malaysia and Thailand remained higher than China with Indonesia about 30% less expensive. However, Indonesia had suffered from sporadic political and social disruptions and the economic crisis was increasing the unrest. Additionally, many inside Mattel felt that the local inflationary forces would continue to narrow any cost advantage.

Montalto concluded that if China made sense in the first place, a presumed short-term shift in real labor costs should not invalidate the location strategy. The Guangzhou location was aligned with Mattel's overall strategy for die-cast cars, it supported Mattel's diversified portfolio of operations, and it remained a cost-competitive option even after the currency shift. So Mattel went ahead with the plant in Guangzhou, breaking ground in June 1998. The first production occurred during the summer 1999. The plant was designed to handle

Exhibit 5.7.   Capacity Management Lessons from Mattel.

Risk	Lesson	Example
Product Supply		
Short Product Life	• Manage product variety with rolling mix	• Building collector markets creates long-life brand and smoothes capacity requirement
Manufacturing Capacity	• Outsourcing strategy	• Outsourcing improves economies of scale and asset utilization
	• Combine off-setting seasonal products	• Snow sleds and swimming pools
Currency Fluctuations	• Financial hedging	• Contracts in stable currency, forward contracts
	• Diversify supply	• Several suppliers in different countries
	• Operational hedging	• Several plants in different countries
Supply Disruptions from Political Issues	• Diversify supply	• Several suppliers/plants in different countries
Control Over Core Products	• Dual sourcing with both internal and outsourced manufacturing provides control while providing risk management.	• Hot Wheels produced both within Mattel facilities and by outsourced partners

65M units with the possibility of adding another 65M. Matchbox production was centralized in the new plant and the rolling mix strategy was initiated in 2000. Bangkok and Shanghai were transitioned to Hot Wheels and other die-cast products (larger scale). In 1998, Penang was able to boost production to 12.5M cars/month covering most of the Hot Wheels demand. Hot Wheels cars that were outsourced were shipped to Penang to be assorted. The subsequent years showed that the decision to go to Guangzhou was a good one.

The Mattel case illustrates many important lessons for those seeking to leverage low-cost sourcing (Exhibit 5.7):[11]

- First, the case shows how toymakers couple their demand management initiatives with strategies to manage supply. For example, the rolling mix strategy was designed to both increase demand and build long-term brand excitement. As it was implemented by Mattel, it also created a smoother, less seasonal capacity requirement by building demand from year-around collectors. It also eased many of the forecasting and logistics challenges of replenishing multiple SKUs from a long-leadtime, Asian supplier base.

---

[11] Johnson, M. Eric (2001), "Learning From Toys: Lessons in Managing Supply Chain Risk from the Toy Industry," *California Management Review*, Vol. 43, No. 3, 106–124.

- To reduce investment risk stemming from short product lifecycles and high-demand variability, toymakers like Mattel use coordinated outsourcing strategies. For toy marketers, outsourcing enables both small and large toy companies to bring products to market without large investments in plant and equipment. By working with a pool of outsourced suppliers, who mitigate their risk by working with many different toy firms, both groups reduce their risks. Contract manufactures can also couple toy production with other counter-seasonal products to reduce swings in their capacity requirements.
- Mattel effectively hedges against political and currency risk by sourcing in many different countries. This operational hedging strategy not only mitigates the risk of currency moves and political upheavals, but also provides toymakers with the opportunity to shift production to take advantage of short-term cost fluctuations.
- By employing a dual sourcing strategy, Mattel achieves high productivity in its own plants while ensuring that changes of customer demand and preferences can be satisfied through outsourced partners.

Powerful lessons like these prove that managers can learn again from toys.[12]

---

[12] Johnson, M. Eric (2005), "How Can North Pole Workshops Better Respond to Shifts in Demand," *Harvard Business Review*, December, 44.

Part II

SUPPLY CHAIN MANAGEMENT IN EMERGING ECONOMIES: CHALLENGES AND OPPORTUNITIES

Hau L. Lee and Chung-Yee Lee (Eds.)
Building Supply Chain Excellence
in Emerging Economies
©2007 Springer Science + Business Media, LLC

Chapter 6

MANAGING SUPPLY CHAIN OPERATIONS IN INDIA
Pitfalls and Opportunities

Jayashankar M. Swaminathan
Kenan-Flagler Business School
University of North Carolina at Chapel Hill, USA

Abstract: In the last decade, India's role in the global supply chains has been steadily increasing. A massive population coupled with a large talented workforce have made India attractive both as a market and a source. In this paper, we discuss the pitfalls and opportunities in effectively managing supply chains in India. We discuss topics that would interest multi-national firms and local Indian firms.

6.1.   Introduction

An extensive area of research in operations relates to creation and delivery of value through better demand supply coordination which is also termed as supply chain management. Supply chain management is often referred to as efficient management of the end-to-end process, which starts with the design of the product or service and ends with the time when it has been sold, consumed, and finally, discarded by the consumer (Lee and Billington 1993; Swaminathan and Tayur 2003). One could think of any product or service supply chain to be consisting of five major value elements – design, planning, procurement, production and delivery. In addition, managing supply chains typically entails efficiently coordinating the flow of information, products and finances (Swaminathan 2001). Managing a supply chain within a single country is complicated due to various types of uncertainties in demand, supply and process. In most developed economies, there are limited uncertainties in availability of basic necessities for any kind of business such as power, roads, water etc. However, in developing economies infrastructure is weaker and that poses several newer types of challenges. It may even cause successful well tested strategies that worked well in developed economies to fail. A classic example is that of Wal-Mart which has an efficient network of cross docking facilities in the US that store minimal inventory in them while simultaneously enabling more frequent supplies to the retail stores. This system relies heavily on the fluid highway transportation system in the country which enables the firm to accurately estimate the travel times of trucks and efficiently coordinate their arrival and departure to the cross-docking facility, and therefore the inventory flow. When Wal-Mart went into operation into South America, it found it very difficult to run a logistic system based on such cross-docking facilities and had to adapt its approach. Therefore operating supply chains in developing nations often requires firms to be able to tailor their existing supply chain strategies or develop newer ones for that environment.

In this chapter, we will first discuss the characteristics of India as both a source and a market for the world economy. Next we will take a process oriented approach of the supply chain and discuss how operating a supply chain in India is different for each of those sub processes and highlight the challenges and opportunities therein. Finally, we discuss some of the recent changes in the Indian economy and how they might affect future operations in India.

6.2.   India: Source and Market

India became an independent nation in 1947. At that time it was among the poorest countries in the world and was put in the category of developing nations. For the next 40 years, India followed a centralized socialistic democracy with close ties with the Soviet Union. During this time, India started plans to improve agriculture and irrigation that mainly were focused on rural India. Simultaneously, the government set up programs and institutions of higher excellence for science and technology to compete with the very best in the world. However, the economy was closed for most part, bureaucracy was powerful and the efficiency and innovation of a capitalistic economy were lacking. In the early 90's with the fall of the Soviet empire, the economy in India was opened up, foreign investment constraints were eased, multi-national firms were allowed in several sectors, bureaucracy was reduced and businesses encouraged. The government during this time also focused on some new sectors such as software development and telecommunication where India could catch up with other world competitors. These changes have positioned India to the world both as a source and as a market.

6.2.1 India: A Source

Firms typically decide to source from an international location when it adds value in terms of cost, quality, time or capability. Whenever a firm decides to relocate its operations or invest in new facilities it needs to answer some key questions such as –

1. Is this move going to reduce or increase our costs?
2. Are we going to get better quality output as a result?
3. How does this affect the firm's ability to quickly respond to changing business needs and thereby promote or hinder business innovation?
4. How does this improve the firm's capability in the global market?

In today's context, cost and quality are important opportunities that give advantage to India. Due to Indian government's focus on education in science and technology after independence, today India graduates 200,000 english speaking engineeering graduates every year, the largest in the world. These and other english speaking graduates are in great demand for operating the world manufacturing and service economy due to cost advantages. Their wages are one fifth or sometimes one-tenth of comparative wages in the western world. For example, a recent graduate of engineering in India might get an annual salary of $5000 as compared to $30000 to $50000 in the U.S. The cost advantage in the labor pool is not restricted to service sector or engineers alone, this advantage spans through manufacturing sector as well. For example, a worker in an Indian textile company could be paid around $500-$1000 per year whereas a similar worker would be paid around $25000 per year in the U.S. This provides an opportunity for a multinational firm to reduce cost when they relocate or outsource their operations to India. That is why we find so many multinational companies are either setting up their own subsidiaries in India (like GE, IBM, Intel, Ford, Glaxo, Google) or have outsourced business functions like software development, software maintenance, call centers and credit processing. Although cost has traditionally been the main driver for shifting operations to India, today the value proposition spans quality as well. For example, in the software sector, initially firms outsourced development to Indian companies for cost alone. Today India boasts of having the largest number of firms that are at CMM process maturity level 5, an international measure of quality of process used for software development, testing and maintenance. Firms choose the best Indian vendors such as Infosys, Wipro, Tata Consulting and HCL, who can not only provide low cost but deliver the highest quality in their work. This has led to a boom in the services and technology sector in India in the last decade. According to one estimate, India's service and technology sector would be at $50 billion in 2008 and at $80 billion in 2010 (Engardio

2005). The time element is very useful in the global service industry because of the ability to work 24 hours a day on a particular task. It is not uncommon for U.S. software firms to work with an Indian partner to create a joint virtual development team where the U.S. engineers work on the project during their day time and the Indian engineers work on the same project during the night time in US while it is day time in India. This provides a tremendous advantage in fast life cycle innovation based industries such as software. The distance between the two teams sometimes creates challenges in terms of communication and integration, but many firms have found that advantages of shortening the development cycle overweigh the challenges.

Due to global outsourcing trends in the service sector and India's key role in it, the Indian services and technology sector has been in the limelight in the last few years. The manufacturing sector in India has traditionally lagged China in both cost and quality. China's manufacturing sector is estimated at $900 billion in 2006 as opposed to India's sector which is closer to $100 billion (Engardio 2005). However, the Indian manufacturing sector is catching up in a big way. Top Indian manufacturing firms such as Reliance, Hero Honda, TVS and Tata Motors have among the very best processes and have won several international quality awards including the coveted Deming Prize. While today India lags China by a huge margin in this sector, going forward, value that Indian firms would bring to the table will be in terms of improving strategic capabilities for multinationals. Indian firms are uniquely positioned to benefit from the diversifying strategy of multi-national firms. For example, the textile manufacturing is mostly concentrated in China today. As western firms move overseas they want to have alternative world class locations in addition to China. Intimate Clothing, a subsidiary of Mas Holdings, a leading Sri Lankan firm that supplies to major brand names all over the world and was once described as "Victoria' Secret's Well Kept Secret" by Forbes Magazine, is playing on this advantage in a very clever manner. They are strategically expanding their operations in India (as opposed going to other lower cost countries in the region) since they want to develop not only a "low cost" operation but also a "world class" operation. They have set up two plants already, one in Chennai and another in Bangalore with the very best processes. Their plan is to expand their operations in India to span the whole production spectrum from design, prototype, assembly and distribution. It is clear such facilities outside China will play an important role in improving the capabilities of multinational firm in the future.

6.2.2   India: A Market

A population of over a billion people makes India an attractive market for selling products and services. The growth in the economy has been above 6 percent over the last decade and the per capita income has almost doubled to $543 in 2004 (Bharadwaj et al 2005). Although India accounts for only 2% of the current world GDP, it is estimated that it could account for 17% of the world GDP in 2050 (Engardio 2005). The consumer market is even more attractive given that India is already in the top 10 markets with a value at $250 billion (Bharadwaj et al. 2005). It is projected that in the next five years India will be in the top five in terms of market potential for consumer goods. It is important to understand that even though in developed nations there are typically three economic strata – upper income, middle income and lower income groups, in India which is in the transition from a developing to a developed nation there are more strata (Joseph 2005). One could call them the upper income group earning $6000 per annum or above, the next is middle income group that earns between $3000 and $6000 per annum, the lower income group that earns between $1000 and $3000 per annum and the very poor group that earns less than $1000 per annum. It is to be noted that purchasing power parity adjusted these numbers need to be multiplied by a factor of 5. It is estimated that 150 million households are in the very poor category, 43 million households are in the lower income group, 3 million households are in the middle income group and 200,000 households are in the upper income group. Clearly, the total number of potential customers is very high in India but prices that the market will bear are very different than those in developed countries. For economies with the demographics and purchasing power as in India, multinational firms will need to innovate in order to successfully satisfy the consumers.

The sheer size is a great opportunity but it is a different story when it comes to execution. Between 1947 and early nineties, India was pre-occupied with agricultural development that it neglected infrastructural development. As a result, the roads are well below the standards in any developed economy and are worse than in some of the other developing countries. It is only in the last couple of years that the government is systematically building highways that connect major cities. The demand for power in India far exceeds the supply. This problem is worse in big cities such as Delhi and Mumbai. Further, India is a heterogeneous nation that has several states, several languages, several cultures and as a result very diverse customer preferences. The income distribution as well as preferences in rural and urban markets are also significantly different. Therefore, cracking the Indian market is not an easy task to say the least, however, the rewards for successful firms could be substantial. A firm that has successfully developed its supply chain for this market environment is likely to be well positioned to be a leader.

6.3.   Supply Chain Management in India

The five major process elements of any supply chain are design, planning, procurement, production and delivery. In the next few passages we will highlight the key challenges and opportunities faced by firms operating in India in each of these processes.

6.3.1   Design

Product design is an important element of any supply chain. A significant proportion of total costs incurred during the life cycle of the product are determined at the design stage (Swaminathan and Lee 2003). In India, an important characteristics that needs to be kept in mind at the design stage is the income diversity in the population. Just the huge differences in the ability to pay for a product, makes many of the successful products of the developed economy unviable in India. The example of shampoo sales in India serves this very well. When multinational firms such as Unilever first started selling shampoos in India they bottled them in packages similar to their offerings in developed countries and priced them around the same price (around $2). With this strategy, they were able to address only the topmost layer of the Indian households and could not penetrate the huge market. In the last few years, they have started selling shampoo in sachets (one time use) that sell for 2 cents. Clearly, by doing so, they have made their product more accessible to the customers and today it is estimated that close to 45% of the Indian population purchases shampoos

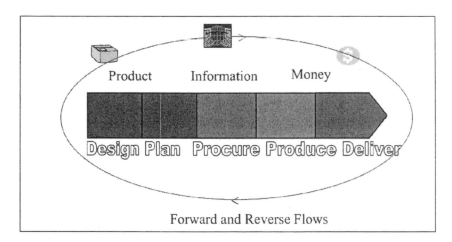

Figure 6.1. Process View of Supply Chain Management.

up from 18% ten years ago (Bharadwaj et al. 2005). In this process, the multi-national firm had to come out of the traditional thinking of large sizes and economies of scale and focus its attention more on innovative packaging that could enable pricing products at 2 cents. Clearly, this involved redesign of both the product and the package and finding local suppliers who could potentially operate profitably at those costs.

Firms are beginning to realize that there is money in markets like those of India at the bottom of the pyramid (Prahalad 2004). Many firms in India are now designing products specifically for the Indian market. For example, Hewlett Packard has developed a solar powered digital camera and a solar powered photo printer that they are piloting in villages that often do not have adequate power supply. Using these products, they have created a new career for some of the village women, who have become photographers that go from village to village and take pictures and help create national identification cards for villagers in the remote parts of the country. This has removed the necessity for villagers to go to the city to get a picture taken and apply for an identification card. Another example is the automobile sector. In the automobile market recently India became the largest small car market only behind Japan with the sales growing at 10% a year. Much of this growth has come from the small car segment. The largest selling car in India (known as Maruti 800) costs around $5000. Recently, Tata Motors, another Indian firm has announced that it will be developing a car for less than $2000. It is important to understand that in order to compete in India in this sector the costs of the car has to be affordable to the masses. Low cost may be enabled by removing safety features that would considered required in a developed nation. For example, most of the small cars do not have any safety airbags. Although it might seem shocking for a multinational firm to try to design products without these features, the need for airbags is limited on Indian roads. For most of the intra-city travel cars speed limits are restricted to 30 miles an hour (in theory) and could be as slow as 5-10 miles an hour (in practice) during peak time. Since Indians typically do not travel inter-state by their personal car, most consumers in that market are not willing to bear the costs of safety airbags. However, as the inter-state roadways improve, the customer preferences are likely to change and they may demand safety airbags in cars. With India's population and economic status, it is not surprising that the biggest market in the automotive sector is the two wheeler (motorbikes and scooters). Hero Honda is the world's largest producer of two wheeler motorbikes whose annual sales exceeded 2.5 million units last year. Two wheelers are not only used for personal transport of one or two people but are often used for transporting families of four and serve both the urban and the rural market. It is important to note the difference between bikes sold in western countries and in India. An important design element of these vehicles is the fuel efficiency and many of these vehicles provide up to 100 miles per gallon. Further, power (or engine size) which is a critical and often differentiating element of bikes in developed countries (such as Harley Davidson) is valued less in this market because of road conditions which hamper the ability of the consumer to relish the thrills of quick pickup and high speeds. As a result, most of the bikes sold in India are in the ranges from 75cc to 225cc, very different from bikes sold in US.

While designing products for India it is important to leave the mentality of tweaking existing product line for a new market. Although the costs associated with development of products specifically for the Indian market may seem prohibitive, firms can exploit the experience in India in developing products for the "bottom of the pyramid" in other countries in Asia, Africa, Europe and South America where a similar demographics exists. HP's initiative in developing affordable digital cameras and printers is likely to have an impact on other parts of the world as well. The bottom line is that firms should assess the market potential and needs accurately and should be prepared to come up with unique products that cater to the needs of the Indian market (and other world market with a similar profile) in a profitable manner.

6.3.2   Planning

Planning function can be for strategic choices such as product launch or for tactical decisions such as sales planning and forecasting. Planning is often quite challenging in India due to lack of visibility of sales at the end customer level. The concept of point of sale systems have caught on in urban areas only in the last decade or so. As a result, a firm selling products has limited integration of information in their supply chain. The rural market is in worse shape in terms of sales information since these shops do not have any point of sales systems. Many of them still rely on paper and pencil. In this environment, it is not prudent to expect that firms will have the capability to capture lost sales and stock-out conditions quickly. In fact, in several firms the operating policies may have serious inefficiencies. For example, a sales executive of one of the large multinational consumer goods firm in India indicated that recently they found that it was quite common for salesperson in rural areas to only take orders for products in-stock at the warehouse. Thereby, items which were out of stock and potentially were in great demand, were never being replenished while slow moving items were being ordered all the time. The lack of SKU level sales and demand information at the retail outlets leads to usage of approximate surrogate measures for customer service such as total coverage of rural retail outlets, number of sales person visits, and the overall product lines sold in a given time frame, which have their own ramifications for the supply chain. The lack of point of sales and information makes it hard to practice innovative concepts such as vendor managed inventory and collaborative forecasting particularly in the retail segment.

In the consumer goods sector much of the sales take place in small local stores that do not have access to bar coding. Further, the lack of such data and information, makes the planning process more of a push strategy primarily driven by sales targets that the firm would like to achieve in a given period. This leads to several unwanted effects like "inventory dump" and "hockey stick" phenomenon that lead to self induced seasonality wherein firms show very high sales towards the last quarter.

Planning new product launches is complicated in India due to two main reasons. The first relates to the diversity of the population and the size of the country. As indicated earlier, India is composed of many states and each state has traditionally has had its own culture, language and preferences for products. Therefore, firms need to plan for a product launch in the spirit as they would do for a multi-lingual region such as Europe. That is the firms needs to develop packaging in different languages, needs to develop customized advertisement campaigns as well as setup distribution network for the large population. Further, the lack of information tracking at the retail level makes it hard to evaluate how a product is doing in the first few weeks after the launch.

Planning is also adversely affected by the over reliance of the rural economy on monsoons. Sales are crucially dependent on when and how much rainfall the country gets each year. Almost 68% of the total sown area in India is vulnerable to draught. If monsoons are good then the agricultural production is likely to be good and the farmers have greater disposable incomes which translates into more disposable spending power for purchase of other goods and services. On the other hand, lack of rain implies lower sales for those goods and services. Therefore, any firm that caters to the rural India needs to carefully take the monsoon predictions into account while planning for a season.

The challenges of firms which are focused on foreign multinational customers is quite unique. The value proposition of an Indian vendors to their client is often either low cost or superior quality or in some cases both. However, many clients have high expectations in terms of response times and communication capabilities from their vendors. Compounded by the fact that they are located thousands of mile apart the need for robust planning capability as well as information technology infrastructure for communication is heightened. Ten to fifteen years ago, Indian firms had a tough time competing along these dimensions. However of late, many Indian firms have invested monetary, human and technological resources primarily catered towards their multinational customer. For example, most of the call centers in India have invested in dedicated telecommunication resources for customer overseas. Similarly, in the manufacturing sector, it is common nowadays to see firms that have made customized investments to improve their planning capabilities while dealing with multinational clients.

Planning can be a very difficult task while operating in India and firms need to have extra flexibility in their execution strategy to be successful. The lack of adequate information systems in the channel makes planning even more difficult. In recent times, many firms are introducing information systems in their supply chains and at the retail level and this is likely to help the planning process.

6.3.3   Procurement

As indicated earlier, one of the opportunities in India lies in the ability to get inexpensive workforce and suppliers. However, there are some critical challenges that firms need to pay close attention to. It is critical for firms to remember that as they grow their business, the ability of the supply base to grow may be limited due to capacity constraints. Further, there is huge difference between the capabilities of big top tier suppliers and the smaller (and often rural) second or third tier suppliers. The best of the first tier suppliers are sophisticated and follow world class best practices. As indicated by Balakrishnan et al. (2005), India had the largest number of firms conferred with the Deming award outside Japan, and most of these have been awarded to large top tier suppliers. However, that does not mean that there are no procurement challenges in India. In fact, the majority of suppliers lack in critical elements of service namely quality, delivery reliability and value added services. Product quality is unreliable mainly due to lack of continual supply of power, water and raw material. In some cases, the quality is adversely affected by the country's over-reliance on the monsoons. Delivery times of suppliers show a huge variance due to road conditions, traffic and distance of the suppliers. It is estimated that only 4% of suppliers in India are within 3 miles of the manufacturing facility, and more than 50% of the suppliers are located beyond 300 miles (Chandra and Sastry 2002). One of the reasons for disparate locations of the supplier base is the presence of industrial zones which were not necessarily optimized for supply chain performance. Uncertainties in delivery lead times make it essential for a firm to make pragmatic changes to their just in time production system or hold very high safety inventory levels in the supply chain. Finally, small to medium suppliers lack major capabilities to design products or services. In most cases a firm finds it hard to collaborate with such suppliers for designs which can impede faster product development and also increase design costs. A common approach employed by multinational firms in India has been to create joint ventures with an Indian supplier. This enables the supplier to ramp up quickly in a cost efficient manner. However, this approach has mostly been successful with medium to large suppliers. Given the uncertainty in the supply process, it is not surprising that a recent survey found that 63% of the firms had more than 100 suppliers and 17 percent of the firms had more than 500 suppliers (Chandra and Sasty 2002).

In addition, the smaller supplier segment is fragmented and mostly rural in nature which makes procurement more difficult. The solution to managing such a diverse base relies heavily on empowering the smallest of the suppliers and creating a mechanism for disintermediation. One such approach is the creation of cooperatives. After independence the number of cooperatives in India has steadily increased from 181000 in 1950–51 to 453000 in 1996–97 while the membership has grown from 15.5 million to 204.5 million. These cooperatives span across a number of industries such as dairy, textiles, cotton, housing, production, processing, finance among others. One successful example of cooperatives in India is the case of a diary products company called Gujarat Cooperative Milk Marketing Federation (GCMMF) that markets milk and other diary products under the AMUL name. The procurement network for AMUL has more than 2 million dairy farmers from whom milk is collected at more than 10000 cooperatives which eventually deliver to the plants. This is comparable to the consolidation of the dairy farmers association which is one of the largest cooperatives in the US consisting of more than 25000 members from 45 states. A more recent disintermediation and supply consolidation effort has been by the multinational Indian Tobacco Company (ITC) in their agribusiness through an initiative called e-choupals. E-choupals are creation of electronic portals in small villages through which farmers can get the most recent information about the market price, the crop, weather conditions, fertilizers among others. This has enabled ITC to form a direct link with the farmers thereby reducing several layers of the supply chain. The net effect has been better prices and profits for the small farmers as well as lower costs and supplier proximity for ITC. This service has reached out to 3.1 million farmers through 5050 internet kiosks in six Indian states (ITC 2005). There are significant cost benefits to procuring products and services in India for multinational firms but they need to be careful in selecting capable partners who can not only meet their current requirements in terms of cost and quality but have the potential to grow with them and add new capabilities over time. While seeking smaller suppliers the best approach maybe to avoid the multiple layers of the channel and utilize the electronic media to link with them in an efficient manner.

Procurement challenges for firms procuring items from outside India are complicated by complex tariff, import and taxation policies. Since the Indian economy was regulated to a great extent until the recent past, the imports and exports were closely related. For example, the government would regulate what percentage of parts that go into the final product could be imported from abroad. This often correlated with factors such as the percentage of final production that is exported and whether the product was a luxury good or not. Further, long delays in customs and port handling often made it very hard for firms to operate under a lean inventory system. With the opening up of the economy and relaxation of some of these limits, firms today have much greater flexibility in what they import (procure from overseas), how they process those parts and what they can (or not) sell within India. The bottlenecks in the ports is far from solved. The government in its most recent initiative is trying improve the transportation infrastructure in the country with focus on rail, sea, road and air.

6.3.4   Production

Tybout (1998) finds that it is typical in less developed countries for a large number of small scale firms to operate together with a few very large firms. India is no different in that context. The number of manufacturing firms in the organized sector (those that are registered with the government) is estimated at 127,000 whereas the unorganized sector consisted of 17 million firms. Small and very small firms which comprise majority of the unorganized sector make products in food, tobacco, lumber, textile and apparel sector. The situation is very similar in the service sector. For example, the retail sector is dominated by small mom and pop stores. The concept of superstores and malls are only beginning to appear and are mainly restricted to the big cities.

Traditional emphasis of Indian firms in both manufacturing and service sector has been on cost. Indian firms have tried to utilize the low cost of operations (both labor and overhead) to compete aggressively in the global market. Take for example, the software sector, where Indian firms for the last twenty years have competed mostly in terms of being able to deliver software professionals and as a result software projects at a fraction of the cost of that in a developed country. It is only in the last ten years that we observe that several of these firms claim the very best software development processes and are competing in terms of quality and innovation in addition to cost and timeliness. As a result, locations like Bangalore, Pune, Hyderabad and Chennai are becoming software and technological hotbeds where the best of the firms such as Microsoft, Intel and Google want to be. This facilitates tapping the best talent for their firms as well as create intellectual property through cutting edge research and development centers. The story is very similar in the manufacturing sector where thirty years ago manufacturers in India produced only for the Indian market and had no competition from multi-nationals. They created products for a captive market. However, today some of the Indian manufacturers are very large in scope and compete with other multinationals in the world economy. For example, Reliance Industries is one of the largest firms in India with annual revenues of $22 billion operates in various sectors including exploration and production (E&P) of oil and gas, refining and marketing, petrochemicals (polyester, polymers, and intermediates), textiles, financial services and insurance, power, telecom and information and communication initiatives. Reliance Industries exports its products to more than 100 countries in the world, itself being a multinational firm. In the last thirty years, firms in the manufacturing sector have adopted better processes thereby producing the very best quality of products as well as have developed competencies in terms of speed of delivery, planning as well as new product development.

One of the biggest impediments related to the production environment in India relates to limited (or lack of adequate) infrastructural support. For example, it is common even for the industrialized zones to suffer from frequent power outages, inadequate water supply and poor roadway connections. This causes unexpected delays in production as well as prevents firms from attaining their best performance. Firms often have to create a contingency plan for basic infrastructural needs. For example, it is not uncommon for firms in the service sector to invest heavily in backup power so that the operations inside receive uninterrupted power supply. Similarly, firms often run their buses to pickup and drop their employees because the public transport system is quite unreliable and many people cannot afford (or do not want to use) to drive to office in their personal vehicles. While running a production or service facility in India it is very important to plan for contingencies that may arise due to poor infrastructure. In addition, firms need to pay close attention to the unique needs of the employee base which could be related to transportation, recreation, household activities- like baby sitting services or traditional economic incentives that will keep the employees loyal to the organization.

6.3.5 Delivery

Delivery of products and services depends on distribution and sales which rely on the transportation possibilities, warehouse choices as well as the retail store coverage. In most developed countries, the retail market is dominated by organized retailing. In US the share of organized retailing is estimated to be around 80%. Europe has an organized retailing share of 70%, Brazil – 40% and China – 20%. The top 50 retailers in the US control almost 36% of the organized retail. India lags behind by a large margin in organized retailing with only 2% of the total sales controlled by the organized retailing although there are 6 million retail outlets in total (India 2005). It is estimated that there is no organized marketing and distribution in 87 per cent of India's villages, which are homes to 50 per cent of the rural population. As a result, large firms often have to create dispersed supply chains to enable them to produce and sell across the country. To give an example, Hindustan Lever Limited, a very large consumer product manufacturer covers an urban population of 1 million retail outlets and a rural market consisting of 50,000 villages through its supply chain network that consists of about 80 factories, 150 outsourcing units, 2000 suppliers and 5000 distributors. Market penetration of products across the country is a daunting logistical challenge due to poor infrastructure and prohibitive costs of customer acquisition, particularly in rural areas.

The dispersed population and the lack of cost effective modes for distribution have influenced firms in India to use multiple tiers in the distribution network. It is quite common for manufacturing firms to first ship products to their depot from where it is taken to a local distributor who then suppliers retailers in a region who finally sell it to the end customer. In some cases, where the firm does not operate its own depots, it employs a carrying and forwarding agent (C&F) who then suppliers to the local distributor. The role of these players in the distribution channel is extremely important because most of the retail stores can stock only 5–10 units of a given stock and often stock competing brands together in a small space. As a result, it is the distributors and sales agents who need to convince the retailer about stocking their products as well as making sure that their items are replenished in a timely fashion at the retail stores. Given the important role of the different parties in the distribution channel, their profit margins are also high. It is typical for a local distributor to have 5–10% margin while a C&F agent may make 2–4%. Multiple layers in the distribution network impact the final price the customer has to pay for the product.

The taxation structure in India is also complex with products typically being taxed twice. Once by the central government and the other by the state government. Such complex interstate tax laws do not enable the firms to optimize the distribution network on pure logistics alone. For example, following the traditional hub and spoke model to distribute goods is not even a viable option in several cases. As a result, firms in India sometimes need to setup multiple warehouses within the same state even though it was not logistically a sound practice. To mitigate these problems the Indian government recently adopted the value added tax structure (called VAT) from the 1st of April 2005. However, only 20 of the 28 states have implemented VAT.

The trucking industry which carries most of business related transportation within India is also highly fragmented with 2.7 million commercial fleet operated by over half a million fleet operators (ET 2005). Although outsourcing of logistical activities to third party firms is on the increase, there are very few organized third party logistics providers in India. Those that are operating are mainly in the trucking industry. A recent survey conducted by a leading Indian business daily found that many of the 3PLs in India lacked the scale and financial support to provide value added services such as inventory management and order processing. As the industry evolves and firms strive to become more efficient in the distribution and transportation, it is clear that such value added services will be in great demand.

For firms exporting out of India another big challenge relates to the lack of modern handling facilities at the airports and sea ports. This could be a major handicap for the firm since in today's global supply chains, speed is of essence. Further, complex governmental tariffs as well as import-export restrictions often delay shipments in and out of the country to a great deal. Some of the ports lack enough space and materials handling capability due to which at peak periods the waiting times could be several magnitudes higher than a modern port in other parts of the world. More recently, the issue of security in global transportation has become very important. As the Indian ports are modernized, close attention needs to be placed on this dimension in addition to improving wait times and increasing handling capacity.

Distribution in India is still a traditional industry and when firms enter the Indian market, their distribution channel strategy can make or break their success. The ability to reach the billion people in the market depends critically on how well distribution is planned and executed.

6.4.   Concluding Remarks

It is well known that supply chain management is challenging even when operating in a developed economy such as the US. It gets even more challenging in an emerging economy like India. It is particularly difficult for multinational firms that may have a successful strategy in their home country that try to utilize the same approaches in India. As indicated earlier, one of the major challenges of operating a supply chain in India is the under developed infra-structure for transportation, power and water. The government policies related to double taxation as well as a poorly developed retail and distribution channel have made efficiently coordinating the supply chain an onerous task. Further, less usage of information technology both at the retail level as well as in the supply chain have made it increasingly difficult for firms to build responsive supply chains. However, these deficiencies are slowly but surely getting tackled. The government recently announced a value added tax system which should enable firms to more efficiently coordinate their distribution networks. Similarly, many large firms have started adopting information technology solutions. A recent survey (ET 2005) found that 31% of the firms had implemented enterprise resource planning (ERP) systems, 56% had adopted customer relationship management (CRM) systems while 42% had adopted warehouse management systems (WMS). Although the numbers above are for larger firms the trend to use technology in supply chain is picking up. The usage of the Internet although low compared to US is also showing growth. A comparison of the numbers by Rahman (2004) and Lancioni et al. (2003) on the use of the Internet in India and the US in 2003 reveals the following numbers. The use of the Internet in purchasing was 49% in India as compared to 86.7% in the US, 30.1% companies that use internet for inventory management in India as compared to 48.5% in the US, 50% of the firms used the Internet for transportation as compared to 84.3% in the USA. The usage of the Internet as a medium in order processing, customer service, production scheduling and relations with the vendors is still very low in India and this is an area that could see some growth as other parts of the supply chain become more developed. Many logistics players in India are now introducing GPS tracking in their vehicles which is common in countries like the US. Such practices give visibility to the in process inventory in the supply chain and enable better decision making.

Another aspect of the supply chain efficiency relates to following best practices such as just in time and vendor managed inventory that remove inefficiencies in the supply chain. Although Indian firms have lagged behind in terms of JIT Implementation in the past, they are aggressively moving towards developing pragmatic inventory systems that lead to lower inventory while taking into account some of the inherent uncertainties in operating a supply chain in India. The government is also working towards building a robust infrastructure to facilitate efficient flow of goods and services. In 2003, the government earmarked a large amount of money for developing an advanced highway system (called the Golden Quadrangle) that would connect the different parts of India. As the infrastructure develops for fast and reliable transportation, many of the inefficiencies related to logistics are likely to be overcome. The power supply situation in India still remains a major issue. The demand for power far exceeds the supply particularly in major metropolitan cities. While increasing power generation and supply, the government is also taking steps to distribute power in an efficient manner. A few years ago a new firm Power Grid Corporation of India was set up to overlook power supply and distribution across the nation. The state of the art information and engineering systems that are being implemented by this organization for power failure detection and correction served as an example for the power outage management in the east of US in 2003. These systems allow the firm to promptly detect the electric failures at a fine level of granularity and reallocate power through the grid with a short time. Such systems bode well for the power distribution in India going forward although the supply and demand gap needs to be reduced substantially.

Indian firms today on average are not at the cutting edge of supply chain practices and usage of information technology. We do see world class excellence and best of the breed practices in few large firms. The real potential that lies in India both in terms of being a world class source and as a world class market is yet to realized. It will be not an exaggeration to say that we are only seeing the tip of the iceberg today. How much potential would eventually be realized depends on several complex related factors including government policies in terms of infrastructural development; the growth rate of the economy; development of the rural economy and eradication of poverty; open economic policies and last but not the least continued growth and high aspirations of the Indian firms.

Acknowledgements

The author wishes to thank executives from several firms including Intimate Clothing, Reliance Industries, Hero Honda, Glaxo, Infosys, Wipro, Shoppers Stop and PowerGrid who shared their views and allowed us to take a close look at their operations in India. This research was funded in part by CIBER grant at the University of North Carolina, Chapel Hill. The author also thanks Sriram Narayanan for assistance in collecting relevant articles.

References

Balakrishnan K., A. Iyer, S. Seshadri and S. Anshul (2005). Indian auto industry at the cross roads. *Working Paper*. Stern School of Business, NYU, New York.

Bharadwaj V.T., G. Swaroop and I. Vittal (2005). Winning the indian consumer. *McKinsey Quarterly*. September, 1–7.

Chandra P. and T. Sastry (2002). Competitiveness of indian manufacturing: Findings of the 2001 national manufacturing survey, *Working Paper No: 2002-09-04*, Indian Institute of Management, Ahmedabad.

Engardio P. (2005). A new world economy. *Business Week*. August 22, 52–58.

ET (2005). *Supply chain and logistics 2005*. Economic Times Knowledge Series.

India (2005). India in business. http://www.indiainbusiness.nic.in/india-profile/ser-retail.htm

ITC (2005). ITC e-chaupal. http://www.itcportal.com/sets/echoupal_frameset.htm.

Joseph T. (2005). *The marketing whitebook 2005*. Businessworld.

Lancioni R.A., M.F. Smith and J.H. Schau (2003). Strategic internet application trends in supply chain management. *Industrial Marketing Management, 32*(3), 211–217.

Lee, H.L. and C. Billington (1993). Materials management in decentralized supply chains. *Operations Research, 41*(5), 835–847.

Prahalad C.K. (2004). *The fortune at the bottom of the pyramid: Eradicating poverty through profits*. Wharton School Publishing, Philadelphia.

Rahman Z. (2004). Use of internet in supply chain management: a study of indian companies". *Industrial Management & Data systems, 104*(1), 31–41.

Swaminathan J.M. and S.R. Tayur (2003). Models for supply chain in e-business. *Management Science, 49*(10), 1387–1486.

Swaminathan J.M. (2001). Supply chain management. *International Encyclopedia of the Social and Behavioral Sciences*. Elsevier Sciences, Oxford, England.

Swaminathan J.M. and H.L. Lee (2003). Design for Postponement. *OR/MS Handbook on Supply Chain Management: Design Coordination and Operation*, edited by Steve Graves and Ton de Kok, 199–228. Elsevier Publishers.

Tybout J. (1998). Manufacturing firms in developing countries: How well do they do and why? *Working Paper No 1965*. World Bank.

Hau L. Lee and Chung-Yee Lee (Eds.)
*Building Supply Chain Excellence
in Emerging Economies*
©2007 Springer Science + Business Media, LLC

Chapter 7

INTEGRATED FULFILLMENT IN TODAY'S CHINA

Jamie M. Bolton and Wenbo Liu
Accenture, China

Abstract: As part of its entry terms into the World Trade Organization, China agreed to
open markets and services that had been protected from global competition.
One such area was logistics – a sector with huge potential to spur business
growth. From a supply chain perspective, however, the road to high performance in China is still riddled with challenges. Underdeveloped infrastructures,
fragmented distribution systems, insufficient technology and onerous regulations are just some of the challenges.

Given these barriers, supply chain mastery will separate success from failure
among companies doing business in China. High Performing Supply organisa-
tions generally strive for an integrated end-to-end presence, with high levels of
coordination, standardization and visibility; essentially managing Logistics as
an "Integrated Fulfilment" process. But as far as China is concerned, traditional
logistics – transportation and distribution – is the most important hurdle.

This chapter provides an overview of China's Integrated Fulfilment sector and
includes practical advice for foreign companies – both small and large – that
are planning to establish an effective presence in China or are looking for ways
to improve their distribution system efficiency. We discuss the impact of World
Trade Organization (WTO) entry on China's distribution and logistics sector,
key trends in the sector, the risks of operating in China that remain even after
WTO entry, and strategic recommendations for companies entering China.

China's attractiveness as an investment environment for multinational corporations (MNCs) in a variety of industries – such as consumer goods, retail, automotive, electronics, and telecommunications – is well known. However, the
country's underdeveloped transportation infrastructure, fragmented distribution
systems, limited use of technology in the Integrated Fulfillment sector, dearth
of logistics talent, regulatory restrictions, and local protectionism combine to
hinder the efficient distribution of domestic and imported products, and thus

reduce returns on investment. The barriers have increased the cost of doing business in China and have restricted consumers' choices.

7.1.   WTO Entry: Integrated Fulfillment

As part of the terms of its WTO entry, China agreed to open market sectors and services that, in the past, were protected from global competition. The opening of the Integrated Fulfillment sector is expected to spur the modernization of the sector over the next three to five years (*see* tables) –

Nevertheless, the Integrated Fulfillment sector remains highly fragmented, with strongly protected local interests. Although foreign companies often possess better management systems and technologies, knowledge of the local operating environment, culture, and customer needs helps local firms create and maintain competitive advantages.

7.2.   Side Bar

7.2.1   Taking Risks Can Reap Rewards in China's Integrated Fulfillment Sector

Contrary to popular belief, China's World Trade Organization (WTO) trading rights liberalization commitments do not encompass distribution rights. According to China's WTO commitments, these are two separate areas of agreement.

7.2.2   Trends in Integrated Fulfillment in China

The Integrated Fulfillment sector has long challenged companies seeking to move and sell products within China. As shown in the Table 7.3, in the past one decade, logistics spending in China amounted to one-fifth of the nation's GDP and twice the proportion spent on logistics in the United States. And, on average, around 90 percent of a Chinese manufacturer's time is spent on logistics while only 10 percent is spent on manufacturing. Selling costs in China are significantly higher than those in the West, too. For many commodities, logistics costs are proportionally 40 to 50 percent higher than they would be in the United States.

Today, selling costs in China are significantly higher than those in the West. Annual working capital turnover (a measurement that compares the depletion of working capital [current assets minus current liabilities] to the production of sales over a specific time) in China is, on average, 1.2 times for manufacturing state-owned enterprises (SOEs) and 2.3 times for commercial SOEs. These

Table 7.1.   Trading Rights and Distribution: Two Different Things.

Trading rights	Distribution services
China's WTO entry documents define trading rights as "the right to import and export goods" and state that "China shall progressively liberalize the availability and scope of the right to trade, so that, within three years of accession, all enterprises in China shall have the right to trade in all goods throughout the customs territory of China except for [state-traded goods]." (Imported goods subject to state trading include grain, vegetable oil, sugar, tobacco, processed oil, chemical fertilizer, and cotton. Exported goods subject to state trading include tea, rice, corn, soybeans, tungsten ore, ammonium paratungstates, tungstate products, coal, crude oil, processed oil, silk, unbleached silk, cotton, cotton yarn, woven fabrics of cotton, antimony ores, antimony oxide, antimony products, and silver.)	China's WTO entry documents define four main sub-sectors of distribution services: commission agents services; wholesaling; retailing; and franchising. "The principal services rendered in each subsector can be characterized as reselling merchandise, accompanied by a variety of related subordinated services, including inventory management; assembly, sorting and grading of bulk lots; breaking bulk lots and redistributing into smaller lots; delivery services; refrigeration, storage, warehousing and garage services; sales promotion, marketing and advertising, installation and after sales services including maintenance and repair and training services."
Commission Agents' Services "...[S]ales on a fee or contract basis by an agent, broker, or auctioneer or other wholesalers of goods/merchandise and related subordinated services."	**Wholesaling** "...[S]ale of goods/merchandise to retailers to industrial, commercial, institutional, or other professional business users, or to other wholesalers and related subordinated services."
Retailing "...[S]ale of goods/merchandise for personal or household consumption either from a fixed location (e.g., store, kiosk, etc.) or away from a fixed location and related subordinated services."	**Franchising** "...[S]ale of the use of a product, trade name or particular business format system in exchange for fees or royalties. Product and trade name franchising involves the use of a trade name in exchange for fees or royalties and may include an obligation for exclusive sale of trade name products. Business format franchising involves the use of an entire business concept in exchange for fees and royalties, and may include the use of a trade name, business plan, and training materials and related subordinated services."

(Source: *Compilation of the Legal Instruments on China's Accession to the World Trade Organization – The US-China Business Council*)

Table 7.2. China's WTO Trading Rights Commitments.

Upon Entry: December 11, 2001	Year One: By December 11, 2002	Year Two: By December 11, 2003	Year Three: By December 11, 2004
Eliminate, for both Chinese companies and FIEs, any export performance, trade balancing, and prior experience requirements. Foreign-invested international logistics companies may offer import, export, and entrustment services for export processors (*see* box, *Logistics Obscure the Way Forward*).	Minority foreign JVs granted full trading rights. Registered capital requirement for domestic PRC trading companies lowered to RMB 5 million.	Majority foreign JVs granted full trading rights. Registered capital required for domestic PRC trading companies lowered to RMB 3 million.	Rights will be granted to all enterprises in China except for sectors reserved for state trading. Such right does not permit importers to distribute goods within China. Trading rights examination and approval system will be eliminated. Registered capital required for domestic PRC trading companies lowered to RMB 1 million.

The latest version of China's Catalogue Guiding Foreign Investment in Industry, revised in March 2002, reflects the country's WTO commitments

JV = joint venture. FIE = foreign-invested enterprise

(*Sources: Compilation of the Legal Instruments on China's Accession to the World Trade Organization; The US-China Business Council*)

Table 7.3. Logistics Expenditure as % of GDP in China.

Year	Transport Cost	Inventory Carrying cost	Administration cost	Total Logistics Cost
1994	11.9	7.0	3.2	22.1
1995	11.0	7.6	3.4	22.0
1996	11.2	7.5	3.3	22.0
1997	11.0	7.7	3.5	22.3
1998	11.0	6.8	3.5	21.4
1999	11.6	6.2	3.6	21.4
2000	11.2	6.5	3.6	21.2
2001	11.2	6.4	3.5	21.1
2002	11.6	6.4	3.4	21.5
2003	12.0	6.3	3.1	21.4
2004	12.1	6.2	3.0	21.3

(Source: *China Federation of Logistics and Purchasing*)

Market — 2001
($Billion)

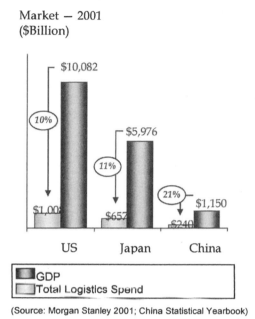

(Source: Morgan Stanley 2001; China Statistical Yearbook)

Figure 7.1. Comparison of Logistics Markets between China, U.S. and Japan.

figures compare with averages of 15 to 20 times in the United States. For many commodities, logistics costs are proportionally 40 to 50 percent higher than they would be in the United States. Accounts receivable – a key measure of inefficient logistics practices – often exceeds 90 days.

Trend 1 – China's Integrated Fulfillment Sector is Growing Rapidly. Despite these weaknesses, China's Integrated Fulfillment sector is growing rapidly. In fact, the logistics industry has reported annual revenue growth rates of 20 percent for 2002, 27 percent for 2003, and 30 percent for 2004 and is forecast to grow 25 percent annually for the next three years. The sector has changed significantly as a result of overall market growth, evolving customer requirements, liberalization of government policies, and China's WTO entry.

China's 10th Five-Year Plan (FYP, 2001–05) mandated massive construction of rail, road, port, and aviation infrastructure, particularly in China's most underdeveloped regions. Table 7.4 shows the current status of China's transport infrastructure and China's government investment plans for further improving China's transport infrastructure.

But figures in government investment plans should be viewed as indicators rather than concrete figures. China may or may not reach the goals, but one

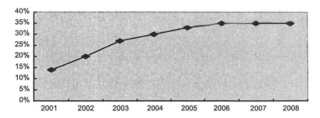

(Source: *China Statistical Yearbook, 2001-2005; China Federation of Logistics and Purchasing; IDC Report 2003*)

Figure 7.2.   Annual Revenue Growth Rate of China Logistics Industry.

(Unit: 000 million tons)

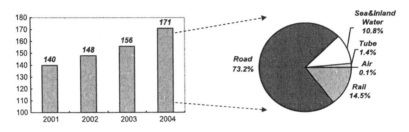

(Source: *China Statistical Yearbook, 2001-2005; China Federation of Logistics and Purchasing*)

Figure 7.3.   China Logistics Industry's Annual Freight Traffic Volume and Breakdown.

thing is certain – these goals indicate that the China government has made infrastructure construction a high priority. The result will be better physical infrastructure that should ease distribution bottlenecks in China.

Trend 2 – The Highly Fragmented Logistics Service Market Starts to Consolidate.   Yet of the more than 18,000 registered companies claiming to offer logistics services in China, not one can offer nationwide distribution services today. No single logistics provider commands more than 2 percent of the market. Because of this industry fragmentation, consolidation of logistics providers and the development of third-party logistics providers are inevitable – especially given shippers' demands for greater efficiencies, scale, breadth of service offerings, and network coverage. Competition is already intensifying in the third-party logistics market, forcing a consolidation of this industry as a whole in China. Foreign companies with strong international networks and better management are gaining market share, while many domestic companies rely on underdeveloped domestic operations. And local and

(Source: *EFT Research 2005; China Federation of Logistics and Purchasing*)

Figure 7.4. China Transportation Infrastructure & Network.

regional distribution systems are replacing state-owned and centrally managed trading and distribution systems.

Trend 3 – MNCs are Relying More on 3PL. To remain globally competitive, Chinese companies must cut costs and expand services. This commercial reality will force local companies to concentrate on their core business, rather than on the traditional SOE goal of establishing a "small and complete" company. The "small and complete" mindset is entrenched, according to a study published in 2002 by China's Development Research Center of the State Council: 70 percent of China's commercial enterprises and 53 percent of industry enterprises own their own vehicle fleet, and 80 percent and 59 percent, respectively, own warehouse facilities.

Though the concept of outsourcing these basic logistics functions is still relatively new to most Chinese companies, many MNCs relying on China as a global sourcing base are inclined to use, and are experienced in using, third-party services – especially those of third-party providers with which the MNCs have established relationships at home. More than 80 percent of MNCs in China currently contract at least a portion of their logistics business to third-

Table 7.4. China's Transport Infrastructure (in 2004).

Types of Trans- port	Length of Transport Routes	Number of Transport Vehicles	Government Investment Plans
Rail	74,408 km (Track in Operations)	528,000 rail containers	• Annual investment will be about $ 8 billion • Half of the rail investment is planned for western China projects, including the world's highest railway, linking Qinghai and Tibet
Road	1.87 million km (Highways)	8.93 million vehicles (average capacity is about 2 tons)	• Annual investment will be about $ 80 billion, by 2010 • Plan to build 0.4 million km express way
Sea	34,000 (shipping berths)	1,500 vessels (capacity of 37 million DWT)	• Annual investment will be about $ 8 billion, by 2010 • Double the number of deep water berths • Specific deepwater port projects include Shanghai, Dalian, Qingdao, Tianjin and Shenzhen
Inland Water	123,300 km	210,000 vessels	• Annual investment will be about $ 1.1 billion, by 2010
Air	2.05 million km (Civil Aviation Routes)	890 airplanes	• Construction or renovation of about 35 airports

(Source: *China Statistical Yearbook, 2004 & 2005; China Federation of Logistics and Purchasing; EFT Research 2005; The US-China Business Council*)

party logistics service providers. Many leading foreign firms in China, such as McDonald's Corp. Dell Computer Corp. and Nokia Corp., have demonstrated great success by using third-party service providers' expertise, capabilities, and assets to offer nationwide Integrated Fulfillment services.

These foreign companies have shown Chinese companies that they do not need to own all of the assets involved in service provision to gain the capability and expertise to offer a full line of services. To reap the benefits of scale, small firms will soon aim to partner with large-volume players in sales and distribution because volume influences the market, strengthens control over the supply chain, and most important, improves efficiency.

According to a report published in October 2003 of the International Data Corp (IDC), logistics outsourcing in China will grow by around 25% annually for the next decade, due to stronger MNC interest and demand for third-party services.

Trend 4 – MNCs are Increasing Control of Down stream distribution. MNCs in China have achieved success through the support of strong, modern distribution networks. Guangdong Honda Automobile Co. Ltd. initiated exclusive four-in-one franchises (sales, repair and maintenance, supply of parts and components, and information service) in China that enabled it to strengthen its brand name and position by better controlling service quality, product price, and market information. Other foreign companies similarly have sought to control more aspects of their distribution system in China. The formation of the SAIC-Volkswagen Sales Co., Ltd. in 2000 is another example of a foreign company seeking to participate directly in distribution business by building a dedicated distribution and after-sales support network.

Trend 5 – Leading Companies Acquire Competitive Advantages Through Alliances. Another major trend in the Integrated Fulfillment sector has been the rise of alliances and joint ventures (JVs) as the top companies in China combine forces to build competitive national distribution chains targeting specific industries. In June 2002, Legend Group Ltd. and APL Logistics (the logistics branch of the NOL Group) announced the formation of a third-party logistics service JV to offer specialized logistics services for the information technology (IT) industry. The JV seeks to capitalize on Legend's position as the leading personal computer vendor in China and APL's globally recognized brand name and expertise in the logistics field. Similar joint ventures have been formed between the TNT Group and the Shanghai Auto Industry Group, among others.

7.2.3   Risks Remain

More than 80 percent of Fortune 500 companies have already invested significantly in China and have gained many years of local market knowledge. At the other end of the scale, small and medium-sized foreign companies are just now entering the country. Regardless of the depth of their experience in China, all companies face risks in the regulatory, political, and market arenas that are likely to persist for the near future.

Risk 1 – WTO Commitment Delays.   China may delay WTO implementation and regulate competition to help local companies compete in the market. As other WTO members have done, China may find ways to work around

Distribution	
1. Allow wholly foreign-owned enterprises (WFOEs) in wholesale, retail, and commission agents' service	1. Regulations issued, but full implementation delayed pending clarification from the PRC Ministry of Commerce (MOFCOM). (Regulations on Management of Foreign Investment in the Commercial Sector)
2. Allow franchising	2. Done. FIEs can expand their scope to include franchising. (Administrative Rules on Commercial Franchising)
3. Allow direct sales	3. Late. Direct sales draft regulation under consideration.
4. Allow retailing and wholesaling of pharmaceuticals	4. Late. Rules for domestic firms exist, but MOFCOM is drafting separate rules for foreign participation.
5. Allow retailing of refined fuel	5. Late. Rules for domestic firms exist, but separate rules for foreign participation are forthcoming.
6. Allow wholesaling of printed matter	6. Done, early (2003). (Rule on Management of Foreign-Invested Book, Magazine, and Newspaper Distribution Enterprises)
Freight Transport Services	
1. Allow foreign majority rail JVs	1. Done. (Revised Catalogue Guiding Foreign Investment in Industry)
2. Allow wholly foreign-owned road enterprises	2. Done. (Second Revision to Administrative Regulations on Foreign Investment in Road Transportation Industry)
3. Allow wholly foreign-owned storage and warehousing enterprises	3. Done. (Second Revision to Administrative Regulations on Foreign Investment in Road Transportation Industry)
Repair, Maintenance, and Leasing	
Allow WFOEs	Unclear. Applications submitted to MOFCOM have yet to receive approval.
Packaging	
Allow WFOEs	Late. No draft reported.

(*Source: The US-China Business Council*)

Figure 7.5. China's WTO commitments due on December 11, 2004.

WTO rules to maintain barriers against imports, for example by erecting WTO-compatible nontariff barriers, such as licensing, health, technical, and packaging standards. As a case in point, China recently released a draft regulation that would require "one license, one product" dealership licenses in the auto sector. Such licenses would prevent newcomers from using existing distribution channels and give existing manufacturers more time to prepare for direct competition. Such requirements will likely create tension between customer expectations and the quality of service available, which may force change faster than the PRC government anticipates.

With the exception of CEPA, for phased WTO liberalization of logistics services continues to delay greater participation of foreign logistics and transportation players.

Table 7.5 and Table 7.6 are summary of the WTO commitments in the Trade and Distribution, the Transportation and Logistics areas, as well as a list of China's implementation efforts to date.

Risk 2 – Regulatory Fragmentation. The distribution and logistics industry has been micro-regulated for years, with different logistics service components governed as distinct sub-sectors by various government departments: the Ministry of Communications governs land and waterway transportation; the Ministry of Commerce administers trading rights and international freight forwarding licenses; and the PRC General Administration of Customs controls brokerage services. Despite central-government efforts to promote coordination – as demonstrated by the August 2004 initiative by nine ministries, including MOFCOM, the State General Administration for Quality Supervision Inspection and Quarantine (AQSIQ), the Ministry of Communications, the Ministry of Railway, the Ministry of Commerce, the State Administration of Industry and Commerce, the General Administration of Customs, the State Administration of Taxation, the State Administration of Foreign Exchange, to issue a plan to promote the country's logistics industry development – the shared jurisdiction system for the logistics sector is unlikely to change in the near future. Companies still must acquire separate licenses through various governing bodies to undertake different activities.

Risk 3 – Lack of Enforcement Capability and Local Protectionism. China's governing structure encompasses multiple layers of central and local governments. Despite central-government efforts to liberalize the market, local-level interpretation and enforcement of laws and regulations can often be arbitrary, favoring local interests and prevent realization of the full benefits of WTO accession to the economy unless they are removed.

MOFCOM issued the Regulations on Management of Foreign Investment in the Commercial Sector in April 2004, specified how foreign-invested commercial enterprises may conduct retail, wholesale, franchise, or commission agency business. The regulation states that new stores opened by foreign-invested distribution companies must suit the urban and commercial development plans of the city in which the store will be located. Given the discretionary latitude possessed by local officials in this regard, this requirement could be used as a market-entry barrier to restrict the number of foreign distribution operations in a given city.

China's licensing regulations require firms to obtain a separate business license for each province in which they operate. In many localities, out-of-province trucks are arbitrarily stopped at city borders and subjected to tolls that local trucks are not required to pay. In some cases, this necessitates expensive unloading and reloading onto local vehicles.

The market-oriented reform in China's logistics industry is hindered by the continued protection of domestic companies. Domestic players still dominate the field with few exceptions. Foreign companies are often left to deal with onerous customs and quarantine policies as well as cumbersome processes for importing goods.

Risk 4 – Capacity vs. Demand.   China has one of the most dynamic markets in the world, making errors in market demand forecasts more frequent and severe than those in more-developed markets. Abundant funds are available in China (both from foreign and domestic sources), and cities are tempted to build up too much capacity rapidly, so that capacity exceeds true market demand. Almost all major cities and regions in China have invested in some form of distribution or warehousing center or logistics park. According to research conducted by the China Storage Association in 2002, 60 percent of logistics centers are empty.

Risk 5 – Social and Political Risks.   Foreign companies planning to enter China through a partnership or JV must be extremely careful about their potential partners' visible and invisible liabilities. For example, protecting local employment is a high priority for PRC governments, so appropriate benefits for excess workers can be a huge issue in contract negotiations.

Corruption also looms large in China – especially in some smaller cities and at local levels – in every industry. Appropriate handling of corruption is essential for success.

7.3.   Recommendations for Foreign Companies

To help foreign companies enter and operate in China successfully, we have identified seven key recommendations for their distribution and logistic operations:

1. Seek deep knowledge of the market

Contrary to what some experts think, there will not be a mad rush by shippers to develop their own sales and distribution channels or networks once China implements its WTO commitments in distribution services – the cost and time needed to establish a network will prevent this. Smart players are building up the capabilities of the PRC distributors with which they already have relationships and are providing incentives and performing audits to improve efficiency of, and control over, distribution and points of sale.

Apart from normal business considerations, such as market size, growth, competition, channel mix (the different paths to market), and regulations, companies must also pay attention to regional and local particularities and relationships (*guanxi*). Every city or investment zone has different policies designed to attract certain types of foreign investment. For example, some zones provide local tax incentives, land leasing, and lower utility fees. Organizations such as The US-China Business Council and the American Chamber of Commerce can help find appropriate locations. A well-connected and trustworthy local partner is also often critical.

Table 7.5. Summary of the WTO Commitments in Trade and Distribution Areas.

Sector	Upon Entry: December 11, 2001	Status	Year One: By December 11, 2002	Status	Year Two: By December 11, 2003	Status
Wholesale & Commission Agents' Services (excluding salt & tobacco)	FIEs can distribute all of their products made in China. Foreign service providers can provide full range of related services for products they produced.	July 2001: First wholesale enterprise with foreign investment approved, i.e., Japan's Maruberi Co. holds 49% of Baihong Commercial Trading Co., which buys products from domestic producers (incl. FIEs) and has limited import & export rights to sell international products	Minority foreign equity permitted except for BMN, pharmaceutical products, pesticides, mulching films, chemical fertilizers, and processed & crude oil	Permitted for international logistics companies as of August 24, 2002, with certain conditions and geographic restrictions	Minority foreign equity permitted. No geographic or quantitative restrictions.	China approved first pharmaceutical JC, Zuelig Xinxing Pharmaceutical Co. in May 2003, Zuelig Pharma holds 49% of company.
Retail (excluding tobacco)	Minority foreign equity permitted in 5 special economic zones and 8 cities. FIEs can distribute all of their products made in China. Foreign service providers can provide full range of related services for products they distribute. Quantitative restrictions apply for each zone/city.		FIEs may do the retailing for BMN.	Permitted as of May 1, 2003, subject to registered capital and other requirements.	Majority foreign equity permitted. All provincial capitals and 2 other cities open to retail JVs.	
Franchising	No commitment		No commitment		No commitment	
Wholesale or retail away from a fixed location	No commitment		No commitment		No commitment	

Table 7.5. (continued).

Sector	Year Three: By Dec. 11, 2004	Status	Year Four: By Dec. 11, 2005	Status	Year Five: By Dec. 11, 2006	Year Six: By Dec. 11, 2007
Wholesale & Commission Agents' Services (excluding salt & tobacco)	100% foreign equity permitted except for chemical fertilizers, and crude oil. May distribute BMN, pharmaceutical products, and mulching films. No geographic quantities or equity restrictions.	Abolished onerous capital requirements (now requiring only RMB 500,000 for wholesalers) Approval from MOFCOM and the provincial authorities is required	No commitment		100% foreign equity permitted for all products. No limits of foreign participation after 2006. May distribute chemical fertilizers and processed & crude oil.	No commitment
Retail (excluding tobacco)	100% foreign equity permitted except for certain products. FIEs may retail Pesticides, pharmaceutical products, mulching films, and processed oil. No geographic restrictions. No quantitative restrictions except for certain products. Majority foreign investment prohibited in chain stores with over 30 outlets.	Abolished onerous capital requirements (now RMB 300,000 for retailers) Approval from MOFCOM and the provincial authorities is required	No commitment	Provisions in the SFDA measures allow local governments to consider broad, subjective standards in reviewing license applications that could justify protectionist at local levels	100% foreign equity permitted, but limits on large chain stores and those selling certain products. FIEs may retail chemical fertilizers, Foreign majority investment in motor vehicle JVs permitted.	No commitment
Franchising	Permitted. No geographic quantitative, or equity restrictions.	Rules to remove all restrictions by Dec. 11, 2004 reported drafted.	No commitment		No commitment	No commitment
Wholesale or retail away from a fixed location	Permitted. No geographic quantitative, or equity restrictions.	Status of new rules uncertain. Nationwide crackdown on pyramid selling underway.	No commitment		No commitment	No commitment

The latest version of China's Catalogue Guiding Foreign Investment in Industry, revised in March 2002, reflects the country's WTO commitments

JV = joint venture. FIE = foreign-invested enterprise. BMN = Books, Magazines and Newspapers

(Sources: Compilation of the Legal Instruments on China's Accession to the World Trade Organization; The US–China Business Council)

Table 7.6. Summary of the WTO Commitments in Transportation and Logistics Areas.

Sector	Upon Entry: December 11, 2001	Status	Year One: By December 11, 2002	Status	Year Two: By December 11, 2003	Status
Rail Transportation	Up to 49% foreign equity permitted.	First license issued in March 2002	No commitment		No commitment	
Road Transportation	Up to 49% foreign equity permitted.	Majority FIEs allowed.	Majority FIEs allowed.	Permitted as of Dec. 1st, 2002, but capped at 75%. Higher proportion permitted in certain sectors and in western areas.	No commitment	
Freight Forwarding	50% FIEs allowed. 5 years waiting period for 2nd JV.	Foreign party needs 3 years' experience; Minimum registered capital = $1,000,000; Branches allowed after 1 year of operation, with $120,000 capital per branch.	Majority FIEs allowed.	75% FIEs allowed from Jan 2003.	National treatment for branch capital (500,000 RMB); Second JV can be set up after two years of operation (reduced from five years).	Second JV allowed after two years from Jan 2003. To some extent, this service overlaps with the services that may be offered by foreign-invested logistics operators
Warehousing & Storage	Minority FIEs allowed.		Majority FIEs allowed	Provided for under road transport, freight forwarding and maritime rules.		
Logistics	No commitment.	50% FIEs allowed for international logistics services; JVs allowed for third-party logistics services from July 2002.	No commitment.			the first round of market-entry mechanisms for foreign-invested enterprises in the domestic transportation and logistics sector

Table 7.6. (continued).

Sector	Year Three: By Dec. 11, 2004	Status	Year Four: By Dec. 11, 2005	Status	Year Five: By Dec. 11, 2006	Year Six: By Dec. 11, 2007
Rail Transport	Majority foreign equity permitted.	Permitted in June 2004 (freight only)	No commitment		No commitment	Wholly owned FIEs allowed.
Road Transportation	Wholly owned FIEs allowed.					
Freight Forwarding	National treatment for minimum capital requirement (2 million to 5 million RMB).		Wholly owned FIEs allowed.			
Warehousing & Storage	Wholly owned FIEs allowed.					
Logistics	No commitment.					

The latest version of China's Catalogue Guiding Foreign Investment in Industry, revised in March 2002, reflects the country's WTO commitments

JV = joint venture. FIE = foreign-invested enterprise. BMN = Books, Magazines and Newspapers

(Sources: Compilation of the Legal Instruments on China's Accession to the World Trade Organization; The US-China Business Council)

2. Focus on value

Companies should bypass inefficient parties and middlemen, thus removing unnecessary layers of bureaucracy and streamlining distribution chains, as is already happening in many industries such as personal computers and consumer electronics. As the middlemen are removed, the capabilities of smaller players are improving. Dell has adopted this strategy through its direct customer model and by outsourcing logistics. Many other companies are quickly following suit. In selected cities in China, Nokia Corp. broke the widely used industry distribution model that follows a "manufacturer, general agent, regional distributor, second-tier distributor, retailer, consumer" pattern and supplied large regional distributors and retail outlets directly, thus cutting distribution costs and raising the company's market responsiveness.

3. Streamline Integrated Fulfillment

Companies and their distributors must integrate, centralize, and streamline Integrated Fulfillment functions, assets, infrastructure, staff, and operations. The goal is for a supply chain to become a separate, yet shared, organization across different business units – one just as important as sales and marketing.

4. Avoid duplicating services

As seen in mature markets, few companies can build or provide the full range of distribution services alone; partnerships and alliances with local distribution service providers are keys to success. By focusing on improving flows, companies' Integrated Fulfillment functions in China can be based more on the transmission of reliable and timely information and less on direct control of the physical movement of consignments.

5. Use technology as a key differentiator

The Integrated Fulfillment sector in China is typically slow to adopt new technologies, partly because of the complexity and seemingly daunting investment requirements of setting up an integrated IT platform. Growth of the sector will require greater sophistication in supply chain planning and will require companies to set up systems that make product movement visible, as a way to track demand in real time. Businesses seeking an efficient distribution chain will increasingly use technology to help make optimal decisions, integrate parts of the chain, and link the beginning of the supply chain to the end.

6. Manage risk effectively

A company's ability to recognize the many risks of conducting business in China, and to plan for them, can spell the difference between success and failure. The companies that understand the baseline cost structure and motivations of their Integrated Fulfillment operation, and track the total cost, will be in a better position to mitigate risks.

7. Build and retain talent for long-term success

According to a survey recently conducted jointly by The Logistics Institute-Asia Pacific in Singapore and the Logistics Institute of the Georgia Institute of Technology, both global and domestic third-party logistics service providers have identified lack of talent as one of the key challenges of operating in China. Value-added services in Integrated Fulfillment require more expertise and skills than most PRC providers currently possess. In general, the concept of functional or service excellence is relatively weak in the Integrated Fulfillment sector. Untrained PRC staff in many domestic distribution firms often have few incentives to perform well; some observers argue that 85 to 90 percent of distribution initiatives fail in China because of workforce errors. For the foreseeable future, training will be an integral part of any company's relationship with a PRC distribution service provider.

7.4.   China's Logistics Rules Obscure the Way Forward

It is important to note that logistics is not treated as a distinct category under China's WTO market liberalization commitments. The non-uniformity of foreign investment regulation in the various sub-sectors within logistics at different times, has led to both limited market liberalization and investment difficulties.

A notice on foreign-invested logistics companies in China, released in summer 2002 by the now-defunct Ministry of Foreign Trade and Economic Co-operation (MOFTEC), meets China's WTO commitment to allow foreign investment in joint ventures for wholesale and commission agent business for most imported and domestic products, but does not address specific products and does not cover retail operations. The geographic area covered by this provisional notice includes the cities of Beijing, Tianjin, Shanghai, and Shenzhen, and the provinces of Zhejiang, Jiangsu, and Guangdong (none of which appeared in China's WTO commitments). Presumably, the geographic scope of the notice will gradually be broadened until the whole country is covered.

The Notice on Relevant Issues Regarding the Experimental Establishment of Foreign-Invested Logistics Companies defines and permits two general categories of foreign-invested logistics operations: international logistics and third-party logistics. International logistics operations have import and export rights but are limited to 50 percent foreign investment, while third-party logistics companies may possess foreign ownership of more than 50 percent. Minimum registered capital for companies of either type is $5 million, and the project must be either an equity or cooperative joint venture.

These collectively constituted a series of restrictive investment criteria that effectively served to discourage foreign investment in this highly fragmented and un-seamless sector.

Either the largest foreign partner or the largest Chinese partner must be a company with "outstanding" logistics expertise, clearly a subjective definition.

The notice raises three difficulties for potential foreign investors in logistics operations:

- China's WTO commitments require the gradual phase-in of trading rights for all FIEs, beginning with minority FIEs in December 2002 and expanding to include wholly foreign-owned enterprises by 2004. The new notice allows two types of logistics companies – one with trading rights and one without. But given the trading rights phase-in (with minority foreign-owned third-party logistics companies being granted trading rights from December 11, 2002), the difference between the notice's two types of logistics companies is unclear
- Existing FIEs seeking to expand their business scopes to include logistics apparently qualify if they have existing investments of at least $5 million and do not need to add $5 million to their existing investments. Nevertheless, FIEs will be discouraged by the requirement that one of the partners have distribution expertise
- FIEs are currently allowed to distribute the products they manufacture in China; within three years, all FIEs will be able to import products. Will FIEs be allowed to ship imported products to their wholesale distributors – as they currently do with domestic products – or will they be required to continue to use a licensed logistics provider or importer to channel the imported products to their wholesalers?

The notice, which took effect August 24, 2002, overlaps with – and contradicts parts of – a Ministry of Communications rule on foreign investment in road transport that was released in November 2002. As more of China's WTO commitments come due, these and other questions will surely be answered, but whether they will be resolved quickly enough for foreign investors is another matter.

(Source: *The US-China Business Council*)

-What does our company have to offer, and want to bring, to China?

-What does the company need to do to establish an effective presence in China?

-Where do we stand today? Is our company developing sufficient knowledge of the market?

-Have we factored in all of the complexities in Integrated Fulfillment that are inherent in selling and moving our products in China?

-Will the service, technological know-how, existing international network, or management skills give our company a competitive advantage over potential international and local competitors?

-Do we know how to lower operating costs to support attractive pricing?

Figure 7.6. Self Assessment Checklist.

7.5. Proceed with Caution, but Do Proceed

As some successful companies are proving, effectiveness in Integrated Fulfillment helps to achieve market share and profitability growth in China. To realize the potential of their China investments, companies need to understand and plan for the market's potential and risks. Companies must also recognize that the Integrated Fulfillment operating model in China is very different from what they would find in the United States, Europe, and other developed markets.

Careful self-assessment and diagnostics should determine the purpose of a company's entry. Below are questions companies should ask themselves that go beyond the supply chain:

Any company investing in China must identify and address these questions. The answers should provide a template for easier market entry and better performance.

References

China Federation of Logistics & Purchasing (2005). *China Logistics yearbook 2005.*

China Society of Logistics and China Federation of Logistics & Purchasing (2005). *China Logistics Academy Frontier Report (2005–2006).*

EFT Research (2005). *China Logistics – Challenges and Opportunities.*

Morgan Stanley (2001). *China Logistics – Spot the Early Bird.*

The American Chamber of Commerce People's Republic of China (2004). *White Paper – Trading & Distribution.*

The American Chamber of Commerce People's Republic of China (2004). *White Paper – Transportation& Logistics.*

The Economist Intelligence Unit Limited (2004). *China Hand – The Complete Guide to Doing Business in China.*

Hau L. Lee and Chung-Yee Lee (Eds.)
*Building Supply Chain Excellence
in Emerging Economies*
©2007 Springer Science + Business Media, LLC

Chapter 8

LOGISTICS MANAGEMENT IN CHINA
Challenges, Opportunities and Strategies

Gengzhong Feng
*The School of Management
Xi'an Jiaotong University, PRC*

Gang Yu
VP, Amazon.com

Wei Jiang
*School of Engineering
Stevens Institute of Technology at New Jersey, USA*

Abstract: China, as the bright spot in the international economic development, has being joined into the global market. The sustained economic growth in China
is the driving force to the rapid development of logistics. In this chapter we
first present a general overview of China's logistics developments. Summarizing the current development of China's logistics industry, we can say that the
market and growth potential is very large, government enthusiasm is very high,
situations change very fast. The rapid growth of the economy and the huge demand for the logistics services sharply contrast the low operational efficiency
of domestic logistics system. Then the obstacles and challenges faced by the
logistics industry in China are analyzed. At last the opportunities and tactics for
developing logistics business in China are discussed.

8.1.   Preface

Today the world economics is being in a profound transformation period
during which the adjustment of global economic structure, the rapid progress

of science and technology, and increasing influence of multinational corporations are the three major directions of economic movements. Some new industries emerge by integrating high technology with modern management techniques. Among them, modern logistics, as an advanced organizational approach and management techniques, has been widely used in different countries and taken great effect in the national economic development.

In the early period of New China, national economy was still in the recovery stage which resulted in the under developed manufacturing and science and technology. Hence the logistics industry was also in its initial stage at that time. During the time of planned economy, the policy of national commodity circulation was to serve the manufacturing and people's living. Manufacturing products were sold, stored and distributed by state-owned enterprises (SOE). Under this mechanism, the circulation system was mainly partitioned into two parts: domestic trade and foreign trade. Domestic trade is divided into subsystems of business, materials, grain, and cooperative units of supply and sales, where each subsystem was further divided into wholesale, retail, warehouse, and transportation sectors. Final products were generally stored and transported by specialized storage and transportation companies in their own subsystem and related wholesale companies. The planning and operation of those storage and transportation companies are generally driven by wholesalers' sale demand.

From 1980's, the market economy system has been gradually established along with the introduction of national reformation and open policies. The traditional structure was broken and a new self-development industrial mechanism was developed quickly. This has resulted in a dramatic change in China's logistics system. On one hand, manufacturing enterprises looked for warehouses and transportation service providers in the logistic market directly, which stimulated the development of non-SOE storage and transportation companies which had been in non-circulation systems. On the other hand, traditional SOE warehouses and transportation companies changed their business from the alliance with wholesalers to the alliance with manufacturers and retailers. These changes gave rise to the importance of logistics services in China.

China's logistics industry is regarded as "the raw land without sunshine" by western countries. The total logistics cost accounts for 16.7% of gross domestic product (GDP) in China, which is higher than the average of 10–15% in developed countries. According to an estimate of the World Bank, if the logistics cost reduces to 15% during the 10th period of "Five-Year Development Plan" in China, manufacturing industry can save US$29B every year. According to a survey to the China's logistics industry by Chinese Association of Storage, "overall, the growth rate of logistics volume is higher than that of GDP, which indicates the healthy development of logistics industry. Under the current economic environment and the market condition, the logistics market is potentially very promising and especially, the third-party logistics has a huge room for development."

Based on the research in China's third-party logistics market conducted jointly by Mercer Management Consulting, Inc. and China Federation of Logistics & Purchasing, the growth rate of third-party logistics sector can be as high as 25% during 2000–2005. The main driving forces to this growth are, first, multinational corporations are outsourcing more and more of their business to China to reduce the supply chain cost. Second, Chinese enterprises are facing pressures of lowering costs and improving competitiveness, which increases the logistics outsourcing demand. Third, Chinese government's incentive policies also stimulate the development of the third-party logistics. Therefore, logistics development becomes an important part of the national economy development.

This chapter is organized as follows. First we present a general overview of China's logistics developments. Then the obstacles and challenges are analyzed. At last we discuss opportunities and tactics for developing logistics business in China.

8.2. Overview of Logistics Market in China

8.2.1 Sustained Economic Growth is the Driving Force to the Rapid Development of Logistics

The recent acceleration of economic reformation has proved a strong force for changes in China. Its direct outcomes, economic development, and market growth are increasing demand for transportation and logistics services. In addition, by changing regulation and encouraging investments in infrastructure, the government is playing a key role in further stimulating the demand for transportation and logistics. More interestingly, logistics innovations bring new business opportunities to traditional warehouse enterprises.

In the last 20 years, China economy keeps continuous high speed increment. Especially in the past decade, the economic growth is far higher than the world average. For example, as shown in Figure 8.1, China's GDP boosts by around three times from US$418.8B in 1993 to US$1650.73B in 2004, and ranks the top seventh in the world in 2000. At present, the output of many industrial and agricultural products such as steel, coal, and grain has been ranked the first place in the world.

Increased economic development and income growth in China have expanded the market of consumer goods to secondary and even tertiary cities. Given the large populations and increasing consuming power of these regions,

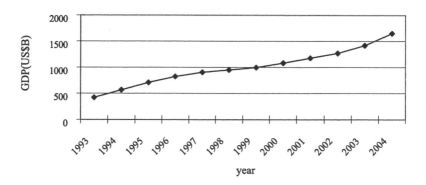

Data source: *China State Statistic Bureau*

Figure 8.1. Annual GDP increment (1993–2004).

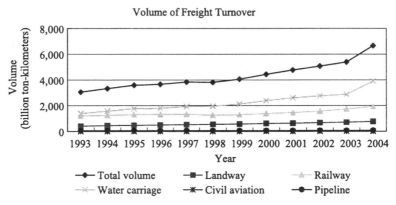

Data source: *China State Statistic Bureau*

Figure 8.2. Volume of freight turnover between 1993–2004.

manufacturers are now looking for practical solutions to reliably move their goods to these attractive markets, without losing control of the goods transported. Along with the economic growth in China, there has been a significant growth of freight forwarding volume, passenger volume, and seaport throughput. As shown in Figure 8.2 and Figure 8.3, from 1993 to 2004, the freight volume increased from 3,052B to 6,667B ton-kilometers, while the passenger volume increased from 785.8B to 1,632.4B person-kilometers. At the same

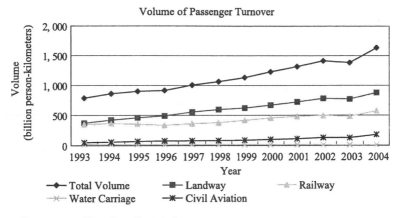

Data source: *China State Statistic Bureau*

Figure 8.3. Volume of passenger turnover between 1993–2004.

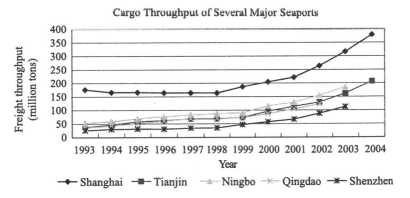

Data source: *China State Statistic Bureau*

Figure 8.4. Cargo throughput of several major seaports in China.

time, the freight throughput of all major domestic seaports also increased dramatically. As shown in Figure 8.4, the cargo throughput increased from 175.96 million to 379 million tons in Shanghai terminal, from 37.19M to 206.19M tons in Tianjin terminal – all expanded about 100% or higher in this decade.

By nature, the development of logistics is generally proportional to the total national economic output and the level of the economic development. During the 10th period of "Five-Year Development Plan", China is expected to maintain a strong growth of GDP at around 7%. The sustained economic growth will unquestionably become the major driver of logistic development and result in a large demand for logistics service.

Logistics development not only closely relates to economic growth, but also strongly depends on the economic output of a country. According to a report of International Monetary Fund (IMF) in 1999, China's logistics cost was US$167B and third-party logistics market share was US$55.8B. Based on China's logistics development projection and IMF statistics, the logistics cost and third-party logistics market share will expand to US$345B and US$145B respectively in 2010, which account for more than 100% increase in around 10 years.

8.2.2 Government Support Helps Boosting the Logistics Development

Along with the increasing demand from the society, China's logistics industry is now undergoing a rocketing development from its initial stage, which has been strongly supported by different levels of the government and policy.

As discussed before, Chinese government has set up a preliminary logistics development plan for the 10th period of "Five-Year Development Plan". The overall plan is to reduce the logistics cost by 2% in GDP by stimulating socialized and specialized logistics companies to achieve a proportion of above 50% in the whole logistics industry at the end of the period, actively exploring the possibility of developing a supply chain network through third-party logistics providers, and building up logistics business unions across different regions, industries, organizations and ownerships. In August 2004, the National Development and Reform Commission, the Ministry of Commerce, the Ministry of Public Security, the Ministry of Railways, the Ministry of Communications, General Administration of Customs, State Administration of Taxation, General Administration of Civil Aviation, and State Administration for Industry and Commerce jointly issued a document – "Opinions on Boosting Logistics Development in China" to further speed up the logistics development. As logistics attracts more and more attentions by various aspects of the society, many local governments also actively plan to develop modern logistics systems. Many big cities such as Beijing, Shanghai, Tianjin, Shenzhen, Wuhan, Dalian, Shenyang, and Ningbo, and major province municipals such as Shandong, Guangdong, and Fujian are planning and implementing the logistics development in their local regions.

The support from the different levels of the government has a great positive effect on the logistics development. In summary, the central government's logistics policy is as follows:

First, actively reform the current administrative behavior to build a healthy market environment for logistics industry. For example, in 2004, the current logistics administration was adjusted, which include standardizing the pre-approval process of the registration and license, reforming the administration in the freight agencies, and properly adjusting the repeat taxation in the logistics industry.

Second, take effective procedures and stimulating policies to promote the logistics development. For example, provide logistics companies with preferential benefit on land and taxation, and finance support. Encourage industrial enterprises to isolate their logistic activities such as material procurement, transportation, and storage from their major business and transfer them to the professional logistics companies which may further motivate the third party logistics business.

Third, strengthen the necessary foundational work to support the logistics development. For example, since the logistic industry has the characteristics of cross-department and cross-industry, in 2003, the administrative departments of the State Council established National Logistics Standardization Committee to try to build up and complete the standardization of logistics technology.

Fourth, strengthen the coordination of logistics administration, which includes building up a national logistics coordination mechanism led by the National Development and Reform Commission and consisted of relevant national ministries and associations. The members include the National Development and Reform Commission, the Ministry of Commerce, the Ministry of Railways, the Ministry of Communications, the Ministry of Information Industry, General Administration of Civil Aviation, the Ministry of Public Security, the Ministry of Finance, State Administration for Industry and Commerce, State Administration of Taxation, General Administration of Customs, General Administration of Quality Supervision, Inspection and Quarantine, Standardization Administration of China, and etc. The main functionality is to provide the policy for modern logistics development, coordinate national logistics development plan, research and solve the problems encountered, and organize and push the logistics development.

8.3. Obstacles and Challenges

No matter how exciting the future will be, several daunting challenges must still be tackled in building a transportation and logistics business in China. Success will hinge on judicious decisions that take into account regulatory uncertainties and unfavorable aspects of economic policy, current limitations in the transportation and logistics sector, and players' own capabilities. The following summarizes problems and challenges faced by China's logistics industry.

8.3.1   Low Efficiency and High Cost Resulted From Self-Logistics by Manufacturing and Commercial Enterprises

Since the logistics expenses account for a large proportion in the total cost of manufacturing and commercial enterprises, the efficiency of logistics operation directly affects the production efficiency and the management performance. As shown in Table 8.1, there is a significant gap between China and developed countries in terms of logistics cost and industrialization. Some estimates put the cost of transporting goods in China at up to 50% more than that in developed regions such as Japan, Europe and North America. Table 8.1 compares China with several countries on the percentage of logistics cost in GDP. The logistics volume generated from every 10 thousand US dollars is 4972 tons/kilometer in China, while it is only 870 and 700 tons/kilometer in USA and Japan. These costs are increased by lack of proper connections in logistics systems, low operations efficiencies, low service levels, and high tolls on roads. Logistics costs (including warehousing, distribution, inventory holding, order processing etc) are estimated to be two to three times the norm and in excess of 20%.

Influenced by the traditional planned economy system, many Chinese enterprises still keep the all-in-one business framework. They often rely on their own departments to accomplish logistics processes from material procurement to product sales. According to a survey done by the Research Institute of Market Economy in the Development and Research Center of the State Council, around 36% and 46% of logistics business on raw materials in manufacturing companies are done by the companies themselves and their suppliers respectively. In commercial companies, the percentage becomes 76.5% and 17.6% respectively. At the same time, most companies hold various logistics facilities which account for an important part of the whole property. This kind of self-logistics considerably increases the business cost of industrial and commercial

Table 8.1.   Percentage of Logistics Cost in GDP.

Country/Region %	China	Taiwan	Hong Kong	Singapore	Japan	USA	England
Percentage in GDP	16.9%	13.1%	13.7%	13.9%	11.4%	10.5%	10.1%

Data Source: International commercial technology, 2002 May

enterprises, limits and postpones the development of highly efficient and professional logistics service. It becomes a bottleneck in rapid development of China's logistics.

In the logistics operations, low exposure to the society and self-service often result in a lack of proper connections in logistics systems, low operating efficiencies, and low service performance. According to recent surveys conducted by the Research Institute of Market Economy in the Development and Research Center of the State Council, around 70% of on-road trucks transport their own goods, with around 37% are empty loaded, and the average speed of transportation is only 50 kilometers/hour. Moreover, poor facilities and management are responsible for high levels of loss, damage and deterioration of stock, especially in the perishables sector. The annual losses due to packaging, loading/transportation, improper storage are US$1.8B, US$6.1B, and US$0.4B, respectively. Part of the problem comes from the insufficient specialized equipments, i.e. proper refrigerated storage and containers, while it is also partly due to lack of training. On the other hand, the floating capital is often over occupied in a company and results in long turn around time. In 1992, the total floating capital in state-owned industrial enterprises was US$122.1B with turnaround rate of 1.65 times/year. In 1999, these numbers became to US$375.4B and 1.2 times/year. Moreover, the turnaround rate was only 2.3 times/year for state-owned commercial enterprises in 1999. However, in Japan, the turnaround rate is 7.5–8 times/year for manufacturing enterprises and 15–18 times/year for non-manufacturing enterprises (including wholesale and retail). The turnaround rate for typical multinational chain groups such as Wal-Mart, Metro, and Carrefour is often as high as 20–30 times/year. This comparison helps explain the large volume of raw material storage and work-in-process inventory in China. Essentially, most Chinese enterprises still adopt purchase-for-inventory rather than purchase-for-order policy. It is generally hard to achieve zero- or low-inventory due to the lack of real-time distribution. The average stock time for raw materials generally exceeds 30 days and large amount of floating capitals are thus occupied by inventory which results in the logistics cost accounts for around 30–40% in the total production cost and excessive shares in total sales. On average, the inventory turnaround time is 45 days for Chinese manufacturing enterprises and 35 days for commercial enterprises while the turnaround time of American automotive, electronics and other retail products is only 12 days in 1997.

8.3.2   Unbalanced Logistics Development Restricts the Business Expansion of Manufacturing and Commercial Enterprises

China's success in joining WTO further opened the national market and attracted many international logistics giants entering China gradually. As different types of logistics companies keep emerging, the competition in the logistics market is becoming more and more intense. According to the existing China logistics market's patterns, there are mainly four types of logistics companies: 1. those supported by the traditional transportation and storage systems, 2. those emerged from the new economy, 3. those originated from the manufacturing enterprises with self-service logistics, and 4. those international logistics companies in China. The characteristics and examples of these types are summarized in Table 8.2.

Although undergoing a fast development with an increase rate of 20–30% in the market volume, most of the third-party logistics companies came from the traditional transportation and storage companies and have not yet achieved superiority in networks, information, and resource utilization regardless of the company's capital size. There are very few logistics companies powerful enough to cover their business in the national market. Due to the limited business scale, no single logistics service provider can achieve a significant market share, usually below 2%. Moreover, most third-party logistics business in China concentrates on some several developed regions due to the unbalanced economic development. As shown in Figure 8.5, nearly 80% of third-party logistics business revenues come from Yangtze River and Pearl River economical zones. Even in these advanced areas, most of the logistics companies can only provide regional logistics service. Logistics companies outside the Yangtze River, Pearl River and Beijing-Tianjin areas are normally much smaller and lack of the ability to provide networked logistics services. In addition, logistics companies' comprehensive abilities in providing logistics services are also relatively weak with about 85% of the total income still coming from the basic services like storage and transportation. The profit coming from the value-added services like repackaging, re-labeling, return, maintenance and financial service only accounts for very little proportion in the total revenue.

Both foreign and domestic companies who are eager to develop their business in China are heavily restricted by the logistics companies' service capabilities, therefore, encounter difficulties in market expansion. For example, the three big retail giants – Wal-Mart, Metro, and Carrefour, who are very good at central purchasing and distribution, have found troubles in their logistics, which result in the failures in their business expansions to some extent in recent years. According to a survey covering more than 100 multinational companies

Table 8.2. Different types of logistics companies in China.

Types	Examples	Characteristics
Traditional transportation and storage companies	China Materials Storage and Transportation Corporation, China National Foreign Trade Transportation (Group) Corporation, China State Post Bureau	(1) Big state-owned enterprises with national networks and a lot of transportation and storage properties (2) Good relationship with central and local governments (3) Relatively high proportion of redundant personnel and low efficiency (4) Internal focused company culture, but not client and performance oriented
New logistics companies	EAS International Transportation Limited, P. G. Logistics Group	(1) Private-owned or joint venture with focus on regions, services and clients (2) Fast growth with relatively high efficiency (3) Only possess limited fixed asset (4) Lack of strong financial support for market expansion (5) The main obstacle for quick expansion is the internal management system itself
Logistics companies originated from manufacturing companies	Haier Logistics	(1) Provide service to internal and external clients with focus on internal clients (2) Specialized in particular areas (3) Have limited assets but good networks (4) Weak at marketing, can't attract more external clients (5) Depend on the parent companies for strategic and future directions
Foreign logistics companies	UPS, FedEx, TNT	(1) Strong oversea networks (2) Rich professional knowledge and business experiences (3) Good business relationship with international clients (4) Advanced information technology (5) Strong financial support from headquarters (6) Still limited business in China with relatively high costs

who have their business in China, these companies have around 90% of their logistics outsourced to third-party logistics companies which are mainly oversea companies. However, the ratio of inventory time between raw materials and final products in China is 30% higher than their oversea companies.

In summary, the unbalanced logistics development can be attributed to the following reasons.

1. Incorrect competitions among manufacturing and commercial enterprises create obstacles for logistics companies, and a bad partnership is typically established in the supply chain. In recent years, many multina-

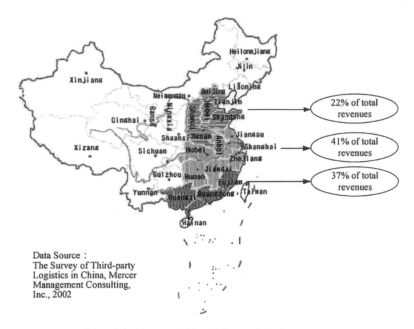

Data Source :
The Survey of Third-party
Logistics in China, Mercer
Management Consulting,
Inc., 2002

Figure 8.5.   Layout of China third-party logistics revenues.

tional retail giants who have entered China brought in an excessive competition in domestic retail industry. Consequently, the average profit rate
of the retail industry is much less than the international level. Because of
based on the strong financial support, the big retail companies are able to
keep the low-price competition in China. For example, the average profit
margin of Parkens' Shop under the Huchson Group is just above 1% in
China, less than 1.5% in Hong Kong. Due to the over competition and the
current economic development of buyer's market, the manufacturing and
commercial enterprises try very hard to make more profits by asking their
suppliers to pay for the logistics cost. Since the agreements between the
manufacturing and commercial enterprises and the suppliers are usually
very tough, the suppliers are under the big risk of losing money. Accordingly, those suppliers usually squeeze the logistics service providers by
reducing and delaying their payment. Even worse, as the logistics companies have to serve thousands of middle to small sized suppliers, they
usually have great difficulties in getting back their money. This kind of
situation constantly causes a lot of headache to different parties involved.
2. Another important reason is the lack of investment in logistics. In China,
logistics is a capital concentrated industry. The logistics companies face

a great pressure on getting proper investment. Their customers such as the manufacturing and commercial enterprises generally require a certain amount of deposit when the logistics companies ask for the logistics services, and even worse their payment for service is generally $3 \sim 6$ months later. This results in the low capital turnaround rate for logistics companies, at most $4 \sim 5$ times annually. Furthermore, the un-smoothness of financing channels adds extra difficulties to the business of logistics companies and their growth.

3. It is difficult to coordinately utilize the resources. Although several big state-owned logistic companies possess good logistics facilities and resources, (for example, China Materials Storage and Transportation Corporation has 63 big logistics centers nationwide with the total area of 13 million square meters and 129 railroads), they cannot take advantage of their resource superiority because of the fragment management and the lack of close business cooperation among their different logistics centers due to various systematic problems. Therefore, they can just usually provide regional logistics services. For the big international logistics companies, as they just started their business in China, their service cost is relatively high and the service scope is relatively small based on their limited storage and transportation resources. Due to the limitation of the logistics network, the domestic logistics companies focus more on the domestic business opportunities while their international counterparts more on the multinational clients.

8.3.3 The Development of Third-Party Logistics is Still Facing Systematic Restrictions

Although different levels of the government have realized that the development of logistics industry is now an important part of China's economic development and are trying to improve the logistics market environment, logistics development is still facing some systematic problems that cannot be solved fundamentally due to historical and systematic reasons and different beneficiaries.

First, regulations are created at a number of different tiers and imposed by national, regional and local authorities in China. Regulations often differ from city to city, hindering the creation of national networks. For example, in transportation management, China adopts the branch gate management system based on different transportation modes and the same structure repeats itself from central to local government. This type of vertical division management has a negative impact and restricts China's logistics development. It not only allows a lot of overlapping and redundant in bureau hierarchy which makes it hard to cooperate and coordinate among different departments and regions, but also fragments logistics processes into segments operated and controlled by different departments and regions. This fragmented system no longer satisfies modern logistics development and considerably constrains the development.

Second, government and business are still not completely separated, hence putting negative effects on logistics industry to be developed in a regulated and healthy way. The phenomenon that government departments directly or indirectly participate in the enterprise management still exists, especially in the railway system. For example, the seaport administration and dock operation are always tied together. This, on one hand, affects the administrative functions to be fairly fulfilled by the government, and on the other hand, prevents logistics companies from improving their market competitiveness.

Third, there is still a lack of clear and effective policy. Under the multi-department, process- partitioned management system, although many policies and regulations have been issued, there are a lot of conflicts of interests among different departments and regions. Take pricing policy for multimodal transport as an example. Cargo container transportation adopts the free pricing policy of "new line new price, hot line high price", while bulk transportation still takes old pricing regulation defined by the government. Therefore, sometimes the former is obviously higher than the latter. This extremely harms the development of container transportation mode and directly affects the other logistics service development based on multimodal transport. Moreover, the taxation policy also has some factors that are disadvantageous to logistics development. For example, the transportation operation tax rate is 3% for transportation enterprises but 5% for logistics or storage enterprises and wholesalers who provide logistics service.

At last, according to a market research of third party logistics conducted by Mercer Management Consulting and China Federation of Logistics & Purchasing, as shown in Figure 8.6, government's restrictions, policies' fuzziness, and lack of professionals are the three major problems faced by domestic and international logistics companies in the China market. Searching for qualified logistics professionals is not easy for both domestic and international logistics companies. Here, the qualified logistics professionals include not only the mid to senior level of logistics management professionals but also well-trained technicians in logistics. Nowadays in China, the logistics professional advancement is far behind the market requirement. According to the analysis, the existing number of logistics professionals is 6 million less than the market demand in 2003. In the next couple of years, beside the shortage of logistics professionals in storage, transportation, distribution, and freight agency, more shortage will be the experts in systematic logistics management, import/export trading operation, e-commerce fields and the advanced international logistics

Challenges faced by domestic logistics companies

0% 10% 20% 30% 40% 50%

Challenges faced by international

0% 10% 20% 30% 40% 50%

Data Source: The Survey of Third-Party Logistics in China, Mercer Management Consulting, 2002

Figure 8.6. Major factors that restrict third-party logistics development in China.

professionals who are familiar with commodity delivery, capital turnover, accounting and related knowledge, and different operational methods.

8.4. Opportunities and Tactics

The rapid growth of the economy and the huge demand for the logistics services sharply contrast the low operational efficiency of domestic logistics system. On one hand, the rapid growth of economy creates huge logistics demands; while on the other hand, the low efficiency of logistics systems provides great opportunities for the new logistics players, which drives the healthy competition and development of logistics companies. In China, the development of the logistics industry faces both challenges and opportunities.

8.4.1   Outsourcing Logistics From Manufacturing and Commercial Companies

Since China is a huge country with complex geography and unbalanced natural resources, which causes an enormous commodity flow span in both time and space. "Northern coal shipped south", "Southern grain shipped north" and "Natural gas transport from the west to the east", and etc result in high logistics costs. At the same time, since small and medium sized companies account for a considerable proportion in China's industries, it is urgently required to lower the circulation cost, increase the logistics efficiency, and adopt the comprehensive logistics management.

Under pressure of low costs, a lot of manufacturing and commercial companies are seeking third-party logistics to outsource their logistics business in order to decrease the logistics cost. However, domestic manufacturing and commercial companies account for a much less proportion than the international companies in outsourcing logistics business. Within the constitution of outsourcing, at present, most of the manufacturing and commercial companies prefer to outsource their storage and transportation business. Among them, manufacturing companies more focus on trunk line service than local distributions while commercial companies more focus on local distributions than storage and than trunk line service. This indicates that the manufacturing companies' logistics demand is different from that of the commercial companies. However, in reality, the proportion of manufacturing and commercial companies who have outsourced their logistics comprehensively is quite small. An importation reason is that it is difficult to find an appropriate logistics company capable of providing the integrated logistics services. From the supply chain point of view, it is easier to outsource logistics in the sale stage than other stages. Logistics of raw material supply is usually undertaken by manufacturing companies themselves or the suppliers. As the raw material logistics directly affects the production itself, manufacturing companies prefer to establish appropriate alliance with their suppliers in managing the logistics.

According to an investigation of the logistics market by Chinese Association of Storage for consecutively five years, over 50% of the manufacturing and commercial companies are willing to look for new logistics companies and to outsource their self-handled logistics business to the third party logistics companies. This, on one hand, indicates that the market demand to the third-party logistics is very large with an upward trend, however, on the other hand, also reveals the unstable relationship between logistics supply and demand parties.

During the searching process for new logistics companies, the first concern of both manufacturing and commercial companies is the service quality, and then their comprehensive logistics capabilities. The criteria in searching third-party logistics are, in the decreasing order of their importance, professional experiences, reputations, networks, prices, strategic assets, coordinating capabilities of logistics resources, information system, and strategic matchabilities. To be successful, third-party logistics companies should improve their professional levels, supply capabilities and service qualities as quickly as possible to enhance the logistics outsourcing trend and establish the strategic cooperation relationship with other parties in the supply chain environment.

In terms of the services provided, besides the traditional logistics services like trunk line service, storage and etc., services such as local distribution, re-design of the logistics system, and loan settlement by demand are also required more and more urgently. Therefore, logistics consulting companies with core business of logistics system design and information consulting will have a big room for development.

8.4.2 Merging and Reorganizing Logistics Companies to Improve the Comprehensive Service Level

Along with the development of information technology and the acceleration of the global market, market competition is more and more intense with new products. Due to the technology advancement and diverse demands, product life cycle is also unceasingly reducing. The competitive environment that the companies have to face now has been changed a lot and the competitive pressure is becoming bigger and bigger, which brings more challenges to the logistics industry. Third-party logistics companies should try to provide comprehensive logistics services to their clients in accordance with their logistics rationalization. To do so, they have to get familiar with the development of the clients' logistics activities and possess capabilities for logistics system's development and innovation. Obviously, these activities are totally different from doing simple transportation, storage and proxy based on client's request.

Due to the low entry cost and bright future of the China logistics industry, not only a lot of traditional storage, transportation and express companies, but also a bunch of newly established logistics companies are trying to grab shares from the logistics market. In China, various companies titled "logistics" keep emerging in recent years. Under the intense market competition and rapid development of the logistics industry, more and more manufacturing and commercial companies are expected to outsource part of their supply chain business to the third-party logistics companies. Due to the large capital investment needed to build a high efficient global or national third-party logistics companies, more and more third-party logistics companies try to expand their services by merge and union. Meanwhile, investment companies who have a lot of fund available are actively looking for new market opportunities and using crosswise conformity as the entering point to the logistics market.

In China, the third-party logistics market has been experiencing many fundamental changes, which basically follow two main directions. One is to accomplish the logistics networking and scaling by horizontal integration among logistics companies to achieve scale expansion. The other is to enhance the ability of service conformity by vertical integration among logistics functions. Vertical integration combines different logistics functions such as storage, handling/carrying, loading and unloading, processing, distribution and transportation into a complete logistics system. It is the base of the horizontal integration which fully utilizes the functionalities of resources from different logistics companies to turn the single company's logistics service into that of alliance.

8.4.3   Commodity Circulations With Chinese Characteristics Provide New Demand for the Development of Logistics Companies

Manufacturing and distribution of staple commodities is crucial to national economy. Based on warehouse and logistics companics, the wholesale market takes a great effect in the staple commodities' circulation process. Different from many developed countries, commodity exchange market such as the wholesale market has been developed and continuously expanded in China during the past decades. It has become the most efficient commodity circulation pattern in promoting the economic development. According to the State Statistics Bureau, there were 93 thousand different commodity exchange markets by the end of 2001 with total exchange volume of US$396.9B, equivalent to 34.8% of GDP in that year. Among them, there were 3,273 large exchange markets whose exchange volume exceeded 12 million USD each year. Moreover, 10 provinces had more than 100 large wholesale markets each with transaction turnover more than US$12M each year. They are Jiangsu (445), Zhejiang (420), Shandong (318), Hebei (265), Guangdong (246), Liaoning (180), Hunan (122), Hubei (144), Anhui (112), and Henan (107). There is a rapid development of the commodity exchange market in recent years, which mainly concentrates on east coast and other relatively well-developed areas.

The development of the commodity exchange market is a continuous process of service innovation, which enormously promotes the advancement of commodity circulation and logistics industry. As the commodity exchange market combines multiple functionalities such as commodity circulation, transaction and logistics together, it becomes the collection and distribution center. Under the support of modern information technology, this type of commodity exchange and market organization format already showed a great vitality and bright future.

Due to historical reasons, the logistics function is still under development in commodity exchange market due to the lack of organization and excessive operational stages in the following aspects. One is that the logistics operations are still traditional storage and transportation functions and have not been integrated efficiently. The other is that the logistics resources have not been fully utilized in the supply chain. The cooperative logistics emerged recently takes alliance strategies such as information sharing and resource coordination in the logistics management to make different parties cooperating and advancing together through their superiority supplement. The key problems in the operational level are how to increase the efficiency of logistics distribution service and reduce the operational cost and how to build cooperative logistics service system. These also bring new opportunities to the integration and development of logistics companies.

In addition, information management is crucial to the success of logistics companies in supply chain. Logistics information management is a process that collect, process, analyze, apply, store, and distribute the logistics information. In this process, optimal resource allocation can be achieved by managing the factors involved in the logistics information activities such as people, technology, and equipment. Since logistics information is location- and time-sensitive, more attentions should be put on preventing the information from missing, distortion, and expiration during collecting, sorting, and processing. Besides technology support, effective management systems are important to information control and distribution.

Recently, E-commerce is emerging as an aid in modern logistics development and has been shown very efficient for the exchange market of staple commodities. It helps fasten and smooth the commodity distribution process to promote the alliance of logistics systems and wholesale markets. The alliance can further be integrated into nation-wide retail network based on Internet with standard service. Moreover, through E-commerce, traditional logistics companies can be improved in logistics solutions of transportation, distribution, and even processing.

Information technology integrates independent logistics entities in supply chain. The two direction of flows – goods flow and information flow – dominates a supply chain. Goods flow generates the information flow and information flow controls the goods flow. It is therefore highly demanded to promptly build logistics information systems and regional or national information exchange platform to facilitate the rapid logistics development.

8.4.4   Business Innovation and Integration Brings Great Potential to Logistics Industry

The integration of logistics with capital flow can greatly improve the logistics operation efficiency and its development. Recently emerged logistics financing is a typical product and innovative field which combines logistics with finance. It fully utilizes the supply chain environment, to bring a new means of financing for small and medium companies, hence producing a positive effect on improving the supply chain operation and increasing the management efficiency.

Different from traditional banking loan business which focuses on real estates or third-party prestige guarantee, logistics financing enables manufacturing and commercial enterprises to get loan service by using their products as mortgage. This new type of financial service is one of the financial derivatives, which gradually changes the relationship between banks and manufacturing and commercial enterprises in loan applications. It is also totally different from third-party prestige guarantee model where the guarantee side has to compensate for the loss. It relies more and more on the third-party logistics companies and thus forms a new cooperative relationship among banks, logistics companies, and loan companies. In consequence, logistics companies become the critical link in the logistics financing business. The corresponding management, control and service level are the major decision factors that determine whether the logistics financing can be developed, how it can be developed, its flexibility, efficiency and risk management level.

Logistics financing is a financial product which brings win-win outcomes to all parties involved. First it helps manufacturing and commercial enterprises get loans from banks to enhance their financing capabilities and capital utilization through temporal pawning of mortgage rights. Second it also helps reduce loan risks to banks since the real products and goods are used for pawns, which are managed by good logistics companies. Third the logistics financing is actually a new value-added business to logistics companies. It not only helps advance their traditional storage business, customize service for different clients, but also brings profits to the companies. Through this business, all three parties can win and share margins brought by reduced costs and promoted logistics business. In the point of view from the whole society, the logistics financing greatly improves efficiency and scale of the whole production and circulation process.

The logistics financing started from 1999, and has expanded fast. Its business has covered various industries such as steel, construction materials, petroleum, chemical engineering, and electrical appliances, etc. Till 2002, for example, credit limits had reached US$300M for Guangzhou Branch of Shenzhen

Development Bank and accredited more than 100 companies. As another example, Guangdong Nanchu Logistics Management Co. expanded its business in 2003 through logistics financing. It has 5 banks as partners, and helps manufacturing enterprises financed 485 million USD and gains breakthroughs in its own storage business. It now starts to develop networks and explore opportunities of storage licensing in different locations. In 2003, there were around 20 logistics centers within China Materials Storage and Transportation Co. who were doing logistics financing business and had credit limits of US$242M. Through six years of development up to 2005, logistics financing has been shown a great potential in China's logistics market.

8.5. Conclusions

In the last 20 years, free trading, growth and integration of global capital market, and the progress of information and telecommunication technology created an increasing global market. The originally fragmented national and regional markets are gradually becoming a united global market. China, as the bright spot in the international economic development, has being joined into the global market. Along with more and more detailed division of labors, and more and more complexity of the economic structure, logistics companies and manufacturing enterprises rely on each other more strongly. Summarizing the current development of China's logistics industry, we can say that the market and growth potential is very large, government enthusiasm is very high, situations change very fast, and both opportunities and challenges exist.

Under this situation, both domestic companies and international companies who have business in China should face the challenges and pressures resulted from the changing environment on time, actively grab any opportunities, and seek development when circumstances permit. In this changing and developing environment, many important social and practical questions and theoretical innovation problems with Chinese characteristics are faced by the researchers. For example, as an important infrastructure, commodity exchange market and the logistics parks have extremely important influence on the development of logistics industry. When they are developed together, they become typical examples of networked supply chain systems, which have great effects in improving the regional economy. However, it also results in some theoretical and practical problems such as the crucial distinction between commodity market and future market, the layout of logistics parks, etc. As the resources shortage problem becomes more and more severe and people's environmental protection consciousness becomes stronger and stronger, the reverse logistics activities such as merchandise returning and repairing, product re-using, and waste processing are gaining a lot of attention. Some policies to be issued regarding manufacturer extension system, etc, will greatly affect the operation and business of manufacturing enterprises. These problems have great social significance and need to be investigated and solved by both researchers and entrepreneurs.

References

Alberts, L.H., Randall, H.L., and Ashby, A.G. (1997). *China Logistics: Obstacle and Opportunity*. MMC Viewpoint.

All China Marketing Research Co., Ltd. (2004). *The Research Report of Logistics Industry in China*.

Armstrong & Associate, Inc. (2004). *China Logistics Report 2004*.

China Federation of Logistics & Purchasing (2003). *China Logistics Year Book 2003*.

Chinese Association of Storage (2002, 2003, 2004). *The Supply and Demand Investigation Report of Logistics Market in China*.

Gengzhong Feng et al. (2003). *Modern Logistics and Supply Chain Management*. Xi'an Jiaotong University Press.

Gengzhong Feng et al. (2004). *Planning and Design of Logistics and Distribution Centers*. Xi'an Jiaotong University Press.

International Business Daily (2002). *The Survey of Logistics Service Demand from Oversea Companies in China*.

The Research Institute of Market Economy, Development and Research Center of the State Council of the People's Republic of China (2001). *The Present Situation and the Future of Logistics Industry in China*.

McKinsey & Company, Inc. (2002). *China's Evolving Logistics Landscape*.

Mercer Management Consulting, Inc. and China Federation of Logistics & Purchasing (2002). *3PL Marketing Survey in China*.

U.S. Consulate (2003). *China Logistics Profile*. Global Agriculture Information Network, GAIN Report.CH3833.

Hau L. Lee and Chung-Yee Lee (Eds.)
*Building Supply Chain Excellence
in Emerging Economies*
©2007 Springer Science + Business Media, LLC

Chapter 9

CONNECTIVITY AT INTER-MODAL HUB CITIES
The Case of Hong Kong

Raymond K. Cheung and Allen W. Lee
The Hong Kong University of Science & Technology

Abstract: As the global economy continues to be increasingly integrated and the transit time required is getting shorter, the inter-modal freight activities around a major hub have become more complex and intense. Although these activities typically represent a very small portion of the global distribution network in terms of distance, they contribute significantly to the total transportation cost. Using Hong Kong, the busiest port in the world, as an example, we illustrate the challenges and issues on managing the connection activities in hub cities. In particular, how cross-border regulatory policies lead to very low utilization of resource is discussed. With partial relaxation of these policies in sight, a new operating model of using a freight relay center for connection activities is emerging. We present a decision model on how freight flow and driver flow can be managed in this setting both at tactical and operational levels and design a solution approach.

9.1.   Introduction

With the process of globalization intensifying, both the volume and the complexity of the logistics flow in the global supply chains have rapidly increased. These factors, together with the ever-increasing customer expectation and speed requirement, have imposed serious pressure on major international hub operators for improvements to efficiency. In the last two decades, there has been a significant productivity improvement within terminal facilities such as container terminals and air-cargo terminals. For example, the designed capacity of Hong Kong Air Cargo Terminals Limited (HACTL) was 2.6 million tons when it commenced operations in 1998, and with a number of stream-lining processes and optimization projects, the effective capacity has recently increased to 3.5 million tons.

At international hub cities, there can be many possible types of inter-modal connections, such as truck-air, truck-ocean, air-water-air (connecting the cargo terminals in a city and those in another city through hi-speed vessels), ocean-barge-truck, and ocean-barge-rail services. These inter-modal activities have created a complex and challenging environment for logistics service providers for increasing the speed in handling freight while keeping the operating costs under control.

The major logistics infrastructures in Hong Kong (such as container and air-cargo terminals) are very efficient. However, the cost of moving freight between Hong Kong and Pearl River Delta (PRD), the manufacturing base in southern China, is very high. For example, it costs $300[1] more to ship a 40-foot container from a production plant in Dongguan, the major manufacturing center in PRD (see Figure 9.1) to Los Angeles via the Hong Kong's Terminals and via the Shenzhen's Yantian Terminals (The Better Hong Kong Foundation, 2004). Given that millions of containers are trucked to Hong Kong from PRD every year, shippers are paying a premium of over a billion dollars a year in order to use Hong Kong's facilities. The cost gap has accelerated the decline of Hong Kong's market share of exporting container in PRD, which has dropped from over 75% in 2000 to less than 50% in 2005 (HKSAR, 2004).

As demonstrated later, while higher land and labor costs are contributing the cost gap, the most significant portion, however, comes from the extremely low utilization of the resources used for the connection activities. This low utilization in turn is caused by various regulatory policies that govern the cross-border freight and driver-tractor flows, and the existing operating models. To increase the utilization, several initiatives to relax the policies are being pursued and new operating models, such as the uses of preloading containers, bonded consolidation facilities and freight exchange centers, are emerging.

The purpose of this chapter is three-fold. First, we provide an overview of the landscape and recent developments in the PRD, with an emphasis on the surrounding area of Hong Kong. Second, we show how the regulatory policies and existing operating models cause the low resource utilization and describe the upcoming logistics operating environment. Third, we demonstrate how decision modeling can be used in the new environment, in particular, the freight exchange centers. We present a decision model for the problem and design a solution approach. This chapter is organized as follows. Section 2 gives the

---

[1] In US dollars unless specified otherwise.

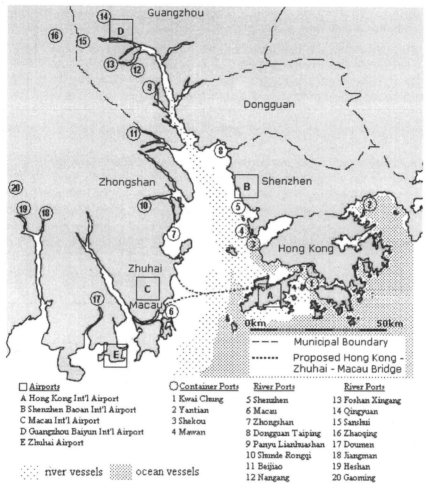

Figure 9.1. The Pearl River Delta Region.

overview of logistics development, Section 3 discusses the sources of low utilization and Section 4 introduces the decision modeling.

9.2. Logistics Development in PRD

Since the beginning of the economic reform in China in 1980, Hong Kong has rapidly expanded its manufacturing capacity in the PRD. Despite the fact that the percentage of manufacturing workers in Hong Kong's total labor force has dropped dramatically in the last two decades from 50% in 1980 to 7% in 2005 (HKSAR, 2005), the actual manufacturing output has mushroomed because over 60,000 Hong Kong-linked companies have established tens of thousands of factories in the PRD, employing more than 10 million workers (Brand Hong Kong Management Office 2005). With the PRD as the hinterland, the exports and imports of Hong Kong were $224 billion and $231 billion respectively. The total amount was three times of the GDP of Hong Kong. In the same year, the exports and imports of Mainland China were $438 billion and $413 billion respectively despite the fact that China's population is 200 times more than Hong Kong's population (International Monetary Fund, 2005). The massive industrial output of the PRD demands a strong logistics infrastructure to support. Due to the lack of deepwater harbors in the central and the western part of PRD, Hong Kong has been the main entrée port to support the PRD area. In 2001, the logistics industry was identified by the Hong Kong government as one of the four industries (the other three are finance, business services, and tourism) that would drive Hong Kong's economy in the next two decades.

Handling over 20 million twenty-foot equivalent container units (TEU) in 2004, Hong Kong has been the busiest container port in the world in 12 out of the 13-year period from 1992–2004. The operations at the cargo hubs were regarded as enormous, efficient, but expensive by world standards (Wang et al. 2005). In the late 90s, the ports in Shenzhen, in particular, Yantian Port (see Figure 9.1), just a few kilometers from Hong Kong started to develop and enjoyed a phenomenal growth. From 2001 to 2004, the annual growth rate of the container throughput in Shenzhen was over 40% while that in Hong Kong was merely 4–5%. As indicated in Figure 9.2, Shenzhen was not even in the top 30 ports in 1994 and was only 17 in 1998. At the end of 2003, Shenzhen's container throughput exceeded the 10 million TEU mark, becoming the 4th busiest port in the world. With Shenzhen's rapid gain of the market shares of the export containers from the PRD to US and to Europe, the leading position of Hong Kong's port was seriously challenged.

For air-cargo, Hong Kong has also topped the world in terms of international air-cargo tonnage since 1996. In 2004, the air-cargo terminals in Hong Kong handled over $140 billion worth of goods a year, which was about 30% of the total trade through Hong Kong. Compared to the situation of container terminals, the air-cargo terminals in Hong Kong currently still enjoy a double-digit growth. The air-cargo throughput in Hong Kong is still very high among all the major airports in China (see Figure 9.3), which was approximately twice the one of Shanghai's terminal, the second busiest in 2003. Following the global trend that an increase percentage of international freight is being handled by air rather than by ocean, more airports are expanding their air-cargo handling capabilities. In the PRD, there are five major airports within 110 km from each

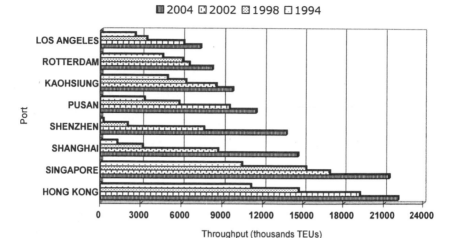

Figure 9.2. Throughput of the top eight container ports from 1994 to 2004.[2]

other. Recently, Guangzhou has opened a new airport in 2004 that has an air-cargo throughput capacity of one million tons, providing a cheaper option for shippers who have been using Hong Kong's terminals. It offers a direct threat to Hong Kong's leading position in air-cargo handling. In addition to air-cargo, integrators such as FedEx, DHL and UPS are setting up their own hub facilities in the region (DHL opened at new hub in Hong Kong in 2004 and FedEx will open a hub in Guangzhou by 2008). On one hand, this represents the geographical advantage of Hong Kong which is right at the center of the 14 major markets (e.g. Tokyo, Seoul, Singapore) in the Far East. On the other hand, the increased volume has made coordinating the freight flow more difficult.

The bottleneck of freight handling in Hong Kong comes from the connection between the terminals and the factories in the PRD. Unlike most major ports, Hong Kong suffers from the fact that there is no rail connection to the container terminals. Therefore, containers are mainly transported across the border by trucks whose productivity, as discussed later, is very low. The land connection to the air-cargo terminals, on the other hand, is relatively long and inefficient due to the airport's location. Furthermore, under the one-country-two-system regime, Hong Kong and the rest of China have different regulations, economic policies, and operating practices. These differences have caused great challenges to the cross-border freight activities. In facing these challenges, the Hong Kong government has worked closely with its mainland

---

[2] Source: Dubai Ports Authority, Port of Hamburg.

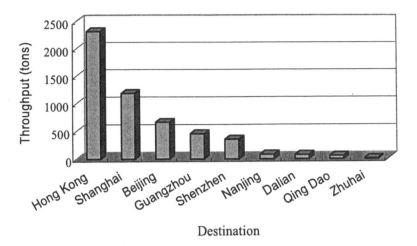

Figure 9.3. Throughput of the major airports in China (2003).[3]

counterparts to co-develop and re-align the logistics infrastructure in the Pearl River Delta (PRD) region. This includes the new container terminals, the logistics centers, and the proposed bridge linking Hong Kong and Macau that will greatly shorten the traveling time between the Western PRD to Hong Kong from several hours to less than an hour. Other than the physical infrastructure, Hong Kong needs to overcome many complicated cross-border issues, including the lowering of operating costs and the development of new services to maintain its competitive edge.

9.3. Challenges on Connectivity and New Operating Models

Handling the connection activities around an inter-modal hub is costly and complex. In terms of cost, although these activities account for less than 10% of travel distance of a shipment, they represent as much as 40% of the total transportation cost (Morlok and Spasovic, 1994). In terms of operations, inter-model hub facilities are quite different from the single-modal hub facilities. Firstly, shipments coming from different modes require different types of entrance points or handling methods. For example, at container terminals, the berths for barges and the berths for ocean-going vessels are different where

---

[3] Source: Dragonair Cargo.

the latter require a deeper water draft. Secondly, at an inter-modal hub, cargos need to be built up, broken down, temporarily stored, and passed through custom clearance. These activities prohibit the more efficient cross-docking operations (moving shipments directly from a dock to another) that typically can be done in a single-modal hub facility. Thirdly, inter-modal operations require multiple handling of shipments, resulting in longer times even though the transits are mostly short-haul. This is because the scheduling of inter-modal shipments is less flexible than single-mode, as shipments have to wait until the scheduled departures.

9.3.1 Impact of Regulatory Policies on Resource Utilization

The $300 cost gap per 40-foot container mentioned earlier comes from two major components. The first is the Terminal Handling Charge (THC) which contributes $100 to the $300 cost gap and the second is the land transportation cost between Dongguan and Hong Kong which contributes $200 to the cost gap. Shipping lines charge THC to shippers for the cost of moving containers between the shore and the vessels. Part of the THC will be paid by the shippers to the terminal operators for this service. It was first introduced by the Far Eastern Freight Conference in 1990. The THC in Hong Kong is among the highest in the world and there has been little transparency on the price-setting of THC (Sun, 2003). The container terminals of Hong Kong are quite unique in the world in that they are fully funded (including the cost of land formation for the terminals), owned and managed by the private sector as compared to other ports where governments are directly or indirectly involved in the capital investment and management. The high private capital investment results in correspondingly high THC. To compensate for the higher THC, the terminal operators provide an accelerated vessel turnaround time – typically within 10 hours (HKSAR, 2003), and provide better services (HKSAR, 2004). Recently, a new terminal opened in mid-2004 and the new operators (one is Singapore-based and the other is Dubai-based) commenced their operations in Hong Kong in late 2004. With overall greater capacity and new competition, the $100 THC gap is likely to be narrowed.

The second component, the land transportation cost gap, is more substantial and is more difficult to reduce. There are tangible factors and intangible factors contributing to the cost gap. The tangible factors are the higher salaries and operating costs in Hong Kong including the license fee for Hong Kong drivers to operate in China. The intangible factors include the productivity of the truck trip and the time required to pass through the border.

Currently, a typical driver can only perform 1.2 trips per day on average (HKSAR, 2004). If we combine the differences between Hong Kong and Shenzhen in terms of tangible factors (i.e. the salary, operating cost, license fees) in a month and divide it by the number of trips per driver, the tangible factors only account for 40% of the $200 land transportation cost gap (The Better Hong Kong Foundation, 2004). In fact, given the relatively short distance (less than 80 km) between the container terminals and the major production sites in the PRD, a truck driver should be able to perform 2 or even 3 round trips per day. In practice, the number of trips is only 1.2. Even worse, in a round trip, the container is loaded one way and is empty on the return. If the number of trips per driver per day can be increased to 2 and the unnecessary movements of empty containers can be greatly reduced, the tangible factors would account for less than 15% of the cost gap.

The low trip productivity (and the low resource utilization) is due to the restrictive constraints in managing resources created by some regulatory policies. For example, there is a 4-up-4-down policy requiring that when a quadruple of driver-tractor-chassis-container going from Hong Kong to PRD, the exact same quadruple has to come back to Hong Kong as the same unit. The origin of the policy can be traced back to the early 90's when empty containers were scarce in China. At that time, empty containers were considered as a commodity rather than as an item of transportation equipment. The policy was set up to avoid illegal import of containers to China. Under this "coupling" policy, the quadruple is considered as a single resource, rather than 4 individual resources. Consider that a loaded container is moving from Hong Kong to a manufacturer in Dongguan (see Trip 1 in Figure 9.4). In addition to the travel time, drivers need to wait at the borders (Hong Kong – Shenzhen and Shenzhen – Dongguan). After it arrives, goods are unloaded from the container at the manufacturing site, which may take hours. During this period, the driver, the tractor and the chassis need to stay there even they are idle. Moreover, unless that there is a load to be shipped back to Hong Kong from the same manufacturer and to board a vessel of the ship liner who owns this container, the quadruple has to be moved back empty to Hong Kong. This leads to a severe under-utilization of the resources. In the case of Trip 1, the productive time for the driver is less than a quarter of his duty hours. Despite the relatively short distance, the driver can hardly make two round trips to Dongguan in a day. Furthermore, due to the need of returning the emptied container to Hong Kong, the unproductive mile is very high.

In addition to the 4-up-4-down policy, there is a 1-driver-1-truck policy requiring that the Hong Kong drivers operating in China need a special license. This license is currently only issued to a particular driver who has to operate on a specified tractor. Whenever the driver is not working, the tractor is idle. Such a policy causes a low utilization of the truck time. For example, if both trips 1 and 2 are handled by the same tractor, then these two trips cannot be done on the same day because the driver who operates this tractor cannot finish

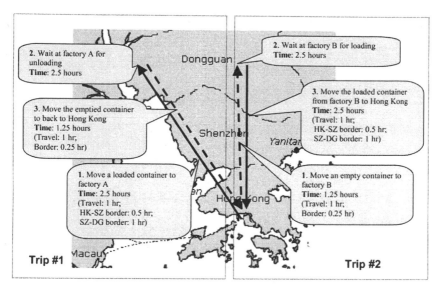

Figure 9.4. Two cross-border truck trips under the 4-up-4-down policy.

two trips within a day. Without this policy, another driver may use this tractor for trip 2. If the above policies were removed, the utilization of drivers and tractors can be drastically increased.

In additional to these policies, the quadruple needs to spend a considerable amount of time at customs clearance check points. For a trip to China, it needs to pass through the Hong Kong – Shenzhen border. After that, it needs to go through the Shenzhen – Dongguan border as they belong to different economic zones.

9.3.2   New Operating Models

As the business environment evolves, we see that several new flow patterns are emerging. This includes the consolidation of the haulage companies, the partial relaxation of the regulatory policies, the development of the barge feeder network, the flow coordination and the information standardization.

New Cross-Border Trucking Operations and Freight Relay Center. Currently, China is the biggest container producer in the world and there is no incentive to illegally import empty containers to China. Thus, the 4-up-4-down policy is an anachronism and outdated. At the present time, there have been rounds of joint meetings between HKSAR officials and their Mainland China counterparts, trying to relax or eliminate these policies. While a complete elimination of them could take many years, a partial relaxation is likely to occur soon. For example, instead of a 4-up-4-down policy, a possible new version could be the 2-up-2-down policy where the driver and the tractor are coupled together while the container and the chassis can be left at the factory for loading and unloading. Figure 9.5 illustrates that under this new environment, a single driver can move more loaded containers per day (instead of one per day shown in Figure 9.4). It is clear that the utilization of drivers can be doubled. Notice that this new environment requires containers to be preloaded at the factory while emptied containers can be stored on site at the factory.

After the regulatory policies, the second most significant contributor to the $200 trucking cost gap is the salary differential of the Hong Kong drivers and the PRD drivers. The difference of monthly salary is around $1200. Under the current situation of 1.2 trips per day, the difference is $50 per trip (with the policy relaxation, the difference can be narrowed to below $30). One way to lower the cost is to develop a freight relay center right across the Hong Kong-Shenzhen border. The container trucking in PRD will be handled by the PRD drivers while that between the center and Hong Kong will be handled by the Hong Kong drivers. From Figure 9.6, we can see that the trip time for a Hong Kong driver can be lowered substantially, making the handling of two or even three round trips possible. Furthermore, with the possibility of stacking empty containers at the relay centers, the Hong Kong drivers can mostly move

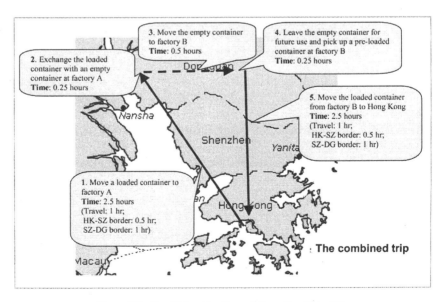

Figure 9.5. Combining trips through the policy relaxation.

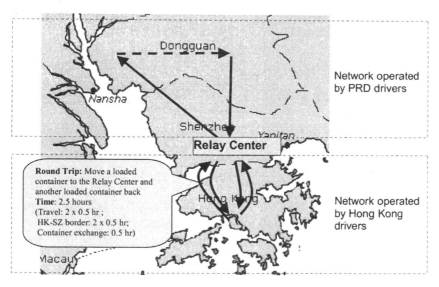

Figure 9.6. New operating network with a freight exchange center.

loaded containers instead of the empty ones. With the implementation of rapid custom clearance at the center, drivers can speed up the cross-border time. A number of new facilities have set up in Shenzhen, such as the South China International Logistics Center located 6 kilometers north of the Hong Kong-Shenzhen border, can function as the relay center.

The optimization of resources as shown in Figures 9.5 and 9.6 can be achieved if there are the economies of scale. Thus, as in the above example, if the driver belongs to a one-truck company, the driver cannot take advantage of the policy relaxation at all as the driver cannot move to another location to pick up a load that belongs to another company. Currently, in Hong Kong, almost all haulage companies are very small. Therefore, with the partial relaxation of regulatory policies, to leverage the economies of scale, there are several options. The first is through merging and acquisition to increase the company size. The second is through strategic alliance with which resources of different companies can be shared to a certain extent. The challenges are how to coordinate and utilize the resources better and how to divide profits among partners in a strategic alliance. The third option is the use of an electronic platform where matching supply and demand for transportation requests is possible. For example, OnePort Limited,[4] formed by the alliance of terminal operations, is an

---

[4] OnePort Limited <www.oneport.com>

example of the resolution that uses an information exchange platform to provide the opportunities for service matching. In addition to matching, such a platform needs to provide various documentation services to harness the cooperation among companies.

Feeder Network by Barge. The high cost differential on land transportation has driven the increase of barge or river trade activities. From 2000 to 2004, while the change of the number of containers handled by trucks is less than 1%, the number of containers handled by barge to and from terminals has increased at an annual rate of 16%. In 2004, around 15% of the local container traffic to and from the terminals was by barge and the share is increasing very quickly. To attract more container movements by barge, terminal operators are setting up satellite bonded terminals along the Pearl River, where customs clearance (for Hong Kong) can be checked. This however, requires the development of a feeder network that can provide regular scheduled barge services (whereas the current barge services are mostly on an ad hoc basis).

Compared to truck transportation, the per-container cost for the local transportation by barge is much lower as one vessel can handle more containers (in the range of 10 to 30) in each trip. In addition, the regulatory policy such as the 4-up-4-down policy does not apply for barge service. Nevertheless, there are still policies governing the activities of barges. For example, the barge service needs to be point-to-point. That is, between a specific local port in the PRD and a specific terminal in Hong Kong, no stop over (stopping at intermediate locations between the origin and the destination) is allowed. This creates a low utilization of barge capacity during the backhaul of a barge trip. If a barge could take multiple stops, the utilization of the barge would be higher. On the other hand, the berth space at the terminals is limited and is primarily for large ocean-going vessels. As barges also need berth space for transferring containers to the terminals, the scheduling and allocation of the berth space becomes more difficult and the waiting time for barge to moor to the terminal is getting longer. For example the situation of a moored ocean-going vessel that is waiting for containers that are being shipped by a barge, but there is no berth space available for this barge.

Bonded Consolidation Centers. Similar to the case of ocean container, the air-cargo terminal charge and aircraft landing fees in Hong Kong are high and the land transportation cost to the terminals is expensive. In particular, there is a toll bridge on the way to the airport. Therefore, air-cargos are typically consolidated before going to the airport (Wan et al., 1998). As for the ocean-cargo, most of them are shipped to and from locations in PRD. Therefore, terminal operators started recently some consolidation centers in the special bonded areas in China, where the goods are treated as not in China. Similar to the relay center shown in Figure 9.6, these bonded consolidation centers allow the use of the lower-cost PRD drivers during the transportation process. Cargo can be picked up from locations in PRD and delivered to these consolidation centers at which the bulk cargos are palletized or consolidated into Unitized Loaded Device (ULD) (the container for air cargo). This type of regional hub-and-spoke network helps reduce the cost for warehouse operations in Hong Kong, saves time in custom clearance, and reduces cost through the use of the bigger cross-border trucks.

Information Standardization and Sharing. In the above new operating models, there are some prerequisites. First, containers need to be loaded at the factory before the driver-tractor arrives. Second, emptied containers can stay at the factory and the relay center. Third, and more importantly, the information about when and where a load is available should be transmitted to the potential haulage companies. Furthermore, the flow coordination needs detailed information exchange, from the bidding or auction of container trucking services to the pre-declaration of goods electronically when using bonded consolidation centers in China. Therefore, information standardization is needed before any strategies can be implemented. Recently, Digital Trade and Transportation Network (DTTN) and Radio Frequency Identification (RFID) are being developed to enhance the efficiency of exchanging data and the business opportunities. DTTN is a designed network providing open architecture to all parties of the business. It is neutral in operation, high in transparency and wide in meeting most accepted industry standards. By allowing prompt services (e.g. trip matching) and accurate data exchange, resource utilization can be improved. Likewise, with the emerging of the RFID technology, custom clearance can be done in the upstream along the distribution network. This can help reduce the processing time of a shipment at the border.

9.4. Managing Freight Flow at Relay Centers

In inter-modal service, high efficiency at individual service providers does not guarantee that the whole process is efficient. In fact, the lack of coordination in the arrival and departure times can be very costly. An example was the chaotic situation regarding the handling of air-cargo at Chep Lap Kok, the new airport in Hong Kong airport when it opened in 1998. At the old airport, there was only one air-cargo terminal operator and one ramp operator (which transferred cargos between terminals and aircrafts). The coordination between the two operators was quite smooth before the airport relocation. At the new airport, there are two air-cargo terminal operators and several ramp operators. When the airport first opened, the coordination of resources and cargos among the multiple players were very poor, resulting in significant delays in cargo-handling. The chaos eased off after the operators set up a new staging area where cargos can be exchanged among operators and allowing resource sharing (for example, the dollies owned by one ramp operator can be used to carry air-cargo pallets handled by another operator subject to some agreed terms). In terms of spirit, the staging area is similar to the relay center and the allowing of resource sharing is a form of policy relaxation. Nevertheless, the six-week chaos led to a loss of over $550 million (Tsang, 1998).

9.4.1   Problem Characteristics

In this section, we use the flow coordination at the upcoming freight relay center to illustrate how decision models can help improve the scheduling. We assume that the 4-up-4-down policy is relaxed to 2-up-2-down while the 1-driver-1-tractor remains intact. The latter policy implies the coupling of a driver and a tractor is unique. For brevity, we simply call the driver-tractor pair a driver.

Two types of flow are involved in the coordination. The first type of flow is container (together with chassis). If the relay center is used, a cross-border container has to pass through both the PRD network and HK network from its origin to its destination. Research on managing flow over a transportation network at the strategic or tactical level has been well-studied in the service network design area. For a given set of demand pattern (transportation requests), the problem of service network design aims at minimizing the total cost by finding an efficient set of routes for the vehicles and for the shipments in the long run (Crainic, 2000). The use of mathematical modeling to formulate the problem and develop solution strategy for less-than-truckload (LTL) network design can be found in Farvolden and Powell (1996). Different from those in the literature, our problem setting is at the operational level. Furthermore, the routes of shipments are almost fixed and the focus is on managing resources that are subject to boundary limitation and regulatory policies.

The second type is the driver flow. There are two distinct groups of drivers: PRD drivers and HK drivers. HK drivers can perform cross-border trucking operations and are more expensive. PRD drivers only work within the PRD but are cheaper. The type difference naturally leads to the multi-commodity flow models where a commodity can be defined as a particular type of drivers. Ahuja et al. (1993) provides an excellent review of the formulations and solution methods of multi-commodity flow models. In these models, different commodities are typically sharing resources in part of the network (e.g. a single truck is used to carry flow of different commodities), leading to the *couple constraint* which says that the total flow of different commodities cannot exceed the resource limitation or capacity. In our problem, however, different groups of drivers do not share resources. The couple constraint comes from the times when a particular container is being handled in different networks. For example, for the container going from PRD to HK, before it arrives at the relay center in the PRD network, it cannot be handled in the HK network.

9.4.2 Formulation

In each individual network, managing the driver flow is similar to the vehicle routing and scheduling problems (see Bodin and Golden (1981) and Fisher (1995)). A common representation of these problems is the time-space representation. As shown in Figure 9.7, a node in the network represents a location at the beginning of a period. In our problem, there are three categories of arcs. The first category represents the movement of a container (a loaded one, or an empty one requested by shippers). Such a move can generate revenue. An example is the solid arc in Figure 9.7 from location j to the relay center starting at time $t = 5$. The arc cost is the negative value of the revenue generated by such movements. The arc capacity, which limits the number of such moves, is the market demand of these movements. The second category of arc represents the pure repositioning of the driver-tractor pair. An example is the arc from location i to location j starting at time $t = 2$. The arc cost is the cost for the repositioning and there is usually no upper limit of such a movement. The last category represents the case that the drivers stay at the current location until the next period. The arc cost is the holding cost for a driver and there is no capacity constraint.

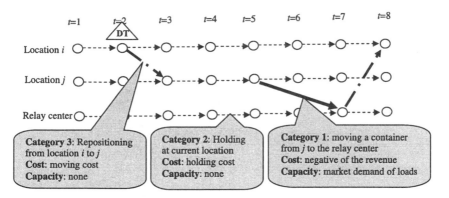

Figure 9.7. Time-space representation for the container trucking problem.

In the framework of time-space network, we now define the parameters and the decision variables as follows. Assume all drivers are available in period 1 and the length of the planning horizon is P. Define

Sets for the network:

N^k	= the set of locations in network k, $(k = \text{PRD, HK})$
A^k	= the set of arcs in network k,

Sets related to container movement:

Λ^k	= the set of containers to be handled in network k, $(k = \text{PRD, HK})$
Λ^{PRD-HK}	= the set of containers to be handled in network PRD first and then in network HK
Λ^{HK-PRD}	= the set of containers to be handled in network HK first and then in network PRD
T_ℓ^k	= the set of possible start times to move container ℓ in network k,

Cost and time parameters

$r_{ij}^{k,t}$	= net revenue for moving a container from i to j starting at time t in network k,
$c_{ij}^{k,t}$	= cost for empty repositioning from i to j starting at time t in network k,
$c_{ii}^{k,t}$	= cost for holding the driver at location i at time t in network k for one period
τ_{ij}	= time to move a container from i to j,
υ_{ij}	= time required for the driver to reposition (without hauling container) from i to j.

Note that $c_{ii}^{k,t}$ represents the holding cost. Since moving a loaded container usually takes more time than a pure repositioning, we use both τ_{ij} and υ_{ij} in the model.

Indicator variable

$\delta_{ij}^{k,\ell}$	$= \begin{cases} 1 \text{ if container needs to be moved from location } i \text{ to location } j \\ 0 \text{ otherwise} \end{cases}$

Decision variables

$s_\ell^{k,t}$	$= \begin{cases} 1 \text{ if container is being moved in network } k \text{ starting at time } t. \\ 0 \text{ otherwise} \end{cases}$
$x_{ij}^{k,t}$	= number of drivers taking containers from i to j starting at time t in network k,
$y_{ij}^{k,t}$	= number of drivers empty repositioning from i to j starting at time t in network k,

The objective is to minimize the total cost (or maximize the total revenue) that can be written as:

$$\min \sum_{k \in \{PRD, HK\}} \sum_{t \in \{1,\ldots,P\}} \sum_{ij \in A^k} \left(-r_{ij}^{k,t} x_{ij}^{k,t} + c_{ij}^{k,t} y_{ij}^{k,t} \right) \tag{1}$$

There are three groups of constraints in the formulation. The first group is the classical network constraints (flow conservation, capacity and non-negativity constraints):

$$\sum_{i \in N^k} \left(x_{ij}^{k,t-\tau_{ij}} + y_{ij}^{k,t-v_{ij}} \right) = \sum_{i \in N^k} \left(x_{ji}^{k,t} + y_{ji}^{k,t} \right)$$
$$\forall k \in \{PRD, HK\}, \ \forall t \in \{1,\ldots,P\} \tag{2}$$

$$x_{ij}^{k,t} \leq \sum_{\ell \in \Lambda^k} \delta_{ij}^{k,\ell} s_{\ell}^{k,t} \quad \forall k \in \{PRD, HK\}, \ t \in \{1,\ldots,P\}, \ ij \in A^k \tag{3}$$

$$x_{ij}^{k,t}, y_{ij}^{k,t} \geq 0; \quad s_{\ell}^{k,t} \in \{0, 1\} \tag{4}$$

Note that in (3), the right hand side represents the total number of containers that are available for the drivers to move. We assume that not all of these containers must be taken by the drivers. In practice, if a container cannot be handled by the drivers managed by the trucking company, the container movement will be outsourced to a third party.

The second group of constraint is about the start time of the container movement:

$$\sum_{t \in T_\ell} s_{\ell}^{k,t} = 1 \quad \forall k \in \{PRD, HK\}, \ell \in \Lambda^k \tag{5}$$

Equation (5) says that the departure time must be within a given time-window.

The third group of constraint relates the containers in different networks. For example, if a container is going from HK to PRD via the relay center, the container arrival time at the relay center in the HK network must not be later than the container departure at the relay center in the PRD network. Thus, we have

$$s_{\ell}^{HK,t} \left(t + \delta_{ij}^{HK,} \cdot \tau_{ij} \right) \leq s_{\ell}^{PRD,t'} \cdot t'$$
$$\forall \ell \in \Lambda^{HK-PRD}, \ t \in T_{\ell}^{HK}, \ t' \in T_{\ell}^{PRD} \tag{6}$$

$$s_\ell^{PRD,t} \left(t + \delta_{ij}^{PRD,\ell} \cdot \tau_{ij} \right) \le s_\ell^{HK,t'} \cdot t'$$
$$\forall \ell \in \Lambda^{PRD-HK}, \ t \in T_\ell^{PRD}, \ t' \in T_\ell^{HK} \qquad (7)$$

The left hand side of the equations (6) and (7) represent the arrival times of the containers at the relay center and the right hand side denote the departure times of the containers at the relay center in the other network. Notice that the location j in these two equations is the relay center.

9.4.3   Design of a Solution Approach

Problem defined by equations (1)–(7) is a large-scale integer programming problem. For example, if we have two networks that have five locations each and the planning horizon is 20 periods, then we have over 2000 integer variables which is too large to solve. To design a solution approach, we look at how to deal with constraints (5)–(7). If $s_\ell^{k,t}$ were not decision variables but were given constants and constraints (6)–(7) were removed, then the problem is separable. It means we can deal with the PRD network and the HK network independently which are the standard minimum cost flow problems. Therefore, we can use the following approach. In the first step, we give the initial values of $s_\ell^{k,t}$ (that is, the departure times) for the containers going to the relay center. In the second step, we set the container departure times of the containers from the relay center as long as they satisfy equations (6)–(7). In the third step, we can solve the separated network sub-problem defined by equations (1)–(4) using efficient algorithms (see Ahuja et al. (1993)). In the fourth step, we use the optimal solutions from the sub-problems to estimate the benefits of changing the start times of the container departure. Based on these estimates, we re-determine the start times of the containers and then repeat the four steps.

To illustrate this approach, let us consider the time-space representation of the PRD network shown in Figure 9.8 where $s_\ell^{PRD,5} = 1$. That is, container should be moved from location j to the relay centre at time $t = 5$. In the optimal solution of the PRD network sub-problem, the driver who is initially available in location i at time $t = 2$ first repositions to location j, waits for two periods, moves the loaded container from j to the relay center, and then returns to location i to finish his duty. Similarly, for the HK network shown in the bottom half of Figure 9.8, in the optimal solution with respect to this network, the driver available in location b at time $t = 3$ moves a container to the relay center and then repositions to location a. In this situation, if we can start the container movement in the PRD network two periods earlier, then we can have the container exchange by the two drivers at time $t = 5$. Thus, the productivity of the drivers is higher (and thus the total cost will be lowered).

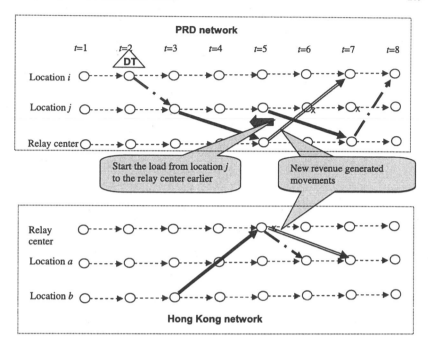

Figure 9.8. Coordinating freight flow in different networks.

Mathematically speaking, let $\hat{c}^k_{\ell,\Delta_\ell}$ be the cost change if the time for starting task ℓ is changed by Δ_ℓ in network k and let $r_{\ell,\ell'}$ be the gain if the two drivers from the two networks can exchange containers ℓ and ℓ' after the changes of start times. The net change in cost is

$$f(\ell, \Delta_\ell, \ell', \Delta_{\ell'}) = \hat{c}^{\mathrm{PRD}}_{\ell,\Delta_\ell} + \hat{c}^{\mathrm{HK}}_{\ell',\Delta_{\ell'}} + r_{\ell,\ell'} \tag{8}$$

One way to obtain $\hat{c}^k_{\ell,\Delta_\ell}$ is by using the dual prices of the arc capacities from the optimal solution in network k. Changing the start time of a container can be viewed as the simultaneous increasing the capacity of an arc and decreasing the capacity of another arc with the same pair of locations but with different start times. Suppose that $\bar{c}^{+,k,t}_{ij}$ is the dual price for the arc from i to j at time t for network k if the capacity of the arc is increased by 1. Similarly, let $\bar{c}^{-,k,t}_{ij}$ be the dual price for the arc from i to j at time t for network k if the capacity of the arc is decreased by 1 (which is defined only for an arc with capacity greater than 0). Then, we can approximate $\hat{c}_{\ell,\Delta_\ell}{}^k$ as

$$\hat{c}^k_{\ell,\Delta_\ell} \approx \bar{c}^{-,k,t}_{ij} + \bar{c}^{+,k,t+\Delta_\ell}_{ij} \quad \text{when } \delta^{k,\ell}_{ij} = 1 \tag{9}$$

To determine how we change the start times, we would like to find the container pairs ℓ and ℓ', and the time changes (Δ_ℓ and $\Delta_{\ell'}$) by minimizing $f(\ell, \Delta_\ell, \ell', \Delta_{\ell'})$ which is defined in (8). This can be done by looping over the container pairs.

Remark: In this section, we have provided a formulation of the problem and discussed the solution framework. The detail implementation of the suggested solution approach may depend on the size of the problem, the types of restrictions imposed on the container exchange, and the number of networks involved. In our discussion, we assume that there are only two networks for simplicity. In practice, drivers belonging to a company can be regarded as a commodity and there can be many commodities involved. This opens up new research opportunities.

9.5.   Looking Ahead

The price pressure and the changing freight patterns have created the need to re-examine the physical logistics network in the PRD. One important new development is the bridge linking Hong Kong with Zhuhai and Macau, in the western part of the PRD, where a new manufacturing base is rapidly developing. Another important infrastructure development is the logistics park near the Hong Kong International Airport. The park will house a number of specialized facilities for high-value added logistics services such as merge-in-transit operations. In terms of service, we will see more scheduled barge services between Hong Kong, new forms of inter-modal transportation services, early clearance, and bonded consolidation centers.

With the operating environment becoming increasingly complex, the ability to effectively coordinate the connecting activities will help determine the success or failure of an international hub. The coordination, however, can hardly be achieved by individual companies alone. An industry-wide information platform concerning container exchange opportunities and an intelligent system for finding the best timing of moving the containers are the necessary ingredients in promoting flow coordination.

Acknowledgement

We thank the invaluable comments from Profs Chung-Yee Lee and Hau Lee which help improve the quality of this chapter. The work was supported by the Research Grants Council of Hong Kong under grant CERG HKUST6291/04E.

References

Ahuja, R., T. Magnanti and J. Orlin (1993). *Network flows: theory, algorithms and applications.* Englewood Cliffs, NJ: Prentice Hall.

Bodin, L. and B. Golden (1981). Classification in vehicle routing and scheduling. *Networks, 11,* 97–108.

Brand Hong Kong Management Office (2005). Pearl River Delta – Factory of the World, July. Website: http://www.brandhk.gov.hk/brandhk/e_pdf/efact12.pdf

Crainic, T.G. (2000). Service network design in freight transportation. *European Journal of Operational Research, 122.*

Farvolden, J. and W. Powell (1996). Subgradient methods for the service network design problem. *Transportation Science, 28*(4).

Fisher, M. (1995). Vehicle Routing, Network routing. In M. Ball et al. (Eds.), *Handbooks in Operations Research and Management Science,* Vol. 8, 1–33.

Hong Kong Special Administrative Region (2003). *Hong Kong Yearbook 2003, 13*(11). Website: http://www.info.gov.hk/yearbook/2003/english/chapter13/13_10.html

Hong Kong Special Administrative Region (2004). Study on Hong Kong Port – Master Plan 2020. Economic Development and Labour Bureau.

Hong Kong Special Administrative Region (2005). Number of Establishments, Persons Engaged and Vacancies (other than those in the Civil Service). Analysed by Industry Sector, Census & Statistics Department.

International Monetary Fund (2005). *Direction of Trade Statistics Quarterly,* March.

Morlok, E.K. and L.N. Spasovic (1994). Approaches for improving drayage in rail-truck inter-modal service. Technical report, University of Pennsylvania.

Sun, C. (2003). Successful transition of Hong Kong exports. *Shippers Today, 26*(2).

The Better Hong Kong Foundation (2004). Restoring Hong Kong's Competitiveness as a Sea-trade Logistics Hub. The Better Hong Kong Foundation, 29th July.

Tsang, Y.K. (1998). *Central News Agency,* 16th July.

Wan, Y., R. Cheung, J. Liu and J. Tong (1998). Warehouse location problems for air-freight forwarders. *Journal of Air Transport Management, 4,* 201–207.

Wang, T.F., K. Cullinane and D.W. Song (2005). *Container port production and economic efficiency.* New York: Palgrave Macmillan.

Hau L. Lee and Chung-Yee Lee (Eds.)
Building Supply Chain Excellence
in Emerging Economies
©2007 Springer Science + Business Media, LLC

Chapter 10

SERVICE PARTS MANAGEMENT IN CHINA

Steven Aschkenase
Deloitte Consulting Overseas Services LLC (Shanghai)

Keith Nash
Deloitte Consulting LLC (Dallas)

Abstract: Aftermarket service parts can be an extremely lucrative business, yet most OEMs only capture a fraction of its profit potential. The situation is even more pronounced in China and other emerging markets, where cultural behavior, infrastructure limitations and market barriers make it hard for foreign OEMs and their dealers to compete with unauthorized service parts providers. This chapter examines the unique challenges of selling and distributing service parts in an emerging market, and offers foreign OEMs specific advice to make their service parts operations more efficient and profitable.

10.1. Introduction

The service parts business is critical to the success of any manufacturing company. It keeps customers happy by providing them with replacement parts they need to keep their equipment up and running. And in mature markets, it is typically a source of enormous profitability. Yet many manufacturers take their service parts business for granted, leaving a lot of potential profits on the table.

Emerging markets are a different story altogether. Customers in developing countries tend to be less sophisticated than their counterparts in mature markets and don't place as much value on service parts. They tend to run equipment until it breaks, then seek out the cheapest possible replacement, often purchasing substandard parts from questionable sources, or fabricating copies themselves.

Under those conditions, it can be difficult for foreign manufacturers and their dealers to command a premium price for genuine OEM service parts – pushing many aftermarket businesses into the red.

In this chapter, we examine the unique challenges of operating an aftermarket business in an emerging economy and explain why it's worth doing well. We look at the differences between a service parts operation and a traditional production-oriented supply chain. We describe the logistical challenges and other factors that make it hard for foreign OEMs to be efficient, price-competitive and profitable. And we offer specific suggestions to help companies improve the financial performance and competitive position of their service parts business. Our primary focus is on China – the world's preeminent emerging market – with real-world examples from leading global manufacturers. However, our insights and recommendations are for the most part applicable to any company in any emerging economy.

10.2. Why Worry About Service Parts?

In mature markets, the service parts business is generally very profitable. Yet few companies even come close to tapping its full potential. According to common wisdom, customers will always need to buy service parts; therefore the aftermarket business can basically run itself. And since most service parts operations are already profitable – at least in mature markets – it's hard to argue the point. The problem with that hands-off approach is it creates a safe harbor for inefficiency and overlooks valuable growth opportunities, leaving a lot of money sitting on the table.

Emerging markets present a greater financial challenge for service parts, but are still worthy of attention for two reasons.

- **Customer satisfaction.** In the short term, service parts are a key driver of customer satisfaction – even for customers who don't realize it. High quality parts reduce unplanned downtime, extend the useful life of equipment, improve operating safety and maximize resale value – helping a foreign OEM establish a reputation for quality and reliability, while building a base of loyal customers. Service parts can also be a significant source of incremental revenue and profit.
- **Long-term profitability.** Emerging markets like China eventually develop into mature markets with tremendous profit potential. Focusing on service parts now positions an OEM to dominate later, when the stakes are much higher.

Improving the efficiency and performance of your service parts operation can deliver significant benefits in both the short term and the long term – transforming money-losers into money-winners, and elevating successful businesses into cash cows.

10.3. What Makes the Service Parts Business Different?

Although traditional production supply chains get most of the attention, the service parts business is every bit as complex and challenging – just in different ways.

Mainline manufacturing typically revolves around high production volumes and careful planning – cranking out a large quantity of identical items, then pushing them onto the marketplace for mass consumption – all as predictably and efficiently as possible. The service parts business, on the other hand, is much less predictable – driven by product failures and other factors that are harder to control. A service parts operation has far more SKUs to manage, and the lifecycle for each part is typically much longer than for finished products. Demand is often unplanned and varies widely, ranging from parts that are extremely low value and high volume (e.g., nuts and bolts that sell by the thousands) to parts that are extremely high value and low volume (e.g., specialized parts that the company only sells once every few years).

These unique service parts challenges require unique processes, systems, facilities and expertise.

The Service Parts Framework

Figure 10.1 depicts the key capabilities for an efficient service parts supply chain. Each of the 12 basic building blocks provides a foundation for the ones above it, creating a pyramid that ultimately drives customer and shareholder value. Although the framework applies both to mature and emerging markets, it is especially critical in emerging markets given the importance of efficiency in a price-driven environment.

Looking at the pyramid from the bottom up:

Foundations: People, organization and systems – along with data and the physical supply chain network – provide the basic *foundations* for all supply chain activity. Major problems in these areas make it difficult or impossible to get anything else done, and can significantly increase overall costs.

Enablers: Visibility, performance management and collaboration build on that foundation, improving execution efficiency and *enabling* the more advanced supply chain capabilities.

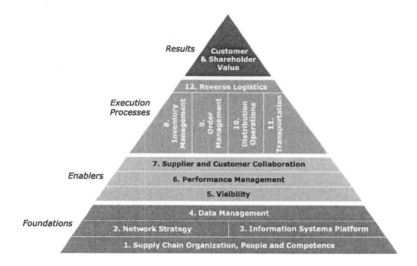

Figure 10.1. The "Ideal" Service Parts Supply Chain.

Execution processes: Inventory management, order management, distribution, transportation and reverse logistics are the core supply chain *processes*. These activities are where the rubber meets the road.

Results: The goal for any service parts business is to create value for the company's customers and shareholders. That goal is both the starting point for service parts strategy, and the ultimate objective for operations.

Most aftermarket supply chains have considerable room for improvement in some or all of the 12 building blocks. Figure 10.2 provides an example of the dramatic results a company can achieve by optimizing its service parts operation.

The manufacturer achieved these results through a wide range of improvements, including: hiring and developing people with the special skills needed to manage a distribution-oriented supply chain. Using advanced statistical analyses, rather than rules-of-thumb. Rigorously applying basic principles of supplier management. Deploying global, integrated systems for managing inventory, orders, and warehouse operations. And cleaning up the company's inventory data.

The results in Figure 10.2 reflect improvements across the manufacturer's entire global business, including both mature and emerging markets. However, the improvement potential for emerging markets alone can be even greater, as operations and infrastructure in developing countries tend to be inherently less efficient than in more established parts of the world.

A closer look at the 12 building blocks

FOUNDATIONS **Organization, People and Competence.** A service parts business requires top-notch people with the skills and experience to run a distribution-oriented supply chain. It's not the same as a manufacturing supply chain.
Network strategy. The physical supply chain network should be continuously optimized for speed and efficiency, not mired in the past.
Information systems platform. Supply chain applications should be fully integrated with other enterprise systems, with the flexibility to adapt to changing conditions. Many of today's systems are standalone applications that have been rigidly coded to a particular supply chain structure.
Data management. For reasonable efficiency, a supply chain needs transaction data that is at least 95 percent accurate and product data that is at least 98 percent accurate. World class supply chains strive for Six Sigma data accuracy.

ENABLERS **Visibility.** Many service parts supply chains are large and global, yet riddled with blind spots – places where inventory temporarily disappears from view. Visibility helps a company overcome problems with the physical network and provides a foundation for advanced capabilities like collaboration.
Performance management. Most service parts supply chains don't react to a problem until it hits the customer – and by then it's too late. Spotting problems sooner makes them easier to resolve and gives a company more options.
Supplier and customer collaboration. Sharing information and co-managing inventory can improve efficiency and decision-making, but even basic techniques like supplier management – consistently applied – can deliver significant benefits.

EXECUTION **Inventory management.** Inventory generally represents a huge portion of the
PROCESSES overall supply chain investment, yet it is often managed with rules-of-thumb and by holding a lot of safety stock. The advanced statistical techniques for optimizing inventory already exist. They just need to be applied.
Order management. A well-designed order management system is the central hub for all supply chain activity, seamlessly integrating with other systems to provide fully automated transaction processing and a single point of reference for all customer inquiries and data.
Distribution operations. Lean methodology is now being used successfully in distribution to simplify operations and drive continuous improvement.
Transportation. Transportation is a major source of supply chain costs – as well as a major source of headaches. Leading companies are using techniques such as inbound logistics and direct shipping to reduce costs and ensure a smooth flow of parts.
Reverse logistics. Most companies – no matter how proficient – struggle with returns. The key to success is viewing reverse logistics as its own fully integrated process.

RESULTS **Customer and shareholder value.** An effective service parts supply chain creates value for customers and shareholders alike.

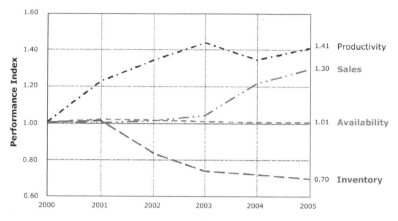

Figure 10.2. Improvements at a Leading Global Equipment Manufacturer.

10.4. Service Parts in Emerging Markets

Two underlying factors define the service parts challenge in emerging markets. The first is *perceived value*. Customers in markets such as China often don't place as much value on service parts. They tend to focus on price, which drives them away from authorized dealers and genuine OEM parts. The second issue is *cost structure*. Foreign OEMs – and their dealers – generally don't have the lean cost structure necessary to compete on price and still produce a profit. Both factors make it difficult for a foreign OEM to capture its usual share of the aftermarket business.

For example, a typical manufacturer might provide only roughly 30 percent of the aftermarket service parts for its equipment sold in China. The remaining 70 percent consists of lower-priced alternatives – *grey market* parts, *counterfeit* parts, and *will-fit* parts – sold through unauthorized channels (Figure 10.3).

Grey market parts are genuine OEM products sold outside of authorized channels – typically through small, independent, locally operated traders and retailers. *Counterfeit* parts are cheap knock-offs fraudulently sold under the OEM brand name. *Will-fit* parts are non-OEM products – essentially anything that fits into the machine – sold under various brand names. They are often inferior quality and may not meet the required specifications.

These lower-priced alternatives enjoy a significant cost advantage at every step of the process – from cheaper labor and materials to reduced logistics costs, import taxes and dealer overhead – making it very difficult for foreign OEMs to compete on price. To make matters worse, foreign OEMs often find themselves stuck with the least desirable part of business (slow-moving,

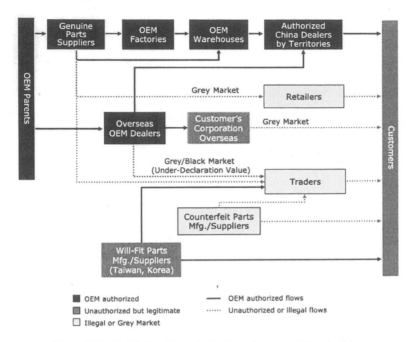

Figure 10.3. Distribution Channels for Heavy Equipment Parts in China.

expensive components) while counterfeiters and other cut-rate competitors skim off the routine, fast-moving parts that generate most of the profit.

10.5. Why Emerging Markets Are Less Willing to Pay a Premium for Genuine OEM Parts

There are a number of reasons customers in emerging markets are less willing to buy premium-priced parts from authorized dealers. The main issue is that it's simply not what they're used to. In China, for example, independent *traders* have historically been the "official" source for service parts from Chinese OEMs (which generally lack a formal dealer network) so they are naturally the source that customers turn to first. Traders also tend to offer lower prices and more personal service. The parts themselves may not be as good, but many customers consider them "good enough" – or simply don't know the difference.

Lack of sophistication. Eighty percent of the heavy equipment in China is privately owned, most of it by small businesses that are first-generation equipment owners. Thirty percent only own one machine, while another 30 percent

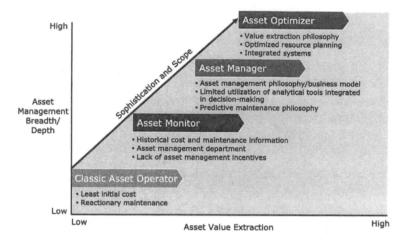

Figure 10.4. Climbing the Asset Management Curve.

only own two or three. These small businesses tend to focus on the basics of operating an asset because that's what they understand. When something breaks, they fix it. And when they shop for a replacement part, they typically look for the lowest price because that's something any business owner can recognize and appreciate.

Although a few Chinese businesses are starting to move up the curve, most still lack the scale and expertise for sophisticated asset optimization and often don't fully understand the importance of high-quality service parts and maintenance. (See Figure 10.4).

Relationships with local traders. As noted, the market for service parts in China is dominated by *traders* – local, independent merchants that offer convenience and low prices on a wide range of parts (usually a mix of grey market or black market, will-fit, and counterfeit). Traders range in size from regional and small (a handful of trucks and retail outlets) to very small (one person delivering parts by car, or even bicycle).

Traders offer a number of compelling advantages: personal service, local convenience, low prices and the security of an ongoing relationship. They visit customers on a regular basis and respond quickly to breakdowns and other urgent requests. They are also the "official" service parts channel for many Chinese OEMs.

In contrast, foreign OEMs and their authorized dealers are generally located much farther way, charge higher prices, and often don't have a relationship with the customer beyond the initial sale.

That last point is particularly important. In Chinese culture, cultivating and maintaining relationships is a top priority – much more so than in the West. The Chinese talk about *guanxi* (pronounced "guanshee"), which reflects their level of trust, commitment and sense of obligation. The closer the relationship, the greater the guanxi. Local traders, with their ongoing customer relationships, have the most guanxi – and thus capture the lion's share of the service parts business.

Authorized dealers selling genuine OEM parts offer higher quality, a factory warranty and the peace of mind that comes from doing business through authorized channels. But in an emerging market like China where price and relationships are paramount – and where liability laws are less of a factor – those advantages simply aren't as compelling.

Limited use of leasing. Companies that lease are more likely to use genuine OEM parts to ensure they get the full residual value for their equipment at the end of the term. In fact, regular maintenance using genuine OEM parts is often explicitly required in a lease agreement.

Leasing is a common practice in mature markets, but is far less prevalent in emerging economies. According to the Shanghai Rental Trade Association, only 2 percent of the heavy equipment purchased in China during 2003 was financed with leasing – compared to 30 percent in the United States and 16 percent in Germany. In fact, until recently, foreign companies were specifically prohibited from creating their own wholly-owned financing subsidiaries in China. GE Capital made headlines in 2004 by being among the first foreign companies to be granted Chinese leasing licenses – giving them the ability to lease not only their own equipment, but equipment from other companies as well.

Low-ball insurance practices. Another factor working against OEMs is that Chinese insurance companies generally insist on the cheapest possible repair, forcing customers to accept will-fit or counterfeit parts instead of pricier OEM options.

10.6. Why Foreign OEMs Have a Hard Time Competing on Price

The intense focus on price – to the exclusion of all other factors – makes it difficult for foreign OEMs to compete in emerging markets. At the heart of the problem is the fact that high-quality parts simply cost more to produce than low-quality parts. They use better, more expensive materials, and they are manufactured to stricter quality standards that require more sophisticated machinery, more involved manufacturing processes, more extensive testing and workers with greater expertise – all of which cost money.

Local suppliers often lack the advanced manufacturing processes and capabilities that have evolved in more developed countries, forcing OEMs to source their parts abroad. The result? Even more layers of complexity and cost, including expensive foreign labor and materials, higher transportation costs and longer lead times.

Trade restrictions and import fees are another important factor, particularly in China where the government aggressively encourages local sourcing. For example, Chinese regulations used to require foreign OEMs to maintain an independent distribution network for goods manufactured outside of China – completely separate from the channels used for domestic products. That restriction was relaxed in December 2004 as part of China's commitment to the World Trade Organization. However, it still applies to foreign automakers and pharmaceutical companies. Car companies, for example, must maintain one dealer network for the vehicles and parts they produce in China, and a separate network for products made elsewhere. That two-tier distribution requirement – along with China's significant import tariffs and customs fees – makes imported service parts extremely expensive.

Will-fit and counterfeit parts don't face the same strict quality requirements as genuine OEM parts, so they can be produced for a lot less money. They also tend to be manufactured locally, providing a distinct cost advantage at every step of the process – from raw materials and energy to labor and distribution. And like their grey market and black market counterparts, locally produced components escape the burden of import taxes and trade restrictions.

10.7. Logistics Problems Add to the Challenge

Intense price competition is only one reason foreign OEMs struggle to make a profit on service parts in China. There are also a wide range of logistics challenges that hamper efficiency and erode profit margins.

In China, logistics costs such as transportation and warehousing typically account for 30 to 40 percent of the final selling price. In western countries, that number is generally well under 10 percent. One reason for the discrepancy is that Chinese labor and other production costs are much lower than in other markets – as are retail selling prices – making logistics costs seem proportionally high. Yet one could also argue that logistics in China should be just as inexpensive as everything else.

The real problem is China's inefficient and unreliable logistics network. The Chinese government is acutely aware of the situation and is investing billions of dollars to upgrade key infrastructure components. Yet the country still has a long way to go.

China's network issues are exacerbated by a lack of information and visibility. Visibility can help a company work around problems in the physical network. Conversely, a reliable network makes visibility less critical because things are generally where they are supposed to be. But when both are lacking – as is the case in China – it's hard to know where anything is.

Roads. China's primitive road system is undergoing significant upgrades. Express highways are being built to link major cities, and a national trunk highway is slated for completion by 2010. But for now, the road system remains a significant impediment to efficient commerce.

Trucking. In China, trucks are the primary means for transporting finished goods – yet there is no national network in place. Protectionist policies by provincial governments impede the flow of material and finished goods, limiting transparency and predictability. Meanwhile, road tolls throughout the country remain extremely high, often accounting for more than 20 percent of total transportation costs.

Rail. The rail system is China's most extensive transportation network, handling more than 90 percent of all bulk shipments. Yet a lack of automation often makes the system unreliable. A shipment that takes two days in one situation might take two weeks in another, leaving companies with little choice but to drop off their cargo and hope for the best. Intermodal capabilities are limited, further reducing efficiency.

Shipping. Container terminals in China's coastal cities are receiving significant upgrades, but for now still lag behind their western counterparts. Within China's borders, barges on the Yangtze, Yellow and Pearl rivers continue to carry a large percentage of the country's commodity cargo.

Air. Although flights between major cities are on the rise, the overall system for air freight remains fragmented. Passenger travel consistently takes precedence over cargo, and processes for handling air freight are often inefficient.

China's network deficiencies are particularly troublesome for foreign OEMs, who are highly dependent on faraway suppliers – making it virtually impossible for them to compete on cost. Chinese companies are less affected for two reasons. First, they tend to operate locally or regionally and are therefore less reliant on the country's transportation systems. And second, they are accustomed to the challenges and know how to work around them.

10.8.   The Profit Challenge

As things stand, it is difficult for foreign OEMs to sell service parts profitably in an emerging market like China. Figure 10.5 shows the profit margins for a hypothetical part that sells for $100 in a mature market, but only $50 in China (due to intense competition from local traders and cheap parts). In this example, the OEM in a mature market enjoys a net profit of 35 percent, while its dealers net 7 percent. In an immature market, both are unprofitable – with the OEM losing 36 percent and its dealers losing 10 percent.

	Mature Market	Immature Market
Market-based price to end-user	100	50
Dealer profit	7	-5
Dealer costs (import duty, VAT, freight, warehouse, marketing)	18	19*
OEM price to dealers	75	36
OEM profit	26	-13
OEM costs (raw material, labor, freight, warranty, warehouse, insurance, marketing etc).	49	49
Dealer profit margin	7%	-10%
OEM profit margin	35%	-36%

* Higher due to import duties

Figure 10.5.   Current Market "Realities" Can Make Service Parts Unprofitable.

10.9.   Improving the Business

Companies in emerging markets often put up with less than stellar profit performance in exchange for rapid growth. But it doesn't have to be that way. Improving the service parts business can have an immediate impact on both profit and revenue, while positioning a company for long-term dominance as the market matures.

This section looks at four things a foreign OEM can do to improve the performance of its service parts business in an emerging market.

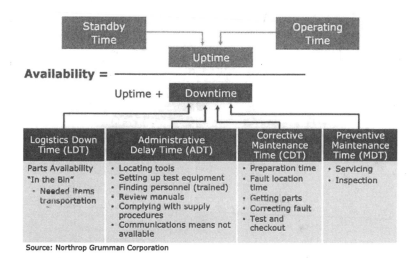

Source: Northrop Grumman Corporation

Figure 10.6. Selling Equipment Availability, Not Equipment.

Increase the perceived value of service parts and maintenance. The biggest problem in emerging markets like China is that customers tend to view service parts as a commodity. Many don't understand the benefits of high-quality parts, and they often don't worry about repairs and maintenance until their equipment breaks down.

The best way for foreign OEMs to boost service parts profitability is to help customers become more sophisticated asset managers – optimizing assets, instead of operating them into the ground.

The first step toward that goal is basic education. Showing customers how OEM parts differ from cheaper alternatives, and how those differences translate into greater reliability and lower cost of ownership.

Another way to foster an asset management mindset is bundling products with service during the initial sale. Product and service bundles help shift the customer's focus from *initial price* to *total cost of ownership*, while ensuring equipment is properly maintained with high-quality, genuine parts.

Taking that idea a step further, leading customers in mature markets are increasingly interested in *performance-based logistics* that focus on business goals and desired outputs, not the equipment itself. Mining companies, airlines and the military are just a few of the organizations looking to buy *power by the hour* – actual operating hours and availability – rather than parts and equipment. With performance-based logistics, the manufacturer and/or its dealers assume full responsibility for repairing and maintaining the equipment using genuine OEM parts, allowing customers to stay focused on the critical needs of their business. (See Figure 10.6.)

Teaching customers to focus on *overall business value* and *total cost of ownership* – not just lowest price – will help OEMs differentiate their service parts from cheap knock-offs. It's a big challenge, and won't happen overnight. But with steady and persistent effort, foreign OEMs can expect to capture an increasingly large share of this lucrative market.

Use relationships and technology to beat traders to the punch. Another way for foreign OEMs to protect the value and market price of their parts is by building stronger relationships with customers. In China, local traders currently own the customer relationship, so they're the ones most likely to get the call when parts are needed. But if foreign OEMs and their dealers can create better ongoing relationships with customers, they could dramatically improve their capture rate and lose a lot less business to traders.

Calling to say hello is not enough. To build a really strong relationship, foreign OEMs and their dealers must provide customers with something of genuine value – a reminder about scheduled maintenance, for example, combined with a discount offer on the required service.

Taking that a step further, a few leading manufacturers have started using sophisticated diagnostics, radio frequency identification (RFID) and advanced telematics to detect equipment failures before they happen. These systems give an OEM and its dealers a chance to warn the customer and offer appropriate repairs before local traders even know there's a problem. A growing number of OEMs are also building advanced diagnostic systems into their machines. These systems help technicians quickly pinpoint mechanical problems, but even more important, they require sophisticated tools to make use of the data – tools that small, independent traders generally don't have access to.

Reduce costs through local sourcing. Although a foreign OEM's primary strategy over the long term should be increasing the perceived value of service parts, companies must also do everything possible to minimize costs. A continuous focus on cost reduction improves competitiveness in the short term, while magnifying any top line improvements.

Local sourcing is one of the best opportunities for foreign OEMs to reduce their cost structure. Purchasing service parts from local manufacturers reduces the cost of parts, transportation and import taxes – and minimizes the need for

	Mature Market	Immature Market	Immature Market with Local Sourcing
Market-based price to end-user	100	50	50
Dealer profit	7	-5	5
Dealer costs (import duty, VAT, freight, warehouse, marketing)	18	19	9
OEM price to dealers	75	36	36
OEM profit	26	-13	7
OEM costs (raw material, labor, freight, warranty, warehouse, insurance, marketing etc).	49	49	29
Dealer profit margin	7%	-10%	10%
OEM profit margin	35%	-36%	19%

Figure 10.7. Restoring Parts Profitability Through Local Sourcing.

a two-tiered distribution network. Local sourcing can also improve responsiveness and agility by reducing lead times.

Figure 10.7 builds on our earlier example, showing how local sourcing can restore profitability for a foreign OEM and its dealers.

Of course, local sourcing is easier said than done. Different countries have different specialties and it's not always possible to find local suppliers with the process capabilities needed to do the job. In China, most manufacturers focus on high-volume production – cranking out large, predictable quantities of a single item with relatively loose tolerances. That's the exact opposite of the service parts business, which generally involves small quantities, tight tolerances and an unpredictable mix of low and high-value parts.

Foreign OEMs may need to invest significant time and money choosing potential supply partners and helping them upgrade their capabilities. Developing a base of qualified suppliers is typically a long-term effort that occurs in stages, starting with vendors that assemble finished products from kits of foreign-produced parts, then gradually expanding the use of locally sourced components until all manufacturing and assembly is done in-country.

Finding and developing qualified suppliers isn't the end of the story. For service parts, one of the most important development steps is establishing a local logistics center to provide key value-added services such as inspection, painting and kitting. OEMs can also expect ongoing conflicts with their local service parts suppliers, some of whom may continuously push for the higher volumes and greater predictability they have come to expect from supplying traditional manufacturing operations.

Despite the challenges and potential pitfalls, local sourcing is a key strategy for any service parts business – and may be the only way for foreign OEMs to compete in an emerging market.

Create economies of scale through shared facilities. In a market as large and complex as China, it's virtually impossible to operate at a profit without achieving a certain scale. Yet, in spite of the market's explosive growth, most foreign OEMs are still in the process of establishing a beachhead and are a long way from achieving that minimum scale on their own.

One strategy a company with multiple brands should consider is consolidating operations – sharing facilities and supply chain networks across the different brands to improve everything from parts availability and inventory management to warehouse utilization, transportation and more.

A company operating on its own can achieve similar benefits by partnering with others through joint ventures and acquisitions. Even companies that compete directly in their main line of business – automakers, for example – may be able to collaborate on service parts distribution since a customer in the market for genuine OEM parts generally won't view another OEM's products as a substitute. Outsourcing is another option, enabling a company to reduce costs by capitalizing on a vendor's scale across multiple customers.

Figure 10.8 shows the specific improvements a leading logistics provider achieved through sharing facilities and optimizing its global supply chain network – cutting overall logistics costs by 15 percent.

	Before	After	Improvement
Inventory Value	$95 M	$65 M	**$30 M**
Facing Fill Rate	87%	94%	**+7%**
Inventory Turns	2.2	4.0	**+1.8**
Warehouse Space	1.6 M Sq. Ft.	1.2 M Sq. Ft.	**-0.4 M Sq. Ft.**
Number of Facilities	6	4	**-2**

The optimal network will help increase your efficiency and lower your logistics cost by 15%

Figure 10.8. The Benefits of Shared Facilities and Optimized Networks.

10.10. Getting Your Service Parts Supply Chain into Fighting Trim

Operating a service parts business in an emerging market like China presents a number of unique challenges. Yet in other ways it's just like any other market. As always, efficiency leads to profitability – and provides the foundation for sustainable competitive advantage.

One of the best ways to boost the performance of your service parts business is to focus on the basic building blocks. Although every company has its own unique challenges and strategic priorities, the opportunities highlighted in Figure 10.9 are almost always worth pursuing.

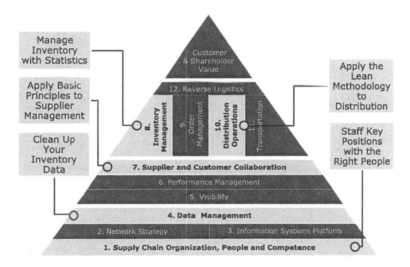

Figure 10.9. Improving Your Service Parts Supply Chain – Five Good Places to Start.

These five areas typically provide a rapid payback. And best of all, they are entirely within your control. With strong commitment and steady determination, positive results are virtually guaranteed.

- **Staff key positions with the right people.** An aftermarket operation isn't the same as a traditional manufacturing business. Look for people with deep experience running a distribution-oriented supply chain, and make sure they have the training and education to tackle complex problems. Putting the right people in a few key positions can really boost your overall performance.

- **Manage inventory with statistics, not rules-of-thumb.** Most companies manage their service parts inventory with best guesses, rules-of-thumb and a ton of safety stock. But it doesn't have to be that way. Inventory management is governed by the laws of probability and statistics, and the ideal operating methods are already well established. Success is simply a matter of putting them into practice.
- **Apply basic principles to supplier management.** When it comes to managing suppliers, a few basic principles go a long way. Establish performance standards that make sense. Set reasonable lead times. Then create a basic scorecard and hold suppliers accountable for meeting their targets.
- **Use lean methodology for distribution.** Lean methodology was originally developed to minimize waste and idle time in manufacturing – but is now being used with great success for distribution. It provides a clear set of principles and tools to improve asset utilization and space efficiency, while reducing cycle times and handling errors. Best of all, it doesn't require a lot of fancy technology or other major investments. You can start with a small pilot in a single warehouse, then expand from there.
- **Clean up your inventory data.** An efficient supply chain needs transaction data that is at least 95 percent accurate and product data that is at least 98 percent accurate. Data quality is primarily driven by employee behavior, not technology. Provide employees with basic training in data quality techniques – then continuously follow-up to make sure those techniques are being applied.

Improvements in each of these five areas will help maximize the efficiency and performance of your service parts supply chain – a good idea in any market, and an absolute necessity in a price-driven market like China.

Emerging markets present a unique set of service parts challenges, and financial success can be difficult to achieve. Yet the potential rewards are substantial – as are the opportunities for immediate improvement. Investing time and effort in your service parts business today will help you strengthen your brand and build a loyal base of customers, boosting short-term profitability while positioning your company to reap even greater benefits as the market matures.

Hau L. Lee and Chung-Yee Lee (Eds.)
Building Supply Chain Excellence
in Emerging Economies
©2007 Springer Science + Business Media, LLC

Chapter 11

DHL IN CHINA
The Role of Logistics Governance

Kelvin Leung
DHL

Paul Forster
Hong Kong University of Science & Technology

Abstract: In developed economies, the flow of information and materials proceeds relatively unimpeded by the logisitcs governance because many of the larger governance obstacles are removed over time in an effort to improve economic performance. In developing economies, the flow is often blocked by deeply embedded obstacles that inhibit the efficient flow of materials and information
among participants in the logistics chain. For the emerging logistics industry in
developing economies, these governance obstacles constrain the design of the
logistics chain, improvement of services, add costs, and hamper the opertional
and financial performance of the logistics chain. For management, consideration of the impact of logistics governance on design, implementation and performance of the logistics chain is part of the strategy. There are several lessons
learnt from DHL's experience with logistics governance in China. Ignorance of
the details of the governance or the role of the governance in the logistics chain
while attending to marketing, investment, or technology strategies can be a fatal
mistake. Logistics governance issues are an essential component of any China
strategy for logistics.

11.1.   Introduction

As a shipment moves from shipper to consignee through the logistics chain
it travels through an invisible obstacle course of rules, regulations, work prac-

tices and customs that together form the logistics governance regime of a logistics chain. Logistics governance both facilitates and constrains the design, implementation and performance of the logistics chain as a whole and of its many participants. In this chapter we draw upon DHL's experience to discuss the role of logistics governance in shaping the logistics sector in China.

In developed economies the flow of information and materials proceeds relatively unimpeded by logistics governance because many of the largest governance obstacles are removed over time in an effort to improve economic performance. In developing economies the flow is often blocked by deeply embedded obstacles that inhibit the efficient flow of materials and information between participants in the logistics chain. For the emerging logistics industry in developing economies, these governance obstacles constrain the design of the logistics chain, improvement of services, add costs, and hamper operational and financial performance of the logistics chain.

Logistics governance is a necessary topic in any discussion of logistics in developing economies. Where logistics governance is harmonized, standardized and stable, as is generally the case in developed economies, the business of logistics can proceed with the knowledge that the rules of the game today are pretty much the same as those tomorrow; that the rules will not interfere with good business decisions; that the rules at the shipment source are similar to the rules at the destination, and; that the rules do not privilege one set of firms over another. In short, in developed economies, logistics governance is not a major source of uncertainty and does not command a great deal of managerial attention. Firms can make long-term investments in the confidence that the governance regime will remain stable. Logistics governance in developed economies provides a "level playing field" for logistics firms and the supply chains they serve. While there still may be obstacles, at least the topography is visible to all.

In developing economies logistics governance is often fragmented, contentious, non-standard, non-transparent and dynamic. As such, not only is the playing field uneven, but the participants are working with different maps. Logistics governance becomes a critical source of uncertainty that affects every facet of the business, requiring managerial attention to understand, anticipate, react, and strategize about logistics governance. Long-term investments are risky as critical assumptions about governance can change before a return is ever seen. The supply chains that are dependent upon efficient logistics bear the costs of higher costs and lower performance.

Logistics governance may not be harmonized in developing economies for many reasons. One primary reason is history and the evolution of logistics. The concept of the finely-tuned integrated global logistics chain is only a recent phenomenon and the individual pieces of the logistics chain have evolved differently. Individual modalities of surface, air, ocean, rail, river have emerged differently with different governance structures, professional bodies and cultures. Nations have evolved different transportation systems that reflect physical topography and political ideology. Such different histories manifest themselves as conflict particularly when they come together at the handoffs between stages in the logistics chain: between international and national interests, between transportation modalities such as surface and air, and at interfaces such as ports and customs.

While differences can be historical and unintentional, disharmony in logistics governance may also be intentional, where regulation is used to protect indigenous firms, protect tax revenues, or favor one group of firms over another. For instance, at the regional level, regulation is often used to protect local interests, attract investment and prevent the flight of industries.

Where regulation is not harmonized, such as is often the case in developing economies, heterogeneity between regulation at the local, regional and national levels can create different operating environments in each sphere. Where information exchange is not standardized, where performance expectations are inconsistent, where work processes vary from region to region, coordination becomes more difficult to manage, more costly, and less effective. Governance becomes a barrier to integrating the components of a logistics chain.

In China, logistics costs are comparatively much higher than in developed countries. The proportion of logistics costs to total production costs is estimated at 20–30% in China, vs. 10% in developed countries (Dekker, 2002). Supply-chain related costs are 30–40% of wholesale prices vs. 5–20% in the U.S. (Tanzer, 2001). In 2001, logistics spending in China was approximately 20% of GDP, twice that of the U.S. An estimated 90% of a Chinese manufacturer's time is spent on logistics, and only 10% on manufacturing.

There are many factors that contribute to this disparity: an immature transportation infrastructure, non-competitive state ownership of distribution, low labor costs and low education levels are some. State owned enterprises such as Sinotrans and Cosco are not challenged to be highly competitive and efficient, low labor costs favor a higher labor component over investments in coordinating technologies such as electronic data interchange, and low levels of education inhibit development of complex service offerings. In this chapter, we focus on another significant factor - how logistics governance shapes the opportunities, design and performance of logistics chains.

The case of China, while representative of issues in logistics governance in developing countries is also a special case in point. Chinese culture, history, and economics are visible in the state of logistics governance. The expectations of Western logistics providers as they enter China may be that the rules of governance in China ought to be similar to those in the developed countries, or that once the China "gets it" they will design a governance structure similar to the West. This is not the case at all, and it is critical that Western managers invest the time to understand the logic of logistics governance in China if they wish to become serious players in logistics in this part of the world.

In the remainder of this chapter, we explain why the logistics governance in developing economies is of crucial importance for management, and we describe some of the more powerful forces at work in logistics governance in China.

11.2.   What is Logistics Governance?

Logistics governance is the set of formal rules, regulations and informal work practices that guide productive work in the business of logistics. There is no single entity that provides logistics governance over all the participants in a logistics chain, rather there is a network of organizations that together directs, administers and controls the activities of the participants in the chain: shippers, transportation providers, warehousers, third-party logistics service providers, air and seaports, customs and consignees.

The term "governance" usually refers to the formal laws and regulations cast by organizations such as governments, companies, and legal industry associations. While formal laws and regulations are undeniably the most powerful governance institutions, we also include in our definition the informal institutions of customs, habits, training and education, and work practices that are products of informal organizations such as social groups, communities and occupations.

Together, formal regulations and informal institutions constitute the governance of logistics. In doing so, governance penetrates logistics from work practices in everyday job tasks to the structure of the industry.

11.3.   Why Logistics Governance Matters

Globalization, competition, outsourcing, new business models, long supply chains, customer expectations, and geo-politics place increasing pressure on supply chain participants to provide more for less to consumers. In turn, as the primary customers of logistics, supply chain participants place pressure on logistics providers to be more cost-effective, efficient, responsive, flexible and secure. In order to do so, global logistics providers must increase their level of coordination and control both within their own internal operations and across the entire logistics chain. Coordination in the context of logistics reflects not only the physical movement of materials, but also the effective movement of information in order to control the flow of materials.

The trend in contemporary logistics has shifted from performance measures such as "whenever it gets here" towards time-specific delivery, and in doing so reducing variability around delivery times, increasing predictability, and removing logistics as a source of uncertainty in supply chains. Reducing variability makes the task of logistics more complex. The higher the level of performance demanded of the logistics chain, the more closely the stages in the logistics chain must be coordinated and work processes integrated. Buffers between stages need to be eliminated, more communication is required both upstream and downstream, and tasks need to be carefully integrated. The tight coupling of activities requires greater cooperation and exchange of information between participants in the chain, making the interfirm relationship a critical factor in the chain performance. Therefore, participants in the logistics chain need to be chosen carefully, and the organizational structure of each participant and the interorganizational structure of the chain needs to be designed to motivate higher levels of coordination.

Logistics governance affects coordination across the logistics chain in several important ways. First, logistics governance constrains the possible configurations of logistics chains. The formal rules and regulations such as ownership and FDI, business licenses, access to physical facilities constrain and enable which firms can participate in the chain, types of services that can be offered and where firms can be located.

Second, logistics governance is not plastic or subject to managerial rationality. In general, logistics governance will outlive most individual logistics firms. The rules and regulations are not under the control or influence of any indi-

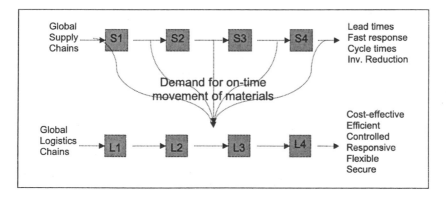

Figure 11.1. Increased coordination across supply and logistics chains.

Table 11.1. Sales growth of DHL China, 2001–2004.

Total	2004 vs. 2003 (% change)	2003 vs. 2002 (% change)	2002 vs. 2001 (% change)
No. of Shipment	25–30%	25–30%	20–25%
Tonnage	15–20%	55–60%	30–35%
Revenue	50–55%	20–25%	25–30%

vidual firm or even an industry. Governance structures emerge slowly, evolve through history, and do not necessarily change in response to the pressing needs of business. Rather, governance structures reflect deeper, more universal interests, and change at unpredictable times. As such, changes in logistics governance cannot generally be considered as a malleable element of business strategy. However, they strongly influence the space of possibilities that can be considered by management.

Third, reporting requirements, bureaucracy, forms, documentation and data exchange are also affected by logistics governance. The volume of documentation required and delays introduced by completing, correcting, transferring and awaiting approval of documentation affect the information flow necessary to coordinate activities, the material flow and in turn, the overall performance of the logistics chain. As logistics is increasingly an information business, and the effective management of information is crucial to the coordination of the flow of goods. Rules and regulations, to the extent that they influence quantity and quality of information exchanged introduce costs and delays affecting the performance of the chain.

11.4.   DHL in China

DHL has been operating in Hong Kong since 1976 and in China since 1991. DHL operates three divisions in China – DHL-Sinotrans, DHL-Danzas Air & Ocean, and DHL Solutions. Maps 1 and 2 show DHL's presence in northern and southern China as of 2005.

DHL's mainland operation has grown in double digits each year since 2000 (Table 1). From 2003 to 2004, shipments rose 25–30%, tonnage rose 15–20%, and revenues rose 50-55%. This growth, while extraordinary, is similar to growth rates throughout the logistics industry reflecting the rapid growth in China trade.

DHL has two main joint ventures in China – one for express and one for logistics. DHL-Sinotrans provides express courier products in competition with

UPS and FedEx. DHL-Zhong Fu offers general air freight, sea freight, road transportation, warehousing and value-added logistics services such as vendor management inventory, cross docking operations, label scanning, pick and pack, vendor and purchase order management.

Joint ventures have been crucial to operating within China. DHL used a 50-50 joint venture with Sinotrans in order to enter the express delivery market

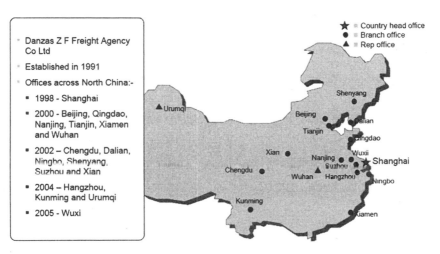

Figure 11.2. DHL in Northern China.

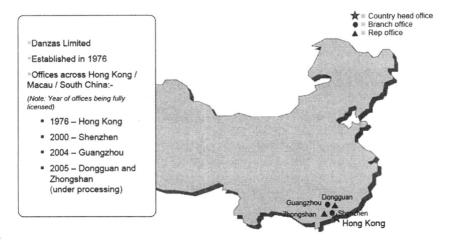

Figure 11.3. DHL in Southern China.

starting as early as 1986. Sinotrans, as a state-owned enterprise had both government support and local knowledge to complement DHL's international experience. Unlike its competitors, particularly FedEx and UPS, DHL also drew upon commercial airlines for lift capacity for its express operations rather than flying its own freighters.

DHL has leveraged the DHL-Sinotrans relationship as a platform from which to offer advanced services such as online tracking and tracing, shipment processing, time definite delivery and services that require higher levels of coordination. The stability of the JV relationship enables both partners to invest in facilities and technologies that are required in order to improve service levels. The relationship has been successful, enabling DHL to capture 37% of the China express delivery market by 2003.

11.5.   National Logistics Governance

At the national level, logistics governance in China is affected by multilateral and plurilateral global trade agreements. While changes in logistics governance can move at a glacial pace, international agreements can be a powerful force for change. There are many such agreements: General Agreement on Tariffs and Trade (GATT), Agreement on Trade-Related Investment Measures (TRIMs), the General Agreement on Trade in Services (GATS), and Agreement on Government Procurement (AGP). While all these agreements have some impact on logistics, we focus our discussion on WTO and CEPA as these agreements are a powerful force in shaping and constraining the logistics sector in China. These agreements also create a challenge for the Chinese government to comply with the negotiated elements of the agreements and yet manage the interests of groups disadvantaged by the changes stemming from the agreements.

11.5.1   World Trade Organization

The World Trade Organization (WTO) affects the macro environment of logistics by shifting more production to China, increasing the movement of raw materials, work in process, and creating opportunities for advanced logistic services. On one hand, the WTO is an entry ticket for firms into China, increasing global trade, and it may bring many positive benefits to China. On the other hand, it might disadvantage indigenous logistics firms by increasing competition from foreign firms with deeper pockets, with greater international experience. In general terms, because logistics costs are such a large component of Chinese exports, increased competition in logistics will push down costs, improve service, and make exports more competitive.

WTO, through several phases, allowed an increase in ownership of Chinese firms by foreign freight forwarding companies from 50% to 75% by January 2004, and to 100% ownership by December, 2005. That is to say, in 2006 foreign firms can establish wholly owned subsidiaries or buy out existing joint venture partners.

On the surface, gradual lifting of ownership restrictions throws open the doors for foreign companies to enter China, increase service levels and lower costs. The phasing is intended to provide the indigenous Chinese firms time to adjust to increased competition. While it is certain that WTO will result in some standardization of regulation and harmonization of internal and external requirements for trade, it is naïve to assume that foreign companies are playing on a level playing field with indigenous firms. As we will see, logistics governance is complex, and while one set of regulations may ease entry, there are many other institutions at work in logistics governance that constrain the opportunities for foreign firms.

11.5.2 Closer Economic Partnership Arrangement

The Closer Economic Partnership Arrangement, or CEPA, is an example of national regulation that influences the emergence of the logistics industry in a certain region. The arrangement is between the Ministry of Commerce in Beijing and the Special Administrative Regions of Hong Kong and Macau. In some sense, CEPA is a gift to Hong Kong and Macau, enabling Hong Kong registered companies to get a head start in mainland China years ahead of the WTO timeline and giving a boost to their stagnant economies in 2003 after the impact of 9/11 and the outbreak of SARS. It also gave Beijing an opportunity to demonstrate that its concern for the economic future of these regions.

The concerns of the Central government extend beyond creating an efficient logistics system but that logistics is a by-product of other overarching interests. We discuss CEPA in more detail below to illustrate how the influence of regional regulation can influence the entire logistics industry.

Under the conditions of this agreement, there are three key changes to the trade environment between the mainland and Hong Kong. First, 273 types of commodities manufactured in China and the Hong Kong Special Administrative Region can be distributed and sold tariff free bilaterally. Second, qualified Hong Kong companies in eight industry sectors are allowed to enter China without the requirement of forming a joint venture company with a local Chinese company and vice versa for the Chinese companies investing in Hong Kong. Third, an official governmental platform was established to facilitate trade between the Hong Kong and the mainland.

CEPA offers a unique advantage to Hong Kong logistics and freight forwarding companies. Prior to the effective date of 1 January 2004 of CEPA, foreign fowarders, including those in Hong Kong and Macau, were required to adhere to the phased relaxation of the WTO. Through CEPA, a Hong Kong company can establish a wholly owned entity in China. In addition, the required capital investment amount for Hong Kong companies was reduced from 1 million USD to 620K USD (5 million RMB).

Regulatory changes such as CEPA lower the barriers to entry to the mainland logistics industry for one select group of companies in Hong Kong and yet these changes have a dramatic impact on the landscape of the whole mainland industry. In the past, Hong Kong companies attempting to enter China were unable to do so directly, but were forced to deal with the airlines and shipping lines indirectly by using a third party local agency. Working through a third party that is not wholly owned creates numerous problems for control and coordination by leaving the Hong Kong company open for opportunistic behaviour by its partner. Through CEPA, Hong Kong companies can now enter the Chinese market with wholly owned partners.

Second, the agreement will have an impact on the structure of manufacturing and the demand for logistics. Efficient logistics will reduce the economic distance to remoter regions. New entrants offering more competitive logistics services can draw manufacturing inland away from the congested urban areas along the coastline. The penetration will depend on several factors including transportation and logistics costs, wage rates, land costs, transportation infrastructure and tax incentives from municipal and provincial governments in the inland areas.

Third, CEPA influences the labor market for logistics professionals. The rapidly growing logistics market requires an experienced labor force. While China is producing high quality university graduates with shipping, trading and logistics educations, they are mostly inexperienced. The expansion of the logistics market as a result of WTO and CEPA continues to contribute to a shortage of experienced personnel in the short term. One by-product of this shortage has been high staff turnover rates. To address this issue, companies are spending more on human resources matters and the related costs such as training, career development and, compensation and benefits are increasing.

While CEPA is a regional agreement, through increasing competition, restructuring the location of manufacturing, creating labor shortages, impacts the structure of a much larger sphere of logistics.

Figure 11.4. Export processing zones.

Figure 11.5. Free Trade zones.

11.6. Economic Zones

China faces a practical problem of sharing economic prosperity across a vast and heterogeneous geographical and cultural landscape. One mechanism for encouraging regional growth has been the creation of economic and development zones. To name a few, the Special Economic Zones (SEZ), the Economic and Technological Development Zone (ETDZ), the Hi-tech Industrial Development Zones (HIDZ), the Taiwanese Investment Zones (TIZ), the Border and Economic Cooperation Zones (BECZ), the Tourist and Holiday Resorts (THR), the Export Processing Zones (EPZ) and the Free Trade Zones (FTZ). The different types of zones have different emphases. Some facilitate technology development, some enhance trade and some develop tourism, and most have political agendas. The zones most relevant to the logistics industry are the Export Processing Zones (see Map 3) and the Free Trade Zones (see Map 4). Both types of zones are supervised by Chinese Customs and are bonded areas. The key difference between them is that the Export Processing Zone is primarily for exports, whereas Free Trade Zones are for imports, exports and re-exports. Within these zones, manufacturers have set up their production bases to gain tax benefits. Similarly, there are a number of logistics service providers who have placed their processing and value-add centres within the zones to provide services such as sorting, pick and pack, quality control, and consolidation for their customers.

These zones enable China to develop both inbound and outbound trade as well as establish new trade rules. For example the Free Trade Zones are export processing zones that facilitate import and export trade within a bonded area, provide a tax rebate scheme to attract manufacturing and allow value-added processes to be done in these areas, whereas these activities cannot be performed outside these areas under the general Chinese Customs' rules and regulations.

The zones create different opportunities for logistics service providers, therefore providers need to be aware of the purpose, design, and regulation of the different zones in order to understand what opportunities lie within each zone. In the Shanghai free trade zone, the zone is considered "outside" China. DHL was the first to establish a local partnership to bring shipments from overseas and to combine with shipments from local production in order to merge, consolidate, sort, relabel, pick and pack, ship to overseas, or re-enter into the domestic market. Elsewhere in China, this cannot be done under the existing rules and regulations, either because such activities are illegal, because high costs are imposed. DHL captured the opportunity in Shanghai by partnering early and developing services and supply chain solutions for this new regulatory environment. DHL was only able to do this by maintaining contact with policy channels and staying abreast of new opportunities that result from changes in governance.

This example illustrates that regulation is a large determinant of the costs of logistics solutions. Logistics providers cannot always provide a cost-effective customer where governance issues prevent the ideal design. Rather providers have to work around the restrictions of regulations when designing solutions. In one sense, governance moderates the relationship between the size of the investment and the performance outcomes. That is, the highest investment in China does not necessarily mean the highest performance when the influence of regulation is considered.

11.7. Pan Pearl River Delta Economic Zone

Another very strong regulatory influence on the growth of the logistics industry in China comes from the economic engine of the Pearl River Delta. Given the dramatic growth in this region in recent years, regulation has been created to share some of the growth with surrounding regions. In 2003, Chang Dejiang, Party Secretary of the Guangdong Province as well as a member of the China Communist Party's Politburo, endorsed the concept of a Pan Pearl River Delta economic zone based on his previous successful leadership in expanding the influence of the Shanghai economic zone into the Jiangsu and the Zhejiang provinces, which led to the creation of the Yangtze River Delta economic zone.

The Pan Pearl River Delta economic zone is intended to speed economic reforms in the eight adjacent provinces: Fujian, Jiangxi, Hunan, Sichuan, Yunnan, Guizhou, Guangxi and Hainan. Together, they form the "9 + 2" economic zone – the nine provinces plus two special administrative regions of Hong Kong and Macau (see Map 2). The economic zone expands the geographical area from 42,824 km^2 to 2,006,129 km^2, and the population from 30.6 million to 456 million. The size of the market GDP will increase from 330 billion USD to US dollar 637 billion USD. The total market GDP of the Pan Pearl River Delta economic zone is projected to exceed 1 trillion USD by 2010.

The zone attempts to harmonize and align the economic roles of each of the provinces and special administrative regions in order to reduce duplication in planning and developing resources, in particular, infrastructure such as airports, seaports and highways. Tens of thousands of kilometres of railways and highways will be constructed to improve the current links among the different areas. The economic zone intends to improve inter-provincial cooperation and to strengthen connectivity among the provinces in the region. The Pan River Delta region will benefit industry in general by relieving Guangdong of the

Figure 11.6. The Pan Pearl River Delta economic zone.

higher costs associated with manufacturing and infrastructure congestion, and achieving the political aims of providing a wider base for economic growth.

For the logistics industry this means, inasmuch as transportation costs from the inner provinces to the Guangdong province and the two special administrative regions are lowered, more plants and factories, especially small and medium-sized, will migrate inland. The management of the total supply chain will then require a larger geographical coverage as well as a more complex and comprehensive inter-modal logistics network.

In general, these policies, including the general "go west" policy, create opportunities for logistics in the coming years, extending infrastructure further inland, strengthening inter-provincial cooperation and interaction, and possibly standardizing regulation, such as customs clearance practices, across provinces. In doing so, the policies change the shape of the "efficient frontier" of logistics shifting it further inland away from the congested and costly coastal cities. Nevertheless, these policies create winners and losers. Those firms that are willing to accommodate the socio-political aims of the policies are also those that are best positioned to gain the benefits.

11.8. Logistics Parks

The rationale of a logistic park is to provide customs supervised sites in strategic locations in close proximity to major seaports or airports with supporting infrastructure. There are roughly 300 logistics parks in China, of which more than half are located in the three main economic zones of the Yangtze River Delta region, the Greater Pearl River Delta region and the Bohai Bay region.

Logistics parks illustrate the difficulty of logistics governance where there are competing interests in regulatory design. On the surface the parks are intended to improve supply chain efficiency by enabling logistics service providers to better serve the trade flow between manufacturers and their customers. However, many logistics parks in China have effectively become property development projects rather than logistics projects. Many of them are small, non-bonded and have not been properly planned. Some of them are far away from the key seaports or airports and lack connectivity between the manufacturers and the ports. While they do serve to develop local interests, and in the long run may reach their intended logistics functions, in the short run, the parks tend to place short-term local interests first. The message for foreign entrants is that they need to spend time to become informed of the interests and incentives of local stakeholders before investing in such projects. Is the park bonded? What are the regulations on tax benefits? What is the location of the park with respect to ports? Is there IT connectivity with other parties? Is there access to trained labor? Are the interests of the stakeholders in alignment with ours?

Even large providers such as DHL can encounter difficulties. At one time, DHL entered into some apparently exclusive arrangements with companies inside a logistics park, but subsequently became aware that the holding company of the intended partners had already invested in another competing logistics park, i.e. the intended partner was planning to become both partner and competitor, which was unacceptable. In another case, a local city government was actively engaged in promoting a bonded logistics park, but through another channel DHL learned that the plans had already been shelved. In these cases, having multiple information channels avoided possibly serious consequences and the significance of "guanxi" in interfirm relationships in China.

11.9. Bonded Facilities

As in other countries, there are also bonded warehouses in China. While zones and parks fall under the administration of the central government, bonded facilities are of a much smaller scale with single locations and are subject to the administration of local customs bureaus and governments.

There are primarily two types of bonded facilities, import and export. Historically these two functions are separate – export bonded warehouses are strictly for export and cannot be used to import and then re-export. This results in procedural inefficiencies for users and higher costs for the facilities and the logistics chains they serve.

The licenses for bonded facilities have been extremely difficult to obtain for foreign companies. Even for local companies the issuance of bonded facility licenses is tightly controlled, ostensibly to control tax collection and smuggling, activities that jeopardize local tax income. Tight control of the licensing process assists management and monitoring of the tax revenue process. While foreign companies are not particularly excluded from obtaining a licence, at this time there are apparently no foreign companies holding a license. To work around this constraint, DHL finds local partners with licenses to indirectly have access to bonded facilities. However, this gives the local partner a great deal of power because it controls the license. It becomes extremely important to maintain a healthy working relationship. Under WTO, a foreign company can buy a local company with a license, creating new possibilities for performance and services.

11.10. Zone-Port Interaction Area (Free Port)

To facilitate exports, one key element is to expedite the tax refund process for manufacturers. Under normal circumstances, tax rebates are given when the export items are on board a vessel or airplane. The Wai Gao Qiao Bonded Logistics Park in Shanghai will be the very first pilot "zone-port interaction area" to further enhance the capability of the Free Trade Zone. When export items enter the park, the shipper can immediately start the tax rebate process, speeding the rebate process by one or two working days. This is crucial for many small and medium sized manufacturers who run on a tight cash flow. The Tax Bureau is the bottleneck for the tax rebates and billions of RMB in rebates are delayed. However, as with much of the regulation in China, the process and timing of the changes is unknown and the Tax Bureau has disclosed little about the details of the new proposal.

The Wai Gao Qiao Bonded Logistics Park in Shanghai is the first step towards the policy of a "Free Port". In China, Customs in the port area and Customs of the Free Trade Zone are under different jurisdictions and different tax income books. Often, it takes a series of tedious handling procedures to move a shipment from the Free Trade Zone to the port-controlled area even if they are located beside each other. The Shanghai Wai Gao Qiao Bonded Logistics Park is a joint cooperation project of the Wai Gao Qiao Free Trade Zone and the port authority with the aim to eliminate this type of unnecessary red tape. It is understood that there will be a second batch of trial ports to follow on this concept. The Yantian Free Trade Zone has already applied to the central government and is likely to be the second zone-port interaction area in China.

11.11. The Issue of Licenses

An illusion of many foreign firms is that WTO creates a "level playing field" in China. While WTO certainly removes some fences and fills in a few gaping pits, it certainly does not level the playing field in logistics. We use a discussion of licenses and how they serve to control the entry into the logistics industry as a case in point.

With the accession of China under WTO, access to the logistics industry by foreign firms is increased. However, ownership does not automatically grant operating rights. While a foreign firm may purchase a local company, it still does not mean it has the right to operate within China if the purchased company does not have the appropriate licenses. Licenses are crucial to the governance of logistics because licensing is a necessary condition for the logistics chain, allowing each member to legally perform their business. This is very different from the situation in developed economies where a business license generally grants the rights to perform all tasks. In China, each new product or service offering may require a new license. This creates huge challenges for both supply chains and the logistics chains that serve them as they attempt to expand their markets and change their ways of doing business.

A few of the most critical licenses:

- The *Certificate of Approval for Establishment of Enterprises with Foreign Investment in the People's Republic of China*. This is the foreign funding assessment and approval process governed by the Department of Foreign Investment under the Ministry of Commerce. All foreign companies require this certificate to operate in China.
- The *International Freight Forwarders License* (commonly, but incorrectly, known as the "A-license"). This is the industrial assessment and approval process carried out by the Department of Foreign Trade under the Ministry of Commerce. This license allows companies to operate business in the area of freight forwarding (which is defined differently from the logistics industry in China).
- Once the International Freight Forwarders License is obtained, the *Air-Transport Qualification Certificate* (the real A-License) can be obtained.

This is necessary if the company wants to deal directly with airlines – freight rates quotation, space handling and issuance of airway bills – instead of dealing via a third party local agency with such a license. The requirements for this license include a 3 million RMB register fund and a recommendation letter from a Chinese airline with international freight services. The applicant company, however, need not be International Air Transport Association (IATA) certified. The official body to approve this license is the Civil Aviation Administration of China (CAAC). There is also the B-license, which covers domestic air transportation.

- Having obtained the International Freight Forwarders License, the company is allowed to carry out intra-city domestic trucking services. However, if cross-city or inter-provincial trucking services are involved, the *Trucking License*, which is to be approved by the Ministry of Communications, is needed.
- In addition to the International Freight Forwarders License, there is also the *Logistics License*. However, the differences between the two types of licenses are very vague. There are some differences in the areas of domestic trucking services and in the Trading License. They are covered by the Logistics License and not by the International Freight Forwarders License in terms of the business scope. Thus far, few logistics licenses have been issued. The register fund requirement for the Logistics License is 23 million RMB.
- Similar to the airfreight license, there is the *Non-Vessel Operations Common Carrier* license. This license is for the sea freight operations if the company wishes to deal directly with the shipping lines. The register fund requirement is 800,000 RMB and the official body to handle this application is the Ministry of Commerce.
- The *Customs Broker License* allows a company to handle customs declaration on behalf of their customers. The applicant company must have customs broker staff certified by the Customs Bureau, and provide evidence that it needs to provide customs brokerage services in day-to-day business operations. This license is very difficult for foreign companies to obtain.
- Finally, there is the *Import/Export Trading License*. With this license, services such as Import of Record, Import License, VAT Invoice, payment collection, foreign currency exchange and bank remittance can be offered to the customers.

There are a myriad of governmental bodies responsible for the issuance of licenses, not a single body. New entrants must learn the purpose of each license, the rules for obtaining each license, the granting bodies for each license, the appropriate channel for obtaining the license, as well as be aware of changes in the ministries and granting authorities. Licensing affects logistics by restricting the expansion of foreign companies in China in spite of WTO access, and by influencing the process of improving logistics efficiency.

The existing licensing structure trades off local, national and international interests. International agreements encourage the entrance of foreign investment, local interests protect the viability of indigenous firms. There is a movement towards simplification of the licensing process, liberalizing of regulations, and shifting of granting licensing authority to semi-governmental bodies. As part of the on-going changes in China to simplify and expedite the process for business investments, the State Council published the State Council of the People's Republic of China Document Publishing Number [2004] 16, which has come to be known simply as Document Number Sixteen. This document introduced several changes to the governance structure. The case of Document Number Sixteen illustrates the dynamism in the governance of logistics and a trend towards liberalization. The Department of Foreign Trade is merged into the Department of Foreign Investment under the Ministry of Commerce. The International Freight Forwarders License is no longer required for domestic and foreign companies. However, foreign companies are still required to apply for the Certificate of Approval for Establishment of Enterprises with Foreign Investment in the People's Republic of China. The application for the Air-Transport Qualification Certificate with the Civil Aviation Administration of China (CAAC) will no longer be necessary, allowing companies to deal with airlines directly. The issuance of the Customs Brokerage License was stopped, allowing companies to set up their own customs broker firms to handle their own customs clearance.

11.12. Conclusions

Our thesis in this chapter has been that logistics governance in developing countries intercedes in the relationship between investments in logistics assets and the performance of individual firms, and of the total logistics chain. There are several concurrent trends in China that raise the significance of logistics governance.

The first is the increasing involvement of specialised third parties. New technologies and custom-made solutions to address the requirements of individual customers make the management of the total supply chain more difficult to handle in-house. The expertise required and the resources that need to be invested into this area of operations often add a tremendous burden, and take attention and resources away from the core business. As more Chinese companies gradually realize this, more are starting to outsource their logistics handling to third party logistics (3PL) or 4PL suppliers. It is estimated that out of the 270 billion USD logistics market in China, only 3% is currently handled by third party logistics suppliers. It is projected that by 2010 the market size will increase to 370 billion USD, and the share of that carried by third party logistics suppliers will increase to around 6%. A large portion of this business falls into the area of contract logistics such as warehouse management, distribution centers, vendor managed inventory, cross-docking, buyer consolidation and domestic distributions. This trend introduces new parties, new requirements, and new technologies and the increases the complexity of logistics chain.

A second trend is the evolution from total supply chain management to total trade chain management. Logistics customers, especially small and medium sized firms that cannot afford their own in-house departments, will look for their third party logistics suppliers to provide them the sourcing, procurement, sampling, quality control and trade financing services. This trend increases the total logistics chain requirements, forcing management to consider not their own position in the logistics chain, but to view their position and relationships with all other parties in order to achieve a total logistics chain perspective.

Third, events such as accession to WTO, CEPA, the establishment of economic zones such as the Pan Pearl River Delta will continue to boost both intercontinental trade as well as the cross border trade with China's neighbors. In turn this will lead to the continuous growth in exports of finished goods and imports of raw materials and spare parts. Logistics suppliers must increase their global network coverage as well as cross-border handling capabilities. This trend is forcing firms to enter new markets, and engage greater competition.

These factors combine to increase the complexity of logistics chains. This is a function of higher performance expectations, the introduction of new technologies requiring greater exchange of information, the increasing involvement of specialized third parties. Together these force closer integration of the stages of the chain, and closer coupling of all parties. In doing so, the role of governance in smoothing the process of integration becomes increasingly significant. In China this applies particularly to rules and regulations that influence the structure of the logistics chain, reducing bureaucracy, standardizing information flows, and the creation of a homogenous qualified labor pool.

Coordination and control of the logistics chain is an issue of the highest priority for logistics managers. Logistics governance strongly influences coordination across the logistics chain and moderates the relationship between investment and performance. In developing economies such as China, we find that logistics governance has a profound impact because of the absence of harmonized regulation, competing interests, and rapid changes in regulation. National trade agreements such as WTO and CEPA while providing some level of standardization they also influence barriers to entry to the industry, the labor market, and the set of possible configurations of logistics chains. Regional agreements such as those that define the Export Processing Zones and Free Trade Zones illustrate how regional logistics governance influences barriers to entry, the structure of operations, and offerings of services. Local logistics governance domains such as logistics parks and bonded facilities illustrate how competing interests affect local operations. The discussion of licensing is an example of how the issue of licenses can be used to control the structure of the industry.

For management, consideration of the impact of logistics governance on the design, implementation, and performance of the logistics chain is part of strategy. There are several lessons learned from DHL's experience with logistics governance in China. The first is that maintaining effective communications with all levels of governance is essential to anticipate changes in regulation. Guangxi, the social network for maintaining relationships is a formidable advantage to those who use it effectively, and an equally formidable barrier for those firms who don't. Using guangxi strategically to appreciate changes in logistics governance, combined with an understanding of how governance influences the logistics chain, is a competitive advantage.

While some elements of governance such as regulation can be very dynamic. Dynamism is shown in such issues as ownership, driven by powerful external forces such as WTO compliance. However, other elements such as educational level and local practices are deeply embedded into Chinese society. Embeddedness is seen in local education levels, a qualified personnel, local behaviour in organizations such as Customs, and language. The result is that there is great heterogencity in governance across time and geography that must to be managed by large organizations. DHL uses Mandarin-local dialect speakers as the "glue" between Asian-China-local levels of operations.

Logistics governance is of strategic-level significance for firms wishing to participate in China. Ignorance of the details of governance or the role of governance in the logistics chain while attending to marketing, investment, or technology strategies can be a fatal mistake. Logistics governance issues are an essential component of any China strategy for logistics.

References

Bolton, J.M. and Y. Wei (2003), "Distribution and logistics in today's China," The China Business Review, Sep-Oct, pp. 8–17.

Dekker, N., (2002), "Gold Rush!" Containerization International, Vol. 35, No. 1, pp. 29–31.

Goh, M. and C. Ling (2003), "Logistics development in China," International Journal of Physical Distribution and Logistics Management, 33 (9/10), pp. 886–917.

Jiang, B. and E. Prater (2002). "Distribution and logistics development in China," International Journal of Physical and Logistics Management, 32 (9/10), pp. 783–797.

Tanzer, A. (2001), "Chinese Walls," Forbes Global, Vol. 168, www.forbes.com/global/2001/1112/091.html accessed 9/1/2005, pp. 74–76.

Part III

BUILDING SUPPLY CHAIN EXCELLENCE: INNOVATIONS AND SUCCESS CASES

Hau L. Lee and Chung-Yee Lee (Eds.)
Building Supply Chain Excellence
in Emerging Economies
©2007 Springer Science + Business Media, LLC

Chapter 12

SUPPLY CHAIN REENGINEERING IN AGRI-BUSINESS
A Case Study of ITC's e-Choupal*

Ravi Anupindi
Ross School of Business
University of Michigan, USA

S. Sivakumar
CEO – International Business Division
ITC Limited, India

Abstract: The main premise of the chapter is that emerging economies are characterized by "broken value chains" that attempt to connect the poor, as sellers and buyers, to markets for products and services. Often these value chains do work (as they need to) but with the help of numerous intermediaries who extract disproportionate value leaving the poor with little residual income. The challenge of fixing these value chains is further exacerbated by factors like fragmentation, dispersion, heterogeneity, and weak infrastructure. Nevertheless, the enormity and the complexity of the task at hand imply that reengineering the farm-to-market supply chain offers a tremendous business opportunity.

In this chapter we describe a large scale agri-business supply chain reengineering effort, called *e-Choupal*, being implemented across various commodities by the ITC Group of India. We argue that this large-scale effort enhances shareholder value, alleviates poverty, lays the foundation for global competitiveness of agriculture, and at the same time sows the seeds of social transformation. We illustrate this through a detailed discussion of ITC's interventions in soybean, wheat, and coffee procurement. Together the three highlight a progression of value provisioning using supply chain reengineering. While the first one focuses on improving the logistics efficiency of a commodity supply chain, the second example illustrates the shift from commodity-based to a variety-based strategy, and finally the third example illustrates the migration from products to services-based strategy in agri-businesses. For each of these cases, we articulate benefits that accrue to the key stakeholders. We close the chapter with a discussion on how ITC's e-Choupal is an innovative business platform to convert agricultural supply-chains into demand-chains or from 'selling what is produced' to 'help producing what is wanted'.

12.1.   Introduction

With a population of about 1.03 billion – and growing – living in an area of 3.2 million square kilometers, India is the second most populous nation after China, the seventh largest country by land mass and the largest democracy in the world. Since its independence in 1947, India followed a mixed-model economy with very modest growth. India finally liberalized its economy in 1991 and has since seen a 6–7% annual growth rate. Today India's GDP is about $600 billion (with over $100 billion in foreign exchange reserves) making it the 12th largest economy in the world (4th largest when adjusted for purchasing power parity). Some experts predict that in 10–15 years, India is slated to be the third largest economy with a share of 14.3% of the global economy. Most of the descriptions and achievements of India written up in the popular press – the India of the third largest pool of scientific and technical manpower, the India of high technology and biotech industries, the India as a destination of choice for services outsourcing – is based on the industrial and services sector largely focused on what we call the urban India.

A sector-wise break up of India's economy illustrates that services (inclusive of government services) contribute about 52% of the GDP; the rest is equally divided between agriculture and industry. At 24% of GDP, agriculture is a significant component of India's economy. Agriculture as a relatively large component of overall domestic economy is a characteristic of most emerging economies, especially when compared to the developed countries; for example, in the United States agriculture comprises just 1.4% of the 2003 GDP. Table 12.1 gives some selected statistics to highlight India's position in the world agriculture. Based on the aggregate metrics of arable land areas and volume

* This document would not have been possible without assistance from several people working at ITC-IBD offices in Hyderabad and Bhopal. In particular, we wish to thank Rajnikant Rai, Shailendra Tyagi, D.V.R. Kumar, M. Srinivasa Rao, C.V. Sarma, and V.V. Rajasekhar of ITC-IBD for their time and patience in numerous discussions and S. Ganesh Kumar and Ravi Naware of ITC Foods Division for their insights into the value of differentiation in the wheat markets. We also thank Professors Hau Lee of Stanford University and Sendil Ethiraj of the University of Michigan for their comments on earlier versions of this paper.

of production of various commodities we see that India's agriculture economy occupies a significant position in the world. Further, with 36 rich and diverse agro-ecological zones,[1] India is considered one of the 12 mega-diversity countries in the world.[2] The overall size, diversity and richness of Indian agriculture imply that India is a potential food factory to the world.

The strong aggregate agricultural statistics belie the plight of the Indian farmer, approximately 25% of who live below the poverty line. While agriculture comprises a quarter of the overall national economy, close to 700 million people or 70% of the population, who live in rural areas, depend on it (see Table 12.1). Consequently, the land holdings are small, averaging about 1.5 hectares; see Table 12.2 for distribution of land holdings. Although about 44% of land was under irrigation in 2001,[3] making it the largest irrigated land mass in the world, 83% of the total area under irrigation is accounted for by only 32% of land holdings.[4] Thus Indian agriculture is characterized by extreme *fragmentation* where a large number of farmers, mostly dependent on monsoons, are engaged in marginal sustenance farming. This lack of scale adversely impacts the production efficiency, productivity and quality of their produce and limits the bargaining power of the farmer in the marketplace.

The second issue relates to geographic *dispersion*. Given the country's physical size, the farmers are dispersed widely increasing the distances and hence complexity of the physical linkages with the market both for their produce and for agricultural inputs. The geographical dispersion also adversely impacts the flow of information in the supply chain. Finally, there is significant *heterogeneity* in Indian farming relating to types of farmers and types of crops. Farmer heterogeneity exists in terms of their investment ability, size of holdings, soil conditions, cash-flow needs, etc. Crop heterogeneity exists on the one hand due to the diversity of Indian ecosystem, and on the other due to the historical cropping patterns and practices employed to mitigate business risks at the individual farmer level. Heterogeneity per se is not a problem except that in combination with fragmentation and dispersion, it poses significant challenges in the delivery of much needed customized solutions for productivity improvement and dissemination of best practice knowledge.

---

[1] Patel, N.R., "Remote Sensing and GIS Application in Agro-ecological zoning", Satellite Remote Sensing and GIS Applications in Agricultural Meteorology, Proceedings of a Training Workshop held 7-11 July 2003 in Dehra Dun, India. Edited by M.V.K. Sivakumar, P.S. Roy, K. Harmsen, and S.K. Saha.

[2] http://www.aaas.org/international/ehn/biod/singh1.htm (Accessed on January 8, 2006)

[3] Source: *Agricultural Statistics at a Glance*, Ministry of Agriculture, Government of India, August 2004.

[4] Source: Agriculture Census Division, Ministry of Agriculture, Govt. of India (Analysis: Authors)

Table 12.1. India's Position in World Agriculture (Selected Statistics).

Item	India	World	India's Position	
			% Share	Rank
1	2	3	4	5
1. Area ** (Million Hectares)				
Total Area	329	13425	2.5	Seventh
Land Area	297	13062	2.3	Seventh
Arable Land	162 F	1364	11.9	Second
Irrigated Area	55 F	272	20.2	First
2. Population (Million)				
Total	1025	6134	16.7	Second
Agriculture	739	3211	23.0	Second
3. Crop Production (Million Tonnes)				
(A): Total Cereals	231	2086	11.1	Third
Wheat	68	583	11.7	Second
Rice (Paddy)	132 *	593	22.3	Second
Coarse grains	30	911	3.3	Fourth
Total Pulses	11	52	21.2	First
(B): Oilseeds				
Groundnut	6 *	35	17.1	Second
Rapeseed	4	36	11.1	Fourth
4. Commercial Crops (Million Tonnes)				
(A): Sugarcane	286	1255	22.8	Second
(B): Tea	0.86 *	3.06 ·	28.1	First
(C): Coffee (green)	0.30 *	7.05	4.3	Sixth
(D): Jute & Jute like Fibres	1.88 F	3.07	61.2	First
(E): Cotton (lint)	1.75 *	21.03	8.3	Fourth
(F): Tobacco Leaves	0.61 F	6.35	9.6	Second
5. Implements (Thousands numbers)**				
Tractors in Use	1525 F	26410	5.8	Fourth

F=FAO Estimates; *Unofficial Figures; **Figures relate to 2000.

Source: *Agricultural Statistics at a Glance*, Ministry of Agriculture, Government of India, August 2004.

   In addition, the country's physical, social, and institutional infrastructure is evolving. *Physical infrastructure* relates to roads, telecommunication, power, transportation, and storage; *social infrastructure* concerns health, education, and social security; and *institutional infrastructure* refers to markets (and access to them), financing, credit, risk management (for production and yield), cost and speed of dispute resolution, etc. While the state of the infrastructure elements identified above is poor for the entire country, they are particularly skewed against the rural areas of India. Many villages in India are not accessible by proper roads, lack any telecommunication connectivity, and have a few hours of power supply per day at best. The poor state of the physical infrastructure increases the cost of the physical and information flows in the

Table 12.2. Distribution of Operational Holdings (Indian Agriculture).

No. of Holdings: '000 Number
Area: '000 hectares
Average size: Hectares

Category of Holdings	No. of Operational Holdings		Average Size of			
			Area Operated		Operational Holdings	
	1990–91	1995–96	1990–91	1995–96	1990–91	1995–96
1	2	3	4	5	6	7
Marginal	63,389	71,179	24,894	28,121	0.39	0.40
(Less than 1 hectare)	(59.4)	(69.6)	(15.0)	(17.2)		
Small	20,092	21,643	28,827	30,722	1.43	1.42
(1.0 to 2.0 hectare (s)	(18.8)	(18.7)	(17.4)	(18.8)		
Semi-Medium	13,923	14,261	38,375	38,953	2.76	2.73
(2.0 to 4.0 hectares)	(13.1)	(12.3)	(23.2)	(23.8)		
Medium	7,580	7,092	44,752	41,398	5.9	5.84
(4.0 to 10.0 hectares)	(7.1)	(6.1)	(27.1)	(25.3)		
Large	1,654	1,404	28,659	24,163	17.33	17.21
(= 10.0 hectares)	(1.6)	(1.2)	(17.3)	(14.8)		
All Holdings	106,637	115,580	165,507	163,357	1.57	1.41
	(100.0)	(100.0)	(100.0)	(100.0)		

Note: Figures in parentheses indicate the percentage of respective column total.
Source: Agricultural Census Division, Ministry of Agriculture, New Delhi.

farm-to-market supply chain. The lack of adequate social infrastructure, particularly in education and social security, limits the ability of the farmer to implement best practices and invest in productivity improvements. Finally, the weak institutional infrastructure shifts the risk inherent in agri-business to the party – the small and marginal farmer – that has the least capacity to bear it. The gaps in physical and institutional infrastructure are partially compensated by the existence of multiple intermediaries in the farm-to-market supply chain. These intermediaries, while delivering critical value, also extract a significant share of the profits from the supply chain leaving little for the small farmer. Fragmentation, heterogeneity, and dispersion of the typical Indian farmer further exacerbate the asymmetry of interaction between the farmer and the intermediaries leading to a vicious circle of low equilibrium (Figure 12.1). Small holdings reduce the risk taking ability of the farmer leading to lower investment which translates to low productivity. Coupled with a weak "market orientation" of these supply chains the resulting value added is low, ultimately leaving very low margins for the farmer and the cycle continues.

The main premise of the chapter is that emerging economies are characterized by "broken value chains" that attempt to connect the poor, as sellers and

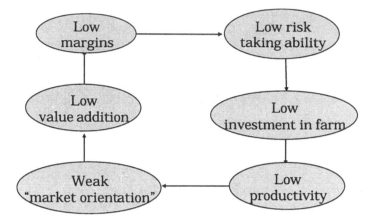

Figure 12.1. Cycle of Low Equilibrium.

buyers, to markets for products and services. Often these value chains do work (as they need to) but with the help of numerous intermediaries who extract disproportionate value leaving the poor with little residual income. The challenge of fixing these value chains is further exacerbated by factors like fragmentation, dispersion, heterogeneity, and weak infrastructure. Nevertheless, the enormity and the complexity of the task at hand implies that reengineering the farm-to-market supply chain offers a tremendous business opportunity. Using the case study of a large scale agri-business supply chain reengineering effort being implemented across various commodities by a business house in India, we illustrate how supply chain reengineering can enhance shareholder value, alleviate poverty, lay the foundation for global competitiveness of agriculture, and at the same time sow the seeds of social transformation.

The ITC Group (www.itcportal.com) is one of India's largest private sector companies with a market capitalization of approximately US $ 11 billion and annual sales of US $ 2.6 billion. ITC has a diversified presence in Cigarettes, Hotels, Paperboards & Specialty Papers, Packaging, Agri-Business, Branded Apparel, Packaged Foods & Confectionery, Greeting Cards and other Fast Moving Consumer Goods (FMCG) products. The International Business Division (IBD) of ITC, started in 1990, exports agricultural commodities such as soybean meal, rice, wheat and wheat products, lentils, shrimp, fruit pulps, and coffee. As a buyer of the agricultural commodities, ITC-IBD faced the consequences of an inefficient farm-to-market supply chain. Increased competition in commodities as India liberalized its economy coupled with low margins made it imperative for ITC-IBD to rethink how it could create a sustainable competitive advantage in the farm-to-market supply chain of which it was only a part. With a mandate to grow its agri-business, in the year 2000 ITC-IBD embarked on an initiative to deploy information and communication technology (ICT) to reengineer the procurement of soybeans from rural India. ICT kiosks (called *e-Choupals*) consisting of a personal computer with internet access were setup at the villages. Soybean farmers could access this kiosk for information on prices, but had a choice to sell their produce either at the local market (called a *Mandi*) or directly to ITC at their hub locations. A hub location would service a cluster of e-Choupals. By purchasing directly from the farmers, ITC significantly improved the efficiency of the channel and created value for both the farmer and itself. The e-Choupal experiment for soybean procurement has been well documented in Prahalad (2005) and Upton (2003). The experiment has been extremely successful for ITC, and by 2004 the network consisted of 5000 e-Choupals and 127 hubs operating in six states of India

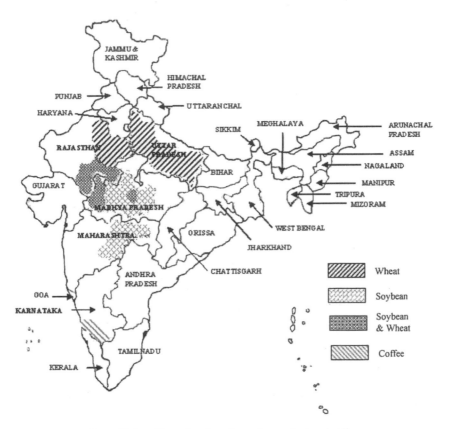

Figure 12.2. e-Choupal regions for soybean, wheat, and coffee.

for procurement of soybeans, wheat, shrimp, coffee, and spices. Figure 12.2 shows the map of India identifying the e-Choupal regions for soybean, wheat and coffee – the subject matter of this chapter.[5]

In this chapter we describe ITC-IBD's rural intervention in reengineering the supply chain. In particular, we discuss three case studies on procurement of soybeans, wheat, and coffee. Each of these case studies highlights a different concept.

1. **Procurement Efficiency in Soybeans**: Using the case study of soybean procurement, in Section 2, we first highlight the inefficiencies that characterize the current (pre-intervention supply chain). Subsequently, we discuss ITC-IBD's reengineered supply chain and its key advantages for all stakeholders involved. We highlight some general issues that pertain to the operation of such supply chains in terms of physical, information, and financial flows.

2. **Variety-based strategy in Wheat**: With ITC-IBD's network in place we then discuss its application to the procurement of wheat in Section 3. While the basic procurement efficiencies of soybean carry over to wheat, wheat raises its own challenges. As compared to soybeans, there are multiple varieties of wheat produced in India (and the world). It is well documented that these different varieties are ideally suited for different end uses for which the buyer and the consumer is willing to pay a premium. However, one of the challenges in wheat procurement has been the loss of identity of the various varieties as they move through the (pre-intervention) supply chain. Ultimately, what gets sold in the market is less differentiated resulting is loss of value. By designing a procurement system to segment and preserve the identity of various varieties of wheat, one could transition from a commodity business to a variety-based strategy that caters to the heterogeneous needs of the consumer base. We will highlight some of the issues involved and discuss how ITC-IBD's network is being leveraged to execute such a differentiation strategy.

3. **Commodity Services in Coffee**: The coffee supply chain offers a different set of challenges as discussed in Section 4. As we will show there is little physical flow inefficiency to be eliminated from the current supply chain that operates. However, some of the other main characteristics of Indian agriculture described earlier – fragmentation and weak institutional infrastructure – still plague this supply chain, in addition to the

---

[5] The map in Figure 12.2 shows five states of e-Choupal operations. This excludes the sixth state of Andhra Pradesh where ITC-IBD's intervention is in aqua-culture.

extreme volatility of the coffee prices in international markets. We will describe ITC-IBD's information technology based intervention to provide a platform to expand the network reach of various participants, increase liquidity, and provide instruments for better risk management.

Together the three highlight a progression of value provisioning using supply chain reengineering. While the first one focuses on improving the logistics efficiency of a commodity supply chain, the second example illustrates the shift from commodity-based to a variety-based strategy, and finally the third example illustrates the migration from products to services-based strategy in agri-businesses. We close the chapter with a discussion on the key learning points from the case studies and articulating the generality of the solution structure being proposed.

12.2. Procurement Efficiency in the Soy-Supply Chain

Soybean is the single largest oilseed produced in the world, contributing over 55% of the total annual oilseed production of 310-320 million tons. Significantly, its production has grown at an annual rate of 5.35% during the last decade. The four largest soybean producing countries include the United States followed by Brazil, Argentina, and China.[6]

Soybeans are an important source of oil and protein to the world. It has several food and industrial uses including usage as a vegetable oil and in the production of margarine. Soya flour from whole beans can be used in baking, as additives and extenders to cereal flour and meat products, and in health foods. With a high content of lecithin, it is also used as emulsifier in the food industry, pharmacy, decorating materials, printing inks, and pesticides. The protein in the extraction meal, in addition to being used for human and animal food, has also been used in the production of synthetic fibers, glues, foams, foam-forming agents. The unripe seeds are eaten in East Asia and the United States. Tofu, soy milk, and soy sauce are some of the other valuable foods obtained from soybeans. Finally, the vegetative portions of the plant can be used for grazing or as hay / fodder / silage or as a green manure. The straw can be used to make paper, stiffer than that from wheat straw.[7]

India is the fourth largest producer of oilseeds in the world and this sector occupies an important position in the agricultural economy covering about 14% of the arable land producing over 25 million tons of oilseeds annually. In 2004-05 India accounted for 7.4% of world oilseeds output; 6.1% of world

---

[6] Source: National Commodities and Derivatives Exchange (NCDEX), India.

[7] Source: American Soya Association

oilmeal production; 3.9% of world oilmeal export; and 9.3% of the world edible oil consumption making it the 4th largest edible oil economy in the world.[8]

Soybean, popularly known as 'miracle bean' in India, is part of the Indian oilseed economy consisting of crops like groundnut, mustard seed, sunflower, etc. from which edible oils are extracted. The market size of soybean in India is over $1.2 billion. Over the last decade production of soybean in the country has been growing at the rate of 5.57%.[9] While aggregate production has been increasing, the yield rate averaging at 870 kg/hectare is still only one-third of the yield rates in United States and Brazil. Soybean is not a very common part of the Indian diet. Consequently 90% of the production is used for extraction of oil and oilmeal. While the process of oil extraction varies by oilseed, broadly it can be thought of as a two-stage process with crushing of seeds in the first phase followed by a solvent extraction in the second. Oil is extracted at each of the stages. The residue or oilmeal is often sold as animal feed. While India imports soybean oil, it is a leading exporter of meal to the Asian region.

Mature beans, oil and meal are the traded forms of soybean and its derivatives – both in the spot as well as in the futures markets. At the international level, the Chicago Board of Trade (CBOT) is the most vibrant futures exchange, whose prices serve as a 'reference' to all other soybean markets across the globe.

12.2.1   Traditional Soybean Supply Chain

An Indian farmer typically has four channels to sell his oilseed produce (see Figure 12.3). The three main outlets include (i) village traders for eventual sale to private mills through wholesalers; (ii) *mandis* or (spot) markets regulated by the state governments; and (iii) a cooperative society for eventual processing by the cooperative mills. The fourth channel comprises of oilseeds kept by farm households as seeds or for direct use often processed in a local small scale crushing plant called the *ghani*. While the fraction of flows through the first three main channels varies across various regions of the country, together they account for close to 90% of all oilseeds sold.

Mandis play a central role in the marketing of commodities in India. They were set up under the Agricultural Produce Marketing Act (instituted by various States) for the purpose of regulating the marketing of agricultural produce. A mandi is like a spot market. While the physical infrastructure is provided by

---

[8] Source: B.V. Mehta, "India – Mover and Shaker of the Global Vegetable Oil Market", Paper presented at the National Seminar on Palm Oil organized by the Indonesian Palm Oil Producers Association (GAPKI), Nov. 17–18, 2005, Bali, Indonesia.

[9] Source: National Commodities and Derivatives Exchange (NCDEX), India.

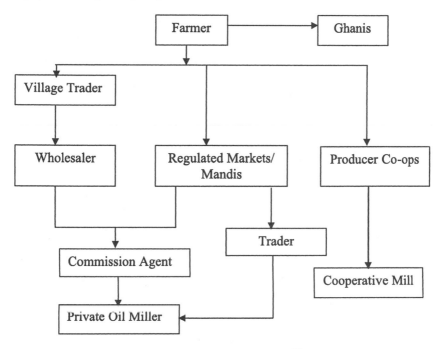

Figure 12.3. Channels for Oilseed.[10]

the government, the actual trading is done by private parties. The mandi prices also serve as a guide to other channels of trade. All flows through the mandi are subject to a tax that is intended for maintaining and upgrading the facilities. Nationwide today approximately 80% of the soybeans are sold through the mandis alone.

We now describe the farm-to-market supply chain for soybeans. Around mid-June at the beginning of the monsoon season, the farmer sows the bean which is ready for harvest by the end of September. The 2–3 months from October – December is considered the peak season for selling soybean. While soybean can be stored, due to the cash cycle of a small and marginal farmer, it is observed that about half of the production is sold during the peak season. Typically, a mandi is located on average between 20–25 kilometers from the farmer's village. With paucity of any communication infrastructure in the village, the only means of price discovery for the farmer is word-of-mouth, based on sales in the previous days by other farmers from the same village.

---

[10] Source: The Indian Oilseed Complex: Capturing Market Opportunities, World Bank Document, Report 15677-IN.

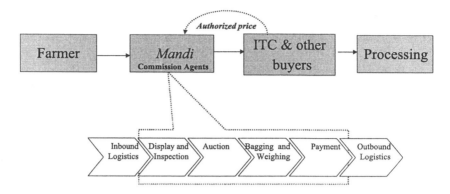

Figure 12.4. The Traditional Soy Procurement Supply Chain.

Figure 12.4 shows the traditional soy procurement supply chain. Once the farmer decides to sell, he transports the produce in bulk to the mandi either in a tractor-trailer or an animal drawn carriage covering the average distance of 20–25 kilometers. This constitutes the *inbound logistics* in the supply chain. The next step in the process is *display and inspection* at the mandi. During peak season, the mandi is usually congested. This implies that it may take a few days before the farmer, who has arrived in the town with his produce, gets an opportunity to display his produce for auction at the mandi. Once at the mandi, he engages a commission agent[11] (CA), who displays the produce in a heap in the mandi yard. Buyers (traders or CAs) move from heap to heap, visually inspecting the produce by picking samples, assessing quality, and making their calculation of the appropriate price for the produce. The produce is then *auctioned*, with auctioning methods being adapted to local practices in each state. Typically, the auction would begin when a government appointed bidder valued the produce and set a minimum bid. Thereafter, like a straight auction CAs bid upwards until the product was sold and the highest bidder is awarded the lot. Buyers from local processing plants also participate in trading at the mandi, either directly or through agents. After the deal is agreed upon, the seeds are sifted manually on sieves of a standard mesh, bagged, weighed, and loaded on the buyer's vehicle. Often due to space constraints at the mandi, the process of bagging and weighing may occur at the premises of the person who won the auction. Payments are made in cash for immediate delivery through the CA / trader who retains the *mandi* fee (paid by the buyer) and other fees as authorized by the market committee. In an ideal situation, the farmer should

---

[11] A commission agent can either be a seller's (farmer's) or a buyer's (e.g., ITC) agent.

get full payment for his produce. In reality, the CA often spreads the full payment over several days forcing the poor farmer to make multiple trips from the village to the town.

Companies like ITC (called buyers) that want to buy soybeans have their own CAs at various mandis. Typically, each day a buyer would authorize its CA to buy a certain quantity at a price not to exceed a maximum limit. Since a buyer has no direct presence at the various mandis, it does not have real-time visibility into the mandi flows and price movements. Once the CA has procured the desired quantity, the CA arranges for the produce to be transported to a location designated by the buyer. The buyer then reimburses the CA for various costs and also pays a commission for the transaction. Payment terms of one week are common in the market with discounts for immediate payment. Some CAs also extend credit terms to processors. Since a large fraction of the beans bought by ITC is processed, it would then get the beans to one of several processing facilities incurring outbound logistics costs.

There are several inefficiencies in this traditional supply chain. Consider the farmer's perspective. At the village, the farmer has no means to get a clear picture of price trends at the mandi before making a decision on when, where, and how much to sell. Price discovery occurs only at the mandi by which time the farmer has already incurred the sunk cost of inbound logistics and it is too late to backtrack should he wish not to sell. With a fixed capacity at the mandi to process flows, peak seasons create a congestion effect. This implies that when the farmer leaves the village with his produce he does not know when he can return, incurring board and lodging costs in the town. The visual inspection of the produce at the mandi is arbitrary and unscientific. As a result, the farmer never gets an appropriate compensation for quality produce. Post-auction often the farmer is charged the cost of bagging. The weighing too is done manually and there is evidence of cheating. Finally, a farmer is not guaranteed to walk-away with full payment for his produce even after several days of this ordeal. While these are physical and financial inefficiencies, the process also extracts a psychological toll on the farmer. Over the decades, the traders and CAs have become a strong financial powerhouse in the rural communities. The small and marginal farmer not only relies on them for selling his produce but also on loans for several agricultural and non-agricultural activities. This dependency means that a farmer cannot alienate the trader / CA which creates a huge asymmetry in all the transactions including selling of the grain. Further, there is asymmetry in social standing between the farmer and the traders / CAs given the disparity in the relative levels of education between the two.

ITC as a buyer is also far removed from the center of real-time transactions. The physical inefficiency of the flows adds to the ultimate cost of procurement for ITC. With no visibility, ITC is at the mercy of the CA who pockets any benefit from intra-day price shifts. Lack of proper quality inspection meant that ITC was not always assured of the quality of produce it bought; furthermore lack of visibility left enough scope for the CA to aggregate and deliver an average quality produce. Finally, ITC being farther removed from the farmer was unable to influence his agricultural practices to improve quality and yield.

The agent, by placing himself right in the middle of the flow between the farmer and ITC, extracted rents from both parties.

12.2.2 ITC e-Choupal: The Reengineered Soybean Supply Chain

Starting in 1990, ITC-IBD operated its commodities supply chain as described above. The supply chain inefficiencies reduced ITC's operating margins. Then in 1996 the opening up of the Indian market brought in international competition in the commodities business. The international players were large established companies with better margin-to-risk profile. Their entry further increased the competitive pressures, limiting opportunities for growth and reducing operating margins. The survival and strategic imperative ITC-IBD faced was to create a sustainable competitive advantage and grow the business. ITC's broader social agenda of nation building as part of their corporate philosophy forced ITC-IBD to consider solutions that could blend shareholder value creation with social development.

There were significant inefficiencies in all of the flows – material, information, and financial – in the traditional supply chain. Material flow inefficiencies included multiple handling of the commodity, unscientific quality grading, and inefficiency of logistics. For the farmer, the information flow (regarding prices) bundled with the physical transaction (at the mandi) was too late to be of any use. For ITC, market signals were not available real-time and its reach was limited to the trader. The intermediary blocked these flows in either direction to his advantage. For financial transactions the farmer was at the mercy of the agent. Any solution had to address all of the flows in the procurement chain.

ITC-IBD decided to "go-direct" to the farmer. Figure 12.5 shows the reengineered supply chain. The first inefficiency that ITC-IBD wanted to eliminate was the bundling of price discovery with physical transaction at the mandi. This meant that price discovery has to happen at the point of origin, viz. the village. This was made possible by establishing an internet kiosk, called an *e-Choupal*,[12] within walking distance of a farmer in a village. The kiosk consisted of a personal computer with VSAT[13] connectivity for internet and a

---

[12] The name is derived from *choupal* which is a traditional village gathering place.

[13] Very Small Aperture Terminal

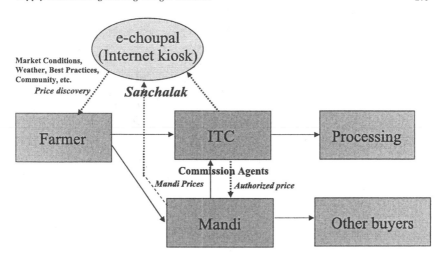

Figure 12.5. The Reengineered Soybean Procurement Supply Chain.

solar power panel to charge the battery to run the computer when no electricity is available. Typically, one e-Choupal would be available per cluster of five villages. A local farmer, called the *Sanchalak* (coordinator) was carefully selected to run the e-Choupal; in fact the computer infrastructure was placed in the farmer's house. Through a web-portal in local language (see Figure 12.6 for an English version of the main page), a farmer could inquire on the prices at the mandi the previous day. Figure 12.7 gives the prices and flows at various mandis. This (and other) information was provided free of cost to the farmer. In addition, to eliminate the inefficiencies inherent in the physical flows through the mandi, ITC-IBD provided the farmer an option to sell the commodity to it directly. This was made possible by supporting a cluster of e-Choupals with a procurement hub[14] where the farmer could deliver the commodity directly to ITC-IBD. These hubs were at a tractorable distance (maximum of 30 kms) from the choupal villages. ITC's commission agent (CA) at the mandi was incorporated into the new infrastructure as a provider of logistical support including cash disbursement at the hub on a (reduced) commission basis (see discussion later).

The reengineered farm-to-market supply chain worked as follows. Every evening, based on the closing prices at local mandi and the commodity futures, ITC determined a fair average quality (FAQ) price for soybeans for direct pur-

---

[14] In some geographic areas, the processing plant acted as a procurement hub. In 2004, ITC has 6 processing plants in the state of Madhya Pradesh.

Figure 12.6. Soya-Choupal Main Portal.

chase at its procurement hub. The price was available to the Sanchalak and the farmer through the e-Choupal web-portal; see Figure 12.8. A farmer could bring a sample of his beans to the Sanchalak who would inspect the produce and give a conditional price quote (adjusting for quality of beans) for direct purchase. Armed with this information, the farmer now had a choice of channels for selling his beans, viz., to ITC directly, to the mandi, or to other private traders. If the farmer chooses to sell to ITC directly, the Sanchalak would provide the farmer with a *parchi* (note) including his name, village, quality of produce, and the conditional price quote, authorizing the farmer to go to the ITC procurement hub.

At the procurement hub, the farmer's produce first underwent a scientific inspection of quality. Unless there was a substantial variation in the quality of the produce between the sample shown to the Sanchalak and actual goods delivered at the hub, the conditional price indicated by the Sanchalak became the final price that the farmer received. Subsequently the produce was weighed

| S.No | Location | SOYA SEEDS - SOYA YELLOW | | | Total Arrivals (Till Date) | Estimated Arrivals (Till Date) | Percentage Arrivals (Till Date) |
| | | Arrival (Bags) | Rate | | | | |
			Low	High			
1	AGAR(MALWA)	1000	1230	1260	131400	300000	43.8
2	ASHOK NAGAR	0	0	0	133500	300000	44.5
3	ASHTA	2200	1200	1235	199900	775000	25.79
4	BADNAGAR	700	1200	1235	147300	400000	36.83
5	BADNAWAR	1000	1210	1240	135000	500000	27
6	BANAPURA	600	1190	1210	59970	300000	19.99
7	BERASIYA	0	0	0	87150	125000	69.72
8	BETUL	1300	1190	1235	44850	200000	22.42
9	BHOPAL	350	1200	1250	91340	300000	30.45
10	BINA	300	1190	1220	66050	100000	66.05
11	BIORA	0	0	0	138700	350000	39.63
12	CHHINDWARA	1500	1175	1208	75150	400000	18.79
13	DAMOH	350	1190	1220	42560	75000	56.75
14	DEWAS	3000	1215	1248	468700	1000000	46.87
15	DHAR	3500	1210	1245	163600	450000	36.36
16	GANJ BASODA	1700	1200	1240	224700	300000	74.9

27-दिसम्बर-2004 तक का मुख्य मंडियों में बीज आगमन रिपोर्ट

| क्र. | प्लांट | सोयाबीन (पीली) | गेहूँ | | | |
			लोकवन	मीन क्वालिटी	शरबती	डब्ल्यू एच 173
1	उकाच्या / इंदौर	1260	650	640	खरीदी बंद रहेगी	खरीदी बंद रहेगी
2	खाडवा	1240	650	640	खरीदी बंद रहेगी	खरीदी बंद रहेगी
3	करेली	1200	खरीदी बंद रहेगी	खरीदी बंद रहेगी	खरीदी बंद रहेगी	खरीदी बंद रहेगी
4	गुना	1220	640	खरीदी बंद रहेगी	726	खरीदी बंद रहेगी
5	शुजालपुर	1225	645	640	खरीदी बंद रहेगी	खरीदी बंद रहेगी
6	टिमरनी	1215	640	खरीदी बंद रहेगी	खरीदी बंद रहेगी	655
	हब / गोदाम					
1	विदिशा	1215	640	खरीदी बंद रहेगी	खरीदी बंद रहेगी	खरीदी बंद रहेगी
2	इटारसी	1235	645	640	खरीदी बंद रहेगी	660
3	सोहागपुर	1210	640	खरीदी बंद रहेगी	खरीदी बंद रहेगी	655
4	बानापुरा	1210	645	640	खरीदी बंद रहेगी	655
5	हरदा	1215	645	640	खरीदी बंद रहेगी	655

भाव (रू.) प्रति क्विंटल

Figure 12.8. e-Choupal Prices of Soybean and Wheat at Plant and Hub Locations.

on an electronic weighbridge. At each of these stages, the farmer was provided with appropriate documentation. Once the transaction was complete, the farmer was paid cash in full.

12.2.3 Benefits

Overall, the reengineered supply chain reduced several non-value added tasks in the traditional chain thereby improving the efficiency of the procurement. Next we describe the benefits – direct as well as indirect – to the various stakeholders.

Direct selling to ITC provided the farmer with several benefits. The first benefit was the monetary savings in transaction costs including savings on commissions and the elimination of costs of handling and other losses. Second, by virtually being guaranteed a same-day transaction, the "sales-to-cash" cycle for the farmer was reduced from several days (when selling at the mandi) to a few hours. Third, ITC eliminated the uncertainty in the price for the farmer by announcing the price for direct purchase the previous evening before he left the village. Fourth, the farmer was assured that his produce will get proper quality evaluation. Finally, logistically, selling direct to ITC involved no greater travel than going to a mandi.

The web-portal, in addition, also acted as a knowledge delivery mechanism. It provided information on farming best-practices, localized weather information, general statistics on commodity production and consumption, links to commodity markets, discussion groups, etc. Such knowledge increases the awareness of the farmer allowing him to make more informed choices.

ITC benefited from the direct-transaction by reduction in freight from the mandi to its processing facility or the warehouse location and elimination of the commissions. Finally, as an incentive to facilitate direct procurement, ITC paid a commission to the village level e-Choupal Sanchalak. Table 12.3 lists the e-Choupal procurement advantage to the farmer as well as to ITC. By building a physical presence in the villages, ITC now had direct contact with the farmers and the ability to influence their farming practices through information and knowledge delivered via e-Choupal. This has potential long-term impact on improving the quality and productivity of Indian soy farming, gains from which will partially be captured by ITC through its direct-procurement infrastructure.

In this reengineered channel, it appears that while the farmer and ITC gained, the CA lost. For the CAs that did the procurement for ITC at the mandi, there were fears of being completely dis-intermediated in this reengineered network. Obviously the role of a CA in the procurement of commodities for ITC has diminished. ITC allayed the concerns of its CAs in several ways. First, ITC still did some buying of commodities at the mandi through its CAs leaving them the usual 3% margins. Second, there were several logistics services needed at the hubs in the e-Choupal network. Since the CA was very resourceful in the local region, ITC co-opted the CA as a provider of these services

Table 12.3. The e-Choupal Procurement Advantage (All prices are in Rupees per metric Ton).

Transaction	Mandi Supply Chain		e-Choupal model	
	Farmer	ITC	Farmer	ITC
Freight	120	120	120	0
Labor / Handling	50	40	0	40
Commission	150	100	0	50
Handling Loss	50		0	
Bagging		75		75
Cash Disbursement Costs				50
Total (for each Stage)	**370**	**335**	**120**	**215**
Total for the Chain		*705*		*335*

which included cash disbursement on a reduced commission (0.5%) basis. Finally, as ITC expanded the use of the e-Choupal for reverse flows into distribution (e.g., of agri-inputs, seeds, etc.) and retail,[15] the CA played a critical role in activities like management of facilities, providing labor resources, buying and selling some of the goods for distribution, etc. earning commissions. Thus ITC provided the CAs an opportunity to expand their role from a 'trader of commodities' to a 'service provider' in procurement and distribution / retail. The CAs were given the option to join ITC in this new role called a *Samyojak* (collaborator). There was initial hesitation among the CAs of the earning potential in their newly defined roles. Presently we observe that the CAs, who opted to become Samyojaks, are earning more money and, due to inherent diversification in the sources of revenue, incur reduced risk. However, as ITC shifts more of its procurement through the e-Choupal and increases its overall market share thereby decreasing the overall mandi flows, the traders as a lot (including the CAs still operating at the mandi for other buyers) would be losing out. Beyond the obvious loss of commissions, the increased transparency that the e-Choupal network brings to the procurement has also increased the bargaining power of the farmer should he choose to sell to a trader.

12.2.4 The e-Choupal Network

Since a procurement hub supports multiple e-Choupals, the reengineered supply chain can be considered as a hub-and-spoke network. In the central state of Madhya Pradesh, which is the prime soy producing region of India, the network consists of 44 hubs and 1750 e-Choupals. Nationwide and across multiple commodities, by 2004, the network consisted of 5000 e-Choupals and

---

[15] See Anupindi and Sivakumar (2005) for details.

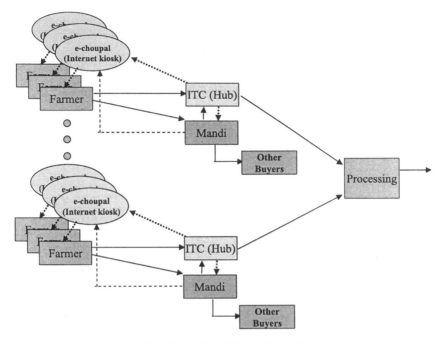

Figure 12.9. The e-Choupal Network.

127 hubs operating in six states of India for procurement of soybeans, wheat, shrimp, coffee, and spices. Additional e-Choupals and hubs are being added to the network continuously. Figure 12.9 illustrates the hub-spoke network and Figure 12.2 (Section 1) shows e-Choupal operating regions within India.

In this reengineered supply chain, while information and knowledge are provided free of cost (at the e-Choupal) physical transactions, and hence revenue, are generated by the strength of the value proposition made available to the farmer. Furthermore, deployment of digital technology reduces the cost of delivery of information and knowledge making it feasible to provide these at the local level. The costs of physical infrastructure for material procurement, however, are relatively higher. Therefore, physical procurement is aggregated at the hub level to leverage economies of scale.

12.3. Variety-based Strategies in the Wheat Supply Chain

Wheat is the most widely grown and consumed grain cereal in the world. With an annual production of 65–75 million tones, India is the second largest producer of wheat in the world after China, accounting for 12% of world production. Wheat production in India has increased almost ten-fold in the last five decades (see Figure 12.10) and accounts for approximately 35% of India's total food grain production.

The roots of this remarkable achievement for independent India were sown in the 1960s. Following a severe food grain crisis of the mid-1960s as well as earlier experiences (e.g., the Bengal famine of 1943 – the world's worst food disaster recorded), there was an urgent need to increase food grain production as well as to develop an efficient distribution system, especially to reach the poor. Under what came to be known as the "green revolution", increased production and yield were made possible by expansion of farm areas, double-cropping, and use of high yielding seed varieties. In addition, the government of India developed a system of institutions with the objective of supporting, controlling and stabilizing food grain prices and seeking to assure basic food availability at reasonable prices to the people. The system included the Commission on Agricultural Costs and Prices (CACP), Food Corporation of India (FCI), and state civil supplies corporations/departments. The CACP studies costs and markets, and recommends minimum support prices (MSP) for procurement. Farmers expected to obtain at least the MSP for their produce. In addition, to improve the distribution and availability of food grains for the masses, national handling – including procurement, transport, storage and release of the grain – is done mainly by the FCI. The distribution is done through the public distribution system (PDS) via state government agencies (state civil supplies corporations / departments) and over 450,000 fair price shops (also called ration shops) spread throughout the country in both rural and urban areas. To improve the country's food security, the central government targets a strategic buffer stock of wheat of about 9.6 million tons; the actual stock stood at 15.4 million tons at the end of 2004.[16]

While the overall production of wheat is quite high, there is a huge concentration of wheat production in the northern parts of India. The three states of Punjab, Uttar Pradesh, and Haryana account for close to 60% of national wheat production. Distribution and consumption of wheat, however, is spread across

---

[16] Source: Annual Report 2004–05, Department of Food and Public Distribution, Government of India.

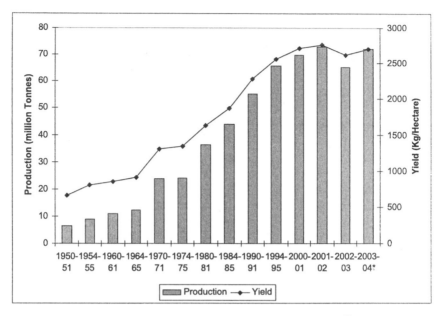

Figure 12.10. Wheat Production and Productivity in India.[17]

the nation. The asymmetry in production and consumption makes distribution of wheat a challenging task.

Environmental factors play a very critical role in wheat production. Temperature, rainfall, and soil moisture are some of the important parameters in the crop production. Due to the year-to-year variability in some of these factors, export trade in wheat based on surplus production has been sporadic. Once production increases substantially over consumption, wheat export becomes a viable proposition. However, export of Indian wheat is today constrained by two factors. The first factor is quality. While some of the wheat produced in India is of highest quality, the traditional farm-to-market supply chain is not capable of segregating wheat by quality. As a result, Indian wheat that comes to market is at best of average quality often containing high levels of extraneous material, poorly filled kernels, broken grain and other infirmities, reinforcing the perception of poor quality of Indian wheat in international markets. The process also has had an adverse incentive effect on the production of higher quality wheat. The second factor is the government regulated MSP. MSP was

---

[17] Source: *Agricultural Statistics at a Glance*, Ministry of Agriculture, Government of India, August 2004.

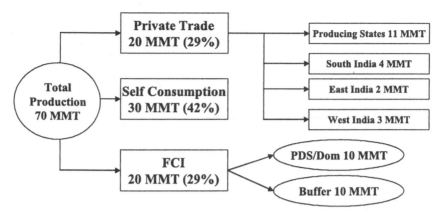

Figure 12.11. Wheat Channels and Markets.[18]

designed to give incentives to the Indian farmer for wheat production; however, often MSP is not in line with the international prices and make Indian wheat uncompetitive.

12.3.1 Traditional Wheat Supply Chain

There are three main channels of wheat distribution and consumption (see Figure 12.11). About 40% of the wheat is consumed locally by the families of the producers and never enters the marketing channels. Half of the remaining production is bought by the government through the FCI for public distribution as well as maintaining a national buffer stock. The FCI procurement is done at the minimum support prices (MSP). The remaining – approximately 20 million tons – is available for private trade. Wheat for private trade either goes to retail consumers directly or is sold to the flour mill industry. The flour mill industry supplies to bakeries, agro-industries, restaurants, as well as retail consumers.

The supply chain for private trade of wheat is very similar to that of soybeans described in the earlier section. A large fraction of the grains move through the mandi system (see Figure 12.4) – a channel whose inefficiencies for the farmer and ITC we already identified in Section 2.

In addition to the various inefficiencies already identified, there is a significant 'quality loss' issue in the traditional wheat supply chain. There are several varieties of wheat produced in India and different quality levels for each variety that depend on hectoliter content, moisture, presence of foreign matter, composition of the kernels, etc. Figure 12.12 lists the main varieties of wheat

---

[18] Source: ITC Internal Estimates.

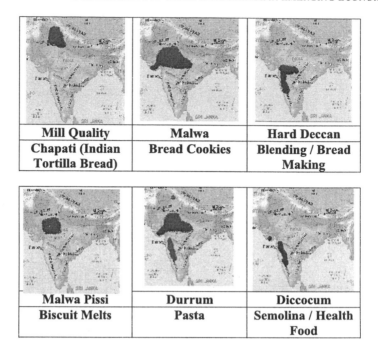

Mill Quality	Malwa	Hard Deccan
Chapati (Indian Tortilla Bread)	Bread Cookies	Blending / Bread Making

Malwa Pissi	Durrum	Diccocum
Biscuit Melts	Pasta	Semolina / Health Food

Figure 12.12. Various Varieties of Wheat Produced: Regions Produced / Wheat Variety[19] / Ideal Usage (Nagarajan, 2004).

produced in India identifying the region of production, the variety of wheat and its intended usage. However, the traditional mandi infrastructure neither allows for scientific analysis of the various quality levels nor permits proper segregated storage. Consequently, the compensation a farmer receives is rarely commensurate with the quality of wheat grown and the grains released in the market are of one "aggregate" fair average quality (FAQ) variety. Often the traders would blend high quality wheat with a lower quality and sell the mix at the price of higher quality wheat keeping the differential.

12.3.2   Wheat Procurement Through e-Choupal

At an annual production of 65–75 million tons, wheat is 14 times bigger than soybeans in Indian agriculture. Given the scale of wheat production and

---

[19] Each of the quality classifications – viz., mill, malwa, etc. – correspond to properties of wheat including the hardness of the grain, hectoliter weight (HLW), grain weight, protein content, etc; see Nagarajan (2004) for details.

trade within India, and buoyed with the success of e-Choupal network in soy-beans, ITC-IBD decided to leverage the opportunity of improved efficiencies in wheat procurement via e-Choupal. The basic processes for wheat procurement via e-Choupal are similar to those described for soybeans (Figure 12.5). Wheat procurement today is concentrated in about 80 hubs spread across three states, viz., Madhya Pradesh, Uttar Pradesh, and Rajasthan. With the e-Choupal net-work, ITC's market share of "non – FCI marketable surplus" wheat in these states is presently about 10%.

Further, in wheat ITC saw opportunities to leverage the e-Choupal procure-ment to compete on variety. Wheat, or for that matter many agricultural prod-ucts like grains, oilseeds, pulses, etc., is fundamentally considered a commod-ity market. Consequently cost is usually the only competitive dimension. Was it possible to go up the value stream in these commodity markets to compete on variety? Surely this was the path followed in many industrial goods, where over time the dimension of competition gradually shifted from cost to quality to variety; see for example the description in Stalk (1988). Could agriculture follow a similar trend? To explore the feasibility of variety-based competition, one needs to look at both demand and supply.

On the demand side, domestically there was sufficient evidence that lack of high quality wheat had stunted the growth of many wheat-based agro-industries like bakeries, biscuit manufacturers, etc. Furthermore, internal con-sumer research by the Foods Division of ITC[20] suggested the existence of differences in consumer preferences for wheat flour in different regions of the country. These differences are primarily driven by the varied usages of wheat flour in regional Indian cooking. Independent research also confirmed the need for variety in wheat production and distribution (Shoran, 2004). Internationally too recent discussions have focused on the need to introduce traceability in the wheat supply chain. While the impetus for traceability issues have arisen from consumer concerns regarding food safety – e.g., to separate out genetically modified (GM) varieties from the non-GM ones – it also allows for competi-tion on variety. But before making investments to provide for segregation and traceability, it is important to know the potential value for segregation. Is the customer willing to pay a premium for higher quality wheat or its products? There is increasing evidence to support the variety premium argument. For example, recent research suggests that wheat with a high protein content com-mands a premium in the international wheat market. Higher protein content in wheat give noodles a short cooking time, firmer texture so that noodles when cooked do not turn out to be sticky or slimy. Frozen dough products require

---

[20] ITC's Foods Division is part of ITC's FMCG (Fast Moving Consumer Goods) segment of the business which includes Cigarettes, Foods, Lifestyle Retailing and others.

Table 12.4. The Value of Segregating Wheat (Johnsson, 2004).

Total Volume	(Avg.) Protein Content	Price	Total Value
Without Segregation			
10,000 Ton	10.95%	$100/Ton	**$1 million**
With Segregation			
2500 Ton	9.5%	$105 / Ton	
3500 Ton	10.5%	$100 / Ton	
3000 Ton	12%	$117 / Ton	
1000 Ton	13%	$126 / Ton	
			$1.0895 million

very high protein content due to denaturizing of some of the proteins when the dough is frozen. Similarly, wheat with higher starch content is ideally suited for baking items as it keeps the cakes and pastries tender and delicate and wheat with high gluten content finds usage in bread for diabetics. Table 12.4 illustrates the value of segregation and identity preservation in wheat; also see Glaudemans (2001), Pauwels (2002), and Mock (2003).

On the supply side, unfortunately, in the Indian context it was impossible to achieve variety differentiation in wheat bought through the traditional farm-to-market supply chain, whether mandi or FCI. The e-Choupal intervention, however, provided ITC-IBD a direct contact with the producer. ITC-IBD first worked with its internal customer for wheat, the Foods Division, to determine the variety needs of the market. Subsequently, it invested in identify preservation infrastructure at its hubs, which included storage systems and testing equipment. Now when the farmer came to sell wheat directly to ITC at its hub, the produce was segregated into lots depending on the quality and variety of wheat determined after scientific analysis. In addition to the physical infrastructure needs to facilitate segregation, ITC also needed appropriately trained manpower. It also needed to determine a variety-based pricing strategy to compensate farmers commensurate with the quality produced. Given the higher costs and the skill set needed to implement segregation and identity preservation, ITC-IBD found it appropriate to implement it at the hub level and leverage some economies of scale.

12.3.3 Benefits

Since the wheat procurement processes are very similar to soybean procurement, efficiency gains similar to those achieved in soybean (see Table 12.3) were also realized in wheat. The benefits to the farmer, ITC, as well as the expanded role of the CA were already discussed in Section 2.3.

Furthermore, unlike soybean, wheat has an internal customer in ITC's Foods division which entered into branded and packaged foods in 2001. It would buy wheat, mill into flour, and then use for its baked goods / biscuits businesses as well as sell the branded wheat flour in retail markets. The e-Choupal procurement network with identity preservation allowed ITC's Foods division to source high quality wheat appropriate for the products it manufactured giving it a competitive advantage over its rivals. In particular, in spite of its delayed entry in the branded wheat flour market ITC's *Aashirvaad* brand became a market leader within a short span of two years. In March 2005, ITC launched a ready-to-cook instant pasta product at price point of Rs. 15 for an 83 gm packet when the only other similar product in the market was imported and priced at Rs. 85. These advantages primarily accrue to ITC due to its ability to segregate wheat and maintain its identity throughout the supply chain, a feature non-existent in the current mandi-based supply chain.

The variety capability allowed ITC-IBD to not only effectively serve its internal customer but also become a supplier of choice with other food manufacturers in the country by meeting their exacting specifications and meeting them consistently. Traceability and segregation also allows ITC-IBD to pay the farmers premiums for higher quality grains. In the longer run farmers will respond to the incentives and increase production of higher quality grains increasing the overall quality of the produce in India. As production of wheat in India grows leading to increasing exports, the system will allow ITC-IBD to compete effectively in export markets.

12.4. Re-engineered Supply Chains for Commodity Services: The Story of Coffee

Next to oil, coffee is the largest traded commodity in the world. In 2004, the total world production of coffee was approximately 112 million bags[21] whereas the consumption was close to 100 million bags leading to a glut in the coffee market. This overproduction phenomenon has been persistent over past several years leading to fall in average coffee prices as shown in Figure 12.13. Furthermore, there is tremendous concentration in both the consumption and the production of coffee. While approximately 65% of world's coffee is consumed between the United States and the European Union nations, together Brazil, Vietnam and Columbia account for 60% of world's production and 55% of world exports. The growing of coffee is thus largely restricted to third world

---

[21] 1 bag = 60 kilograms.

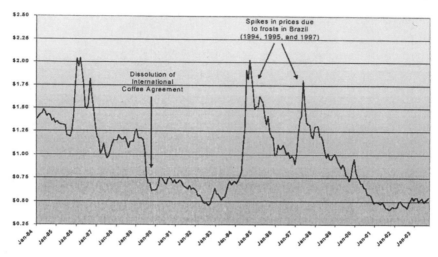

Figure 12.13. Price per pound of green coffee.[22]

countries or the developing world. About 20–25 million families – mostly marginal farmers – in more than 50 developing nations produce and sell coffee. For several of these countries, coffee accounts for at least 20 percent of export earnings. A number of them are facing considerable difficulties due to the dramatic decline in the price of coffee to 100-year lows in real terms (see Figure 12.13). Needless to say, given the significance of the crop in the economies of these countries, price destabilization has caused a severe crisis precipitating in bank failures, public protests, and dramatic falls in export revenues.

Coffee is a perennial eco-friendly in nature but labor intensive crop. Labor constitutes approximately 70 percent of the cost of coffee cultivation. Broadly there are two types of coffee – *Robusta* (lower quality but robust to environmental factors in cultivation) and *Arabica* (premium quality). It takes about 8 years for an economic yield in case of Arabica and 10–12 years in Robusta.

Legend has it that coffee was introduced to India by a pilgrim by the name of Baba Budan around 1650 A.D. who brought with him seven beans, hidden next to his belly, while returning from a pilgrimage to Mecca. He planted these seeds in Chandragiri Hills of Karnataka (also known as Budan Hills) which proved to have the ideal climate and altitude for coffee (Janssen 2004). From here the Indian coffee industry developed centered around the southern state of Karnataka and its neighbors Tamilnadu and Kerala. Today, India's share in

---

[22] Source ICO 2003; extracted from Barry (2004).

Table 12.5. Number, Area and share of production of coffee under different coffee holdings in India 2001–2002.

Size of Holdings In Hectares	No. Of Holdings		Area under Coffee	
	Number	% of Total	Area in Ha	% of total
Small Holdings				
< 2	138,209	77.5	114,546	33.0
2–4	26,549	14.9	67,155	19.4
4–10	10,717	6.0	65,386	18.8
Sub-Total	175,475	98.4	247,087	71.2
Large Holdings				
10–20	1,734	1.0	28,808	8.3
20–40	537	0.3	14,505	4.2
40–60	208	0.1	10,025	2.9
60–80	126	0.1	9,136	2.6
80–100	61	0.0	5,863	1.7
Above 100	167	0.1	31,571	9.1
Sub-Total	2,833	1.6	99,908	28.8
Total (India)	*178,308*	*100.0*	*346,995*	*100.0*

the world coffee market is approximately 3.6%, producing about 300,000 metric tons of coffee 80% of which is exported. The southern state of Karnataka accounts for 70% of India's total production. Around 250,000 growers are in the business of coffee cultivation and 98% of these are small growers; see Table 12.5. Close to 1.5 million families (Titus and Pereira, 2004) are directly dependent on the industry as a means of livelihood.

Given that majority of the coffee production in India is primarily done by small and marginal farmers, the Government of India decided to regulate the export of coffee to protect the farmers from price volatility. The Coffee Act VII of 1942 established the Indian Coffee Board, a name later shortened to the Coffee Board, and run by the Ministry of Commerce. Through the Coffee Board, the government pooled the coffees of its growers and then took control of exports itself. This rather restrictive system ensured that the government could market Indian coffees as a lump unit. However, the very process of pooling meant that the coffees of distinctive qualities would be lost amongst lesser beans. This reduced the incentives to producers of good coffee and the quality of Indian coffee stagnated.

Following the economic liberalization in 1991, the marketing of coffee in India also underwent major changes. Beginning in 1993 and over the next three years, coffee growers were gradually allowed to sell part or all of their produce in open markets both domestically and internationally. The Internal Sales Quota (ISQ) of 1993 entitled growers to sell 30% of their production within

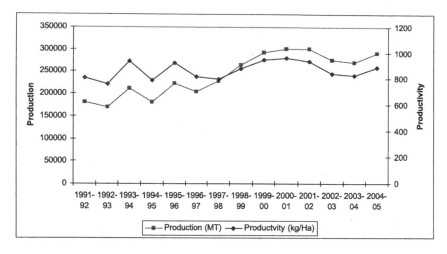

Figure 12.14. Coffee Production and Productivity in India.

the country. A year later under the Free Sale Quota (FSQ) of 1994, large and small coffee growers were entitled to sell, respectively, 70% and 100% of their production within the country and/or anywhere in the world. Since September 1996, a 100% FSQ has been extended for all the growers in the country. Implicitly, the FSQ has brought the integration of domestic markets with the global markets. From this viewpoint, the FSQ may also be considered an important step towards the economic globalization of the coffee sector. While these measures brought about growth in overall coffee production in India, productivity has not improved significantly; see Figure 12.14.

The liberalization of coffee marketing also subjected the growers to the price variations of global markets. Data suggests that the coefficient of variation of global coffee bean prices in the New York Coffee Exchange has been rising since the 1980s and is between 0.20–0.25; see Fitter and Kaplinski (2001). Furthermore, the same study illustrates that the coefficient of variation of producer prices is between 0.30–0.35. Thus a coffee grower is now subject to price risk as well as weather and yield risk. While the overall coffee market has been growing in value, the grower's share has reduced to approximately 10–11% of the final retail price.

12.4.1 Traditional Coffee Supply Chain

In its journey from plantation to cup, the coffee bean goes through the following basic steps. At harvest season, planters pick coffee cherries and usually sell it to traders / agents receiving a *farm-gate price*. The cherries are then processed to give green coffee beans. The beans destined for international markets then go to an intermediary for export and then sold to an international buyer or roaster. The roasters then sell the coffee to retailers.

To get a better understanding of the product and information flow in the context of the coffee supply chain in India, let us look at the process in detail (see Figure 12.15). Unlike soybeans or wheat, coffee is a tree crop entailing a long lead time, typically 4–5 years, for the plantation to bear fruit. In a volatile market the long lead time makes appropriate planning extremely hard. Once planted a tree will bear fruit for 20–25 years. The time span between the blossom and the harvest generally covers eight to nine months, depending on the altitude and prevailing weather conditions. In India the harvest season lasts 3–4 months from December – March. When the berries ripen, pickers need to go to every tree and manually pick the berries.[23] This process makes coffee one of the most labor intensive crops. Unlike the farm-to-market supply chains for soybeans and wheat, there is no physical market like the mandi. The planters (or their agents[24] usually work with a few traders on a relationship-basis. Sometimes, a planter may "promise" his crop in advance of the actual harvest to the trader. After the harvest the planters and the traders agree on a farm-gate price at which point the trader takes possession of the beans (called *raw coffee*) and moves it to the curing works for treatment. There are about fifty curing works in the state of Karnataka, which accounts for 70% of India's coffee production. While there are two broad categories of raw coffee – Robusta and Arabica – after curing one typically gets almost twenty (quality) grades of coffee. Cured coffee is often referred to as *clean (or green) coffee*. Clean coffee can be held in storage for about a year without appreciable deterioration in quality. Clean coffee of up to a year old is designated 'current crop'. Longer storage times alter the quality of the coffee as the beans become drier. About half of the grades of clean coffee are of high quality and trade very quickly in the market. Sometimes the planters (especially the larger ones) sell raw coffee directly to the curing works; the traders buy the clean coffee from the curing works for trade with the exporters. Coffee for domestic consumption is sold to domestic roasters from the curing works. A trader's ability to take inventory of clean coffee allows him some leeway in playing the volatile coffee market by delaying the sale. An exporter typically trades with the foreign buyer for delivery of coffee at a future date, usually a few months from the harvest time. However, the exporter does not want to take possession of the coffee. So he/she strikes a deal with the trader for delivery of coffee at a designated future

---

[23] In larger plantations in countries like Brazil strip-picking or mechanical harvesting is used resulting in labor savings.

[24] An agent is an intermediary between a planter and a trader and works on a commission basis.

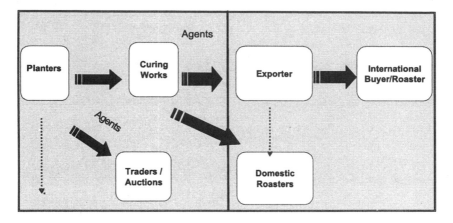

Figure 12.15. The Current Coffee Marketplace in India.

date. The volatility of the market can play havoc on the fortunes of the various players in the supply chain – the planters, traders, and exporters.

On the other hand, unlike the case of soybeans and wheat, the physical flow of coffee from the planter to the international buyer is usually more efficient. Therefore, the opportunities to further increase efficiency of physical flows are very limited; cost of removing these inefficiencies is as much as the value of inefficiency in the physical flow. However, there are inefficiencies in the information and financial flows. Recall that about 98 percent of planters have small holdings (Table 12.5). Overall, while the planters have global trade awareness, they do not know the details of transactions that take place between the trader and the exporter. So any opportunities to profit from interim price movements are captured by the trader / exporter. A trader usually aggregates purchases from various planters. Trader transactions are very relationship-based, both on the supply-side with the planters as well as on the demand-side with exporters. Consequently, the "reach" of any trader is limited to a few players on either side of the transaction. The trader has two main sources of price information – the Bangalore auction and other commodity exchanges. Approximately 15,000-20,000 tons of coffee per year (less than 10% of overall exports) is traded in the Bangalore auction, which primarily allows for price benchmarking. Furthermore, the Bangalore auction (and for that matter all international trade) is restricted to clean coffee. Since the quality of the coffee cannot be easily judged until it is cleaned, there is no "efficient" price discovery process for raw coffee. Yet, clean coffee trading within India only accounts for 40 percent of the overall coffee volumes traded.

There are several problems with the current supply chain. The first problem relates to the price discovery process. The Bangalore auction is an open auction, much like the auction at the mandi for soybeans and wheat. To maintain anonymity, buyers and sellers refrain from taking large positions at auctions and restrict themselves to trading small lots. Consequently, post-auction trades, also known as *kerb* (curb)-*trading*, are common. Domestically, this is the only price discovery mechanism available to traders / exporters in the spot market; futures exchanges have only recently become active. However, the weekly nature of the auction implies that within-week price variations cannot be captured. The second problem relates to liquidity. While half of the grades of clean coffee are of high quality and trade well internationally, liquidity of the remaining grades (referred to as off-grades) destined for domestic consumption has been a problem. Finally, a large majority of the selling in the chain – between the planter and the trader and the trader and the exporter – is relationship-based; the Bangalore auction accounts for less than 10% of the coffee market. Relationship-based selling obviously limits the reach of each party. Further, given the volatility of the export market, defaults are common and contract enforcement has become a big challenge.

12.4.2 Re-engineering Supply Chain to Deliver Commodity Services

ITC commenced its coffee trade operations in 1990 when coffees were sourced mainly from the auctions held by the Coffee Board. After the liberalization of coffee trade in 1994, the planters were allowed to sell directly to interested buyers without the intervention of the Coffee Board. It was imperative for ITC-IBD to quickly adapt itself to the changing scenario as raw coffee had to be procured from the estates to sustain the operations. However, it had to ensure that adequate systems were in place to assure that shipments of quality coffees were affected without hiccups. ITC-IBD setup procurement centers to facilitate buying of raw coffee from the planters through a network of agents. It also made arrangements with select curing works to process raw coffee. A dedicated team from ITC-IBD was stationed at these curing works to oversee the operations and to ensure compliance to the quality specifications as required by the customers. The integrated quality approach and system continues to be in place with constant improvements in each of the processes involving receipt, curing and dispatch of the coffees.

In a market where the daily domestic prices are pegged to the constantly volatile London and New York futures market, a committed delivery with an assured quality is only possible with a reliable network of procurement agents having deep-rooted access across the coffee growing areas and quality coffee curing techniques. Today, with over 10 years of being in the trade, ITC has exported well over 50,000 tons of coffee valued at over US $100 million and is now among the top ten exporters in India.[25]

While successful in its trading operations, ITC-IBD saw opportunities in the coffee supply chain. ITC-IBD understood that the ground realities of coffee are a little different from those found in soybeans and wheat. From an infrastructure perspective, the southern part of India – especially the coffee growing state of Karnataka – is much better off compared to the soybean and the wheat belts in the central and northern parts of India. In particular, better road and communications infrastructure allows for smoother logistics, better information flow, and better access to knowledge and agri-inputs. The average education level of a coffee grower is also higher compared to the soybean or wheat farmer implying a better capacity to absorb knowledge and best-practices. Consequently, ITC-IBD's intervention to provide value in the current coffee supply chain had to be different from its experiments in soybean and wheat sectors.

ITC-IBD followed a two-pronged approach. First it developed a software platform, called **Tradersnet**™, to provide commodity services to the various stakeholders in the coffee supply chain. Second, it developed **Plantersnet**™, to bring information and various commodity services – including Tradersnet – within easy reach of the small and marginal coffee planter.

Tradersnet is an online marketplace to buy and sell both raw and clean coffee on a spot delivery basis (as opposed to "Futures"). It replaces the current relationship-based marketplace with the advantages of an online mechanism. It brings together all the key players in the coffee supply chain (see Figure 12.16). Tradersnet is available to the participants for an annual subscription fee. It helps the participants unlock the value through

- Better, increased reach
- Ensuring price discovery and better returns through various trading mechanisms
- Counterparty flexibility
- Margin model to ensure risk management and factor counterparty adherence.

The increased reach comes from the fact that there are many more participants to trade with, than in the traditional relationship-based selling. Further the participants are guaranteed anonymity in transactions. There are four different mechanisms through which a participant can take part in Tradersnet including *Standard Auction*, *Special Auction*, *Direct Trade*, and *Negotiation*.

---

[25] Source: http://www.plantersnet.com/

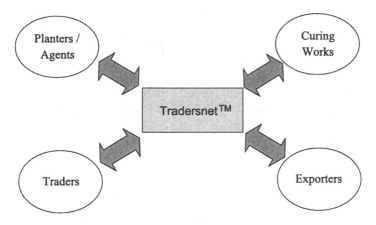

Figure 12.16. Tradersnet – The Key Players.

Standard Auction is a forward auction and the equivalent of the physical weekly Bangalore auction for clean coffee. In the Tradersnet model, there is a daily online auction at 10:30 in the morning. There would be between 3–5 auctions depending on the number of transactions scheduled for the day. Special Auction can either be a forward or a reverse auction customizable by the initiating party. Customization can be along various attributes like the quality of the coffees, auction rules of trade, start and end time of the auction, payment and delivery rules, etc. Direct Trade provides a fixed-price trading mechanism and the Negotiation module allows for trade settlement through online negotiations. Presently, Standard Auction is the only mechanism available. The other three will be offered shortly. Figure 12.17 provides a screen shot of a Standard Auction's auction floor in Tradersnet.

Tradersnet also allows for risk management. Participants in Tradersnet, when they register, rate the counterparties into three categories based on their relationship with them. The three categories include those with whom they may trade without any guarantee, those with whom they would trade but need some guarantee and finally those with whom they do not wish to trade. Tradersnet can facilitate margin money when guarantees are needed. The identity of the trading party, however, is always anonymous until such time as the trade is closed. ITC-IBD's business model on Tradersnet is based on subscription fees as well as commissions for transactions settled through the platform.

Plantersnet, is a web-based initiative of ITC-IBD that offers the planters information regarding coffee prices in various markets, products and services they need to enhance their productivity, improve price realization and cut transaction costs. Planters can access latest local and global information on weather,

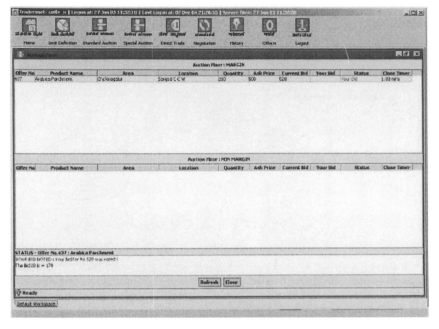

Figure 12.17. Tradersnet – Standard Auction's Auction Floor Home Page.

scientific farming practices as well as market prices at the village through this web portal available in English and the local language Kannada. ITC-IBD has about eighty Plantersnet kiosks concentrated in the coffee growing regions of the state of Karnataka. Figure 12.18 shows the main home page of the Plantersnet portal. A few notable features of the portal include *Parity chart* and *Purchasing Schemes*. All international trade in coffee is based on clean coffee and yet planters only trade in raw coffee. The parity chart tool is a matrix that reflects the corresponding raw coffee prices based on the prevailing international prices. The calculation factors into account variables including yields and grades, prevailing price differentials, and various costs of procurement, processing and other (in-)direct selling costs. A user can also customize the parity chart based on these variables to reflect their own specific conditions.

Under *Purchasing Schemes*, ITC-IBD offers various alternative selling options to the planters. For example, currently they provide a deferred pricing scheme for coffee purchase whereby a planter can "deliver" the crop today, get a portion of the payment, but settle the pricing of the transaction at a later point in time. The scheme enables planters to fulfill their urgent financial needs and

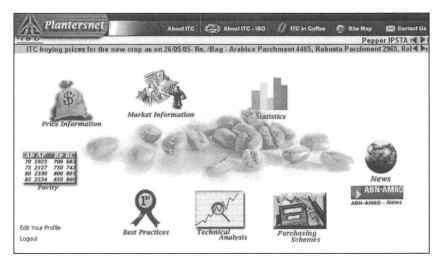

Figure 12.18. Plantersnet Home Page.

at the same time allows the flexibility of selling at a later date to take advantage of possible better prices.

While the focus of Plantersnet is the small and marginal planter, Tradersnet is designed for all the stakeholders in the coffee supply chain. Initial usage of Tradersnet has been restricted to traders and exporters. In the long-run, ITC-IBD's objective is to get planters to participate directly in Tradersnet. To achieve this, the two issues of access and aggregation need to be resolved. Planters can get access to Tradersnet through Plantersnet. The Indian coffee planter, however, is a small and marginal farmer. Consequently, individually, he will not have the volumes to take advantage of Tradersnet. Therefore planters need to aggregate their lots. This is being facilitated by aggregating groups of planters, called *Self Help Groups*.

Thus using Plantersnet and Tradersnet, ITC-IBD is re-engineering the Indian coffee supply chain by providing more accurate and timely price discovery, expanding reach, improving liquidity, facilitating contract enforcement, and offering better risk management tools for all parties involved. One could call this re-engineering a *smart re-intermediation*. These features will not only be particularly useful for the small and marginal coffee planter, but also further enhance the efficiency of the coffee value chain as a whole.

Both Tradersnet and Plantersnet were launched recently and it is too early to document measurable benefits at this point in the experiment.

12.5. Summary and Conclusion

The basic premise of this chapter is that emerging economies are character-ized by "broken value chains" that attempt to connect the poor, as sellers and buyers, to markets for products and services. Often these value chains do work (as they need to) but with the help of numerous intermediaries who extract dis-proportionate value leaving the poor with little residual income. In the context of agricultural sector, the challenge of fixing these value chains is further exac-erbated by the nature of the rural heartland of many of these economies, espe-cially India, which is characterized by geographical dispersion, heterogeneity, marginal farming, and weak – physical, social, and institutional – infrastruc-ture. This creates a vicious low equilibrium cycle where low incomes reduce the risk taking ability of the farmer leading to lower investments and hence low productivity bringing the incomes down further. Private businesses can build a profitable business model out of breaking the vicious cycle by smart supply chain reengineering and bringing the poor closer to the markets. In this chapter we have outlined the case study of experiments by the International Business Division (IBD) of the ITC Group in India. Using the examples of farm-to-market supply chains for soybeans, wheat, and coffee, we have illustrated how ITC-IBD has reengineered these supply chains for economic and social benefit to various stakeholders, viz. farmers, consumers, and ITC.

The nature of the intervention across the three commodities varies and is a function of the type of inefficiency – physical, information, or financial flows – that exists in the current value chain, the level of physical infrastructure (or lack thereof), and the needs of the marketplace. In the current environment in India, while procurement efficiency was the main value driver for soybeans, variety provisioning through segregation and identity preservation, in addition, was the key value driver for wheat. Coffee, on the other hand, required more sophisticated intervention in terms of "commodity services" that included pro-viding liquidity, broader reach, and better risk management instruments. We believe that the three key value drivers of efficiency, variety provisioning, and commodity services are universally applicable to most farm-to-market supply chains. Variety provisioning and commodity services in soybeans and com-modity services in wheat will be the next natural steps in evolution of the re-spective supply chains. ITC envisages three primary types of commodity ser-vices. First is information (on historical spot and futures prices) dissemination and tools for interpretation and analysis to help people make better decisions. Second, ITC can provide a gateway to India's commodity exchange like the National Commodities and Derivatives Exchange (NCDEX) for the small and fragmented farmers via bundling, lot pooling, network of delivery centers, etc.

Finally, the e-Choupal network can allow handling of physical deliveries nationwide.

A key feature of the supply chain reengineering executed by ITC-IBD is to "go direct to the farmer". This is facilitated by the use of information and communication technology (ICT) to build a hub-and-spoke network. The spokes are the e-Choupals (internet kiosks) managed by the *Sanchalak* (coordinator) and the hubs are direct procurement centers managed by the *Samyojak* (collaborator). Within this physical structure there are some key decisions that need to be made regarding information, transactions, and knowledge. For example, how does one handle information? Issues include access – cost (free or charge), location (centralized or decentralized), timing (real-time or not) and whether or not it should be bundled with the transaction (i.e., "I will provide you information if you promise to sell to me"). Next issue deals with the physical transaction itself. At what point in the network should procurement take place – at the village or at the hub? Where should quality checks occur? Finally, key questions in knowledge are whether it should be generic or customized and whether it should be bundled with transaction. The issue is not always either-or. For example, it is reasonable to expect that some types of knowledge will be generic (e.g., how to take soil samples) and others customized (e.g., "My crop appears to be infested. What should I do?"). The resultant network organization empowers the community through real-time information and customized knowledge. Furthermore, with freedom of choice and local management with self-interest through the Sanchalak, it facilitates development of community-responsive grassroots organization. The system maintains efficiency through competition inherent in choice. Finally, it overcomes the challenges of rural landscape through virtual aggregation giving it the power of scale.

A casual reader may draw some parallels with the Dell-Direct model in the computer industry (Margetta, 1998). Dell's model involves selling computers directly to the consumers bypassing traditional intermediaries. Similarly, the e-Choupal model by going direct to the farmer for procurement of soybeans and wheat dis-intermediates the agents at the mandi to gains efficiencies. The similarity ends there. Dell's customers are computer savvy and do not require much assistance in the sales process; any assistance necessary can be provided remotely. The context in which the e-Choupal model operates is very different. The ground reality of rural India is that an average farmer is not very well educated let alone be computer savvy and there are severe gaps in infrastructure. In this environment, ITC has to first provide the infrastructure.[26] Further the transaction has to be facilitated through physical assistance (provided by the

---

[26] See Khanna and Palepu (1997) for a discussion on the need for private enterprises in emerging markets to invest in filling institutional voids.

Sanchalak) at the village level in local language. ITC-IBD also recognized that the agent at the mandi had skills that can be leveraged at ITC's hub operations. So instead of complete dis-intermediation, the agent is co-opted as a service provider at the hub redefining and expanding his role in the value chain.

The e-Choupal infrastructure is clearly proprietary to ITC. By improving the efficiency of farm-to-market supply chains, it becomes one of the main corporate beneficiaries creating sustainable advantage for itself. The e-Choupal network, however, also benefits other businesses through their interaction with ITC as a supplier. For example, ITC-IBD is not always the end-consumer of the commodities it procures. In soybeans, while it processes beans, it also sells the beans to other processors including its competitors; it sells wheat to both internal and external customers; it trades coffee in international markets but also facilitates trades between various parties. So in a sense, ITC-IBD is now the new intermediary in farm-to-market supply chains. By having a direct channel to the farmer, this new intermediary is able to reduce the costs of procurement and, in addition as seen in wheat, satisfy the variety needs of the marketplace. This not only requires a firm understanding of the consumer demands but also the ability to orchestrate the supply chain through appropriate knowledge transfer to the farmer, scientific quality grading, appropriate pricing, and identity preservation. Thus through its e-Choupal network, ITC-IBD is able to connect the customer requirements to the farmer in what could be called "virtual integration".[27]

Lee (2001) has argued that the next competitive battleground for the 21st century will be demand-based management integrating the demand and supply side of the total value chain. While this has been posited for industrial supply chains in the developed world, the concept has an equally strong appeal for agri-businesses in the developing world. Worldwide increased agricultural production and productivity has primarily been driven by the use of high-yielding varieties of seeds and institutional support mechanisms in terms of subsidies and price supports. While the gains in agricultural production has been impressive overall – India increased its staple cereals production from 42 million tons in 1950/51 to 188 million tons by 2000/01 – further gains will be limited as long as agriculture remains a commodity. As societies develop, consumers become more discerning and demand for variety increases. The key challenge for agriculture worldwide then becomes how to respond to the variety needs, which may include not only providing what the customer wants but also provide assurances as to the integrity of that product with respect to its origin and conditions under which it was produced with varying levels of traceability.

---

[27] In this role, ITC-IBD performs an intermediary role similar in spirit to Li & Fung (Magretta 1998), the Hong Kong's largest export trading company.

The answer lies in converting agricultural supply chains into demand chains or from 'selling what is produced' to 'help producing what is wanted'.

In the Indian context, the ITC e-Choupal model facilitates the conversion to demand-driven value chains. As illustrated in the chapter, ITC-IBD has implemented this transformation in the wheat value chain. Going forward, however, opportunities exist in both soybeans and coffee and in other commodities. As we see it, there are two types of opportunities in a demand-driven value chain. The first is simply by providing different varieties of the commodity ideally suited for specific needs; e.g., soybeans specifically bred for industrial uses, wheat with specific protein levels for bakeries, or shade coffee for specific markets. Meeting these needs would require capabilities in grading, segregation, and traceability throughout the value chain. Potential areas for expansion include horticulture, organic products, aquaculture, etc. Some of these may require intermediate processing facilities which can be strategically located at the hubs in the e-Choupal network. The second opportunity in demand-driven value chains is related to provision of enhanced logistical services for customers who demand the right product, in the right quantity, at the right location. ITC-IBD using its e-Choupal hub-and-spoke network with its own storage locations is able to meet the specific logistical needs of its customers which include varying lot sizes, location specificity and timing of delivery, appropriate quality, and desired pricing mechanism. Essentially, ITC-IBD transforms itself from trading commodities to a provider of 'sourcing services' to its customers. The opportunity here for Indian agriculture is immense. It is not hard to see that the global competitive position of Indian agricultural produce is not consistent with the potential of the sector given its rich agro-climatic zones, abundant natural resources and a large pool of labor force. This potential can be realized with efforts like that of ITC.

In this chapter, we have focused on the use of the e-Choupal infrastructure for sourcing commodities, i.e., a flow of products from the farmer to the market. However, the infrastructure can also be as easily used for reverse flows, for example, in distributing products – agri-inputs and consumer goods – to the farmers. In fact, it need not be restricted to products but can be expanded into provision of services. The e-Choupal infrastructure establishes a physical and information 'pipe' all the way to small villages in India. The power of this efficient channel in moving products and services back and forth is immense. ITC-IBD has already started tapping into this potential as a network orchestrator facilitating the provision of various products and services between the farmer and the market. A few examples include provisioning of agri-inputs and consumer goods to the farmers (via direct distribution or retail at hubs), financial products (e.g., insurance, credit, etc.), marketing communication services (e.g., product demonstrations, sampling, exhibitions, etc.), and knowledge delivery (e.g., best practices). Notice that ITC-IBD is often not the producer of these products or services but orchestrates its provisioning in partnership with private and public sector corporations, government agencies, and nongovernmental organizations. As a network orchestrator, ITC-IBD's revenue stream is driven by the fees and commissions that it can charge for facilitating the various transactions. Essentially, as articulated in Anupindi and Sivakumar (2005), the e-Choupal network can be considered as a business platform orchestrated by ITC connecting rural India to markets effectively dealing with the three fundamental characteristics of rural India while sidestepping rural infrastructure deficiencies and facilitating rural transformation.

References

Anupindi, R. and S. Sivakumar (2005). "ITC's e-Choupal: A Platform Strategy for Rural Transformation". Paper presented at the *Business Solutions for Alleviating Poverty (BSAP) Conference.* Harvard Business School, December 1–3.

Barry, M.J. (2004). *Brewing a sustainable future – A study of institutional arrangements and supply chain impacts of sustainable coffee certifications.* Master's Thesis, School of Natural Resources and Environment, University of Michigan.

Fitter, R. and R. Kaplinski. "Who Gains from Product Rents when Coffee becomes more differentiated? A Value Chain Analysis". Institute of Development Studies Bulletin Paper, May 2001.

Glaudemans, H., "Identity Preservation – Dealing with Specificity". Food and Agricultural Research, Rabobank International, February 2001.

Khanna, T. and K. Palepu, "Why Focused Strategies May be Wrong for Emerging Markets". *Harvard Business Review*, July-August 1997.

Janssen, R., "Onward to India". *Fresh Cup Magazine*, May 2000.

Johnsson, Ola, "Tools for Testing and Segregating Grain". FOSS Presentation at the Seminar on Wheat Quality Revolution, October 30, 2004.

Lee, Hau, "Ultimate Enterprise Value Creation Using Demand-Based Management", SGSCMF-W1-2001. Stanford University, Stanford, CA.

Magretta, Joan, "The Power of Virtual Integration: An Interview Dell Computer's with Michael Dell", *Harvard Business Review.* March-April, 1998.

Magretta, Joan, "Fast, Global, Entrepreneurial: Supply Chain Management Hong Kong Style – An Interview with Victor Fung". *Harvard Business Review*, September-October, 1998.

Mock, Jim, "Identity Preservation Systems for Food Processors". Remarks presented to the National Food Processors Association, February 2003.

Nagarajan, S., "Quality Needs of Indian Wheat for Domestic and International Market". Presentation at the *Workshop on Wheat Quality Revolution in India,* October 30, 2004, New Delhi, India.

Prahalad, C.K. *Fortune at the Bottom of the Pyramid: Eradicating Poverty Through Profits*, Wharton School Publishing, 2005.

Pauwels, J. "Replacing Supply Chains with Value Chains". Food Traceability Report, CRC Press LLC, December 2002.

Shoran, Jag. "Diversification of Wheat – Meeting New Demands". Presentation at the Seminar on Wheat Quality Revolution, October 30, 2004.

Sriram, M.S. "Leveraging Information Technology to Access the Poor – A Study of ITC's e-Choupal Experiment". Working paper. Indian Institute of Management, Ahmedabad, India.

Stalk, George. "Time – The Next Source of Competitive Advantage". *Harvard Business Review*, July-August, 1988.

Titus Anand and Geeta N. Pereira. "Shade Coffee at the Altar of Sacrifice". www.INeedCoffee.com, 2004.

Upton, D. "ITC's e-Choupal Initiative". *HBS Case N9-604-016*, October 2003.

Hau L. Lee and Chung-Yee Lee (Eds.)
Building Supply Chain Excellence
in Emerging Economies
©2007 Springer Science + Business Media, LLC

Chapter 13

ESQUEL GROUP
Going Beyond the Traditional Approach
in the Apparel Industry

Barchi Peleg-Gillai
Stanford University, USA

Abstract: The garment industry has gone through enormous changes in recent decades.
Once concentrated in the U.S. and other industrialized countries, the garment
industry has gradually spread to countries with lower production costs, becoming a worldwide industry whose geographical distribution is constantly changing. One of the countries that grew to become a major player in this industry
is China, with US$ 45.76 billion worth of apparel goods exported from China
worldwide in 2003. It is predicted that once quota restrictions on the annual
quantity of goods that can be imported from China and other developing countries are lifted at the beginning of 2005, China and India will dominate world
trade in the global textile and clothing market.

As in other industries, technological advances, globalization, and changing
business practices are affecting the textile and apparel industry. In an industry characterized with over-capacity, companies must keep seeking for ways
to improve their productivity and differentiate themselves from competitors.
Even though the textile industry is already highly automated, textile and apparel companies continue to seek to increase productivity through the adoption
of computerized equipment, introduction of laborsaving machinery and the in-
vention of new fibers and fabrics that reduce production costs. In addition, the
apparel industry is becoming more service-oriented, with increased emphasis
on value-added services and quick response to customer demand.

This paper provides a brief overview of the apparel industry, as well as how it
is affected by various trade barriers, and in particular the elimination of quota
restrictions at the beginning of 2005. In addition, the paper presents a case study
of Esquel Group, an apparel manufacturer headquartered in Hong Kong, which
holds a significant portion of its operations in China. With vertically-integrated
operations, a focus on providing customers high-quality products and services,

adoption of advanced technologies and supply chain management strategies, and emphasis on ethical values, environmental protection, and the well-being of its employees, Esquel's story is quite unique in an industry that is usually old-fashioned and highly focused on cost reduction. We describe Esquel's goals, values, and operations, and discuss some of the pros and cons of their strategy. In addition, we discuss some of the challenges the company is faced with in the changing environment characterizing the apparel industry.

13.1. Introduction

Since the late 1960s, the global MultiFiber Agreement (MFA) has restricted the flow of apparel and textile goods from developing countries like India and Korea to developed countries like the U.S. The restrictions took place in the form of quotas, which set upper limits on the annual quantity of goods that can be imported to each of these countries from individual apparel and textile producing countries. The quota system was established as a mechanism to protect domestic manufacturers in developed countries from low-wage competition.

The landscape of the apparel industry is expected to see some significant changes in the near future. As a result of the Uruguay Round Agreement on Textiles and Clothing (ATC), a full integration of textiles and textile apparel manufactured in countries that are members of the World Trade Organization (WTO) has commenced on January 1, 2005. On the aforementioned date, all textiles and textile apparel manufactured in a WTO country and exported on or after January 1, 2005 will no longer be subject to quota restrictions (U.S. Customs & Border Protection , 2004). It is expected that the elimination of the quotas will accelerate the shift of apparel manufacturing to efficient manufacturers in countries with exceptionally low labor costs. In fact, a report released in August 2004 by the WTO has predicted that China and India will dominate world trade in the US$400 billion global textile and clothing market following the lifting of import quotas in 2005.

In 2003, when the restrictions posed by the quota system were still in place, China exported worldwide US$ 45.76 billion worth of apparel goods, which represented more than 10 percent of China's total exports for that year, and were an increase of more than 25 percent compared to China's worldwide apparel exports in 2002. Out of this amount, US$9.16 billion were exported to the U.S. (U.S.-China Business Council, 2004). The same WTO report mentioned above predicts that once the import quotas are lifted, China alone will supply more than 50 percent of the U.S. clothing market. Moreover, some U.S. executives have publicly forecast that China could grab up to 80 percent of the U.S. clothing market. Supporting these estimates is a report released in 2004 by the U.S. International Trade Commission, which said that China is "expected to become the 'supplier of choice' for most U.S. importers . . . because of its ability to make almost any type of textile and apparel product at any quality level at a competitive price" (Blustein, 2004). Many investors have already started building apparel plants and increasing capacity in China, in anticipation of the end of the quota system.

However, while the economic growth significantly increased the per-capita income in developing countries and lifted millions of people out of poverty, it has also resulted in a range of devastating consequences for the environment. China's overwhelming reliance on coal for almost three-quarters of its energy needs has made its air quality among the worst in the world. Unregulated economic development has also contributed to the devastation of China's forests, which, along with the overgrazing of grasslands and over-cultivation of cropland, has also dramatically changed the geography of the country, contributing to the rapid desertification of China's north and west. The most serious environmental challenge China confronts, however, is access to water. This stems from both growing demand and rapidly increasing levels of pollution (Economy, 2004). Another important challenge facing China is that of unbalanced development across the country, which resulted in widening differences in rural-urban income over the last two decades. More importantly, economic opportunities have continued to dwindle in rural areas and in the inland western provinces. One other concern is labor conditions in China. Numerous articles published in recent years have reported of workers who are grossly underpaid, lack basic protections such as pension and health insurance, are provided with almost no breaks during the day, and are required to work up to 12–16 hours per day, seven days a week. In addition, many workers are exposed to chemical toxins and hazardous machines, and may be required to work in unventilated facilities where temperatures can become very high.

This paper tells the story of one apparel manufacturer, Esquel Group, which is headquartered in Hong Kong and holds a significant portion of its operations in China. Esquel's story is unique in more than a single way. Internally, unlike the widespread practice of outsourcing all non-core operations, which is common in the apparel world as well as in many other industries, Esquel chose to become vertically integrated. Its current internal operations cover the entire apparel supply chain, starting from cotton farming all the way to the manufacturing of the final garments and their accessories. In addition, while the company no doubt strives for growth and financial success, it is determined to achieve these goals without compromising its ethical values and while minimizing damage to the environment. To ensure this, the company has developed a strong internal culture, which focuses, among other things, on ethical business practices, and took multiple steps to ensure its widespread adoption throughout the entire organization. In addition, the company continuously seeks for ways to modify its internal operations so as to make them more environmentally friendly, even if it comes at the price of higher operational costs.

Furthermore, the company takes very good care of its employees, under the philosophy that if it treats its employees well, the employees will in turn care about the company. Another unique characteristic of Esquel is that it does not keep all its profits internally, but rather takes multiple initiatives to give back to the community and improve its well-being. These initiatives include donation to unprivileged areas, donation to local schools and to needy students, and agricultural cooperation with local farmers. More details about Esquel can be found in Peleg-Gillai (2005).

13.2.   Esquel Group – Company Background

Esquel Group (Esquel) is one of the world's leading producers of premium cotton shirts, and among the most dynamic and progressive global-scale textile and apparel manufacturers. The company was founded in 1978 by Y.L. Yang, the first chairman of Esquel. In 1995 his daughter, Marjorie Yang, was appointed deputy chairperson of Esquel and later became the group's chairperson. Y.L. Yang was part of a stream of talented entrepreneurs who fled from China to Hong Kong shortly before the communists took over. At that time the business, which operated out of a small corner office in the Tsimshatsui business district in Hong Kong, pioneered the use of compensatory trade to export garments from China to the United States. The company focused on selling men's cotton shirts, which was Y.L. Yang's area of expertise.

Over the years Esquel, which was part of an old-fashioned industry, gradually grew to become a larger and more modern organization. The company transformed its operations as well as its scope of product and service offerings in multiple ways. Geographically, the company expanded its operations by opening manufacturing facilities outside China. In an effort to get closer to its customers and establish a more direct contact with them, Esquel also established multiple representative offices in the U.S., the U.K. and Japan. By 2005, with 47,000 employees, Esquel had production facilities in multiple locations in China, as well as in Vietnam, Malaysia, Sri Lanka, the Philippines, and Mauritius, and a network of branches servicing key markets worldwide.

As from the late 1980s, Esquel gradually expanded its scope of operations, adding to garment manufacturing the production of woven and knitted fabric, spinning operations, and eventually cotton ginning and farming. Being vertically integrated provided Esquel with the means to better control the entire manufacturing process, and to ensure high quality of the end products. Esquel's textile and apparel production was complemented by strong product development capabilities. The group's design and merchandising teams worked closely with its research and development (R&D) center to create unique fabrics and garments that gave Esquel the cutting edge in the apparel industry.

As for the company's product offering, initially the company made men's shirts of both cotton and blended fabrics. With time, product lines were expanded to include ladies shirts as well as casual pants. Esquel entered the pants business since it believed its customers would value the ability to buy shirts and pants from the same supplier. In practice, however, the added value of this business was small, especially since Esquel did not make bottom weight fabric. Consequently, and since Esquel started facing rapid changes in the external environment while at the same time going through a corporate transformation, the company decided in 2003 to make a temporary exit from the pants market in order to focus on its core products. On the other hand, ladies' shirts proved to be a more promising market, and so even though in 2004 it represented a relatively small portion of the company's sales, this business was expected to grow over the following years. With time, Esquel also decided to move away from blended fabrics, and focus on 100 percent cotton woven and knit shirts.

To ensure that internal operations were conducted in an efficient manner, Esquel developed a series of supporting IT (Information Technology) and supply chain management applications, which ensured a high level of information sharing across the entire organization, and improved production-planning processes. In addition, the company started applying advance RFID (Radio Frequency Identification) technology to capture information that was later used in local production monitoring and planning, inventory control, and overall optimization planning. In parallel, a strong corporate culture was established, with uncompromising emphasis on ethical business practices, creativity, and continual improvement. A new human resources organization was put in place, with a mission to improve the effectiveness of the Esquel organization through people practices.

All these initiatives provided Esquel with the means to offer high-quality, innovative products and services, and to secure a loyal customer base of some of the world's best known and most highly respected brands, including Tommy Hilfiger, Polo Ralph Lauren, Banana Republic, Hugo Boss, Brooks Brothers, Abercrombie & Fitch, Nike, J. Crew, Lands' End and Muji, and major retailers such as Marks & Spencer, Next, Nordstrom, Ito-Yokado, Aoyama and Jusco. As of 2004, 61 percent of Esquel's 50 million pieces of garment sales were to the U.S., 15 percent to Europe, 11 percent to Japan, 4 percent to China, and 9 percent to other markets. Total revenues of the Esquel group reached US$462 million in 2004, with operating profit of US$17 million.

While striving to run a successful business, Esquel also took multiple steps to ensure the well- being of its employees. For example, Esquel provided good working conditions to all its employees, and organized multiple activities, such as educational sessions on HIV/AIDS, to improve the employees' quality of life. Furthermore, Esquel was making an effort to have a positive impact on society. As the largest foreign investor in the province of Xinjiang, in the north-west part of China, Esquel took multiple initiatives to improve the well-being of its people, such as giving donations to strengthen education in the area and to conserve the infrastructures that provided drinking water and electricity. In addition, the company was devoted to protecting the environment in areas where it operated.

The remainder of this chapter discusses in more detail some of these initiatives taken by Esquel over the years.

13.3. Strive for Quality and the Transition to Vertical Integration

Esquel prided itself on being the best-quality fabric producer in China, an achievement that was gained through multiple initiatives, such as the implementation of the ISO9001 quality standard as well as the ISO14001 standard for environmental management, the use of external consultants to provide employees with training on quality issues, and the adoption of the 5S principles of maintaining an effective, efficient workplace.[1] In addition, to better address customer needs, Esquel made a strategic decision to focus on quality, merchandising service, product development capability and speed-to-market rather than on cost. One indication of the high quality level of Esquel's yarn and fabrics is the company's certification in 2005 by USTER®, which gave it the right to use the quality brand USTERIZED® as a seal of quality for its yarn.[2] Esquel's spinning mill was the first in China to have this honor, and at that time there were only a dozen more mills worldwide that had earned this certification.

One of the major transformations the company has gone through in its strive for quality was the gradual expansion of its scope of operations. While Esquel was not opposed in principal to outsourcing and preferred to work with reliable existing suppliers rather than develop the same capabilities internally, it

---

[1] 5S refers to the five Japanese words **seiri**, **seiton**, **seison**, **seiketsu**, **shitsuke**, which have the following meanings: **seiri** – eliminating everything not required for the work being performed; **seiton** – efficient placement and arrangement of equipment and material; **seison** – tidiness and cleanliness; **seiketsu** – ongoing, standardized, continually improving *seiri, seiton, seison*; **shitsuke** – discipline with leadership.

[2] For more details on this certification, see www.usterized.com

would do so only as long as the quality of its products was not compromised. Therefore, when over the years specific needs for quality, reliability, and sufficiently short lead times could not be satisfied from existing external resources, Esquel moved to develop these capabilities in-house. Still, even after having a complete vertically integrated supply chain, Esquel kept on purchasing some commodities in the open market, and when possible, developed strategic partnerships with its major suppliers and educated them on how to achieve the required quality level. Following is a more detailed description of Esquel's supply chain structure and scope of internal operations.

13.3.1 Supply Chain Structure

By 2005, Esquel's internal supply chain covered all the main steps in the garment manufacturing process, starting from cotton farming, all the way to the manufacturing of the final garments and their accessories (see Figure 13.1). All operations, except for cotton farming and spinning of gray yarn, were conducted based on actual customer orders. If needed, portions of the manufacturing process were outsourced when internal capacity was insufficient to meet all customer demands. This unique supply chain helped Esquel to provide its customers with high-quality products and services.

Following is a more detailed description of each part of the company's supply chain, the reasons that led the company to bring each of these activities in-house, and some of the means the company put in place to ensure high quality.

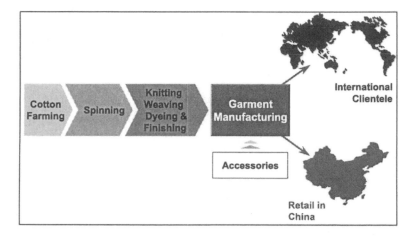

Figure 13.1. Esquel's Internal Supply Chain.

Cotton Farming: High-count fine cotton woven fabric for dress shirts is made from Extra Long Staple (ELS) cotton such as Egyptian Giza or Peruvian Pima. To obtain the highest quality of fabric, Esquel used for its products mostly ELS cotton. Once Esquel discovered that Xinjiang province grows the same quality cotton as the Egyptian Giza or Peruvian Pima it decided to source the majority of its ELS cotton supply from that province, and purchase additional ELS cotton from the U.S. or Egypt only when local supply was insufficient or when there were specific requirements. To gain sufficient knowledge that would ensure successful collaboration with local farmers, Esquel expanded its operations in 1998 and entered cotton farming by forming a joint venture cotton farm in Xinjiang. As of 2005, about 10 percent of the company's total ELS cotton supply was provided from its own farm, with the remainder being purchased from other local farmers. Esquel gave purchase commitments to the farmers in advance, before the cotton was planted, so as to secure the cotton supply and to ensure that the cotton is grown according to the company's specifications. In addition, Esquel educated the farmers on seed selection, cotton farming practices, and on ways to improve yields in order to reach the desired quality, and to eliminate impurities during the picking and storage process. The company also provided incentives to the farmers based on cotton quality rather than quantity, to encourage them to grow ELS cotton that met the required quality.

Once the cotton is picked, it is transferred to ginning mills, where the seeds are separated from the cotton fiber. By 2005 Esquel operated three ginning mills, all of them located in Xinjiang, in close proximity to the cotton farms. The first mill, with an annual output of 1,200 tons, was used for the cotton from Esquel's own farm. Two other mills, which started operating in September 2003 and September 2004 respectively, were used for collecting cotton from other farmers. Together these two mills had the capacity to process 6,000 tons of lint cotton, with the remaining cotton requirement bought from other ginners. Since quality control starts from the ginning process, bringing this activity in-house was valuable as it provided Esquel with in-depth knowledge of the process and the means to better control its quality.

Being a natural product, the quality and characteristics of cotton vary across farms and over time. High Volume Instrument (HVI) tools were therefore used in the ginning mills to inspect each bale of cotton, and an RFID (Radio Frequency Identification) Cotton Bale Management System was used for keeping track of the quality of the cotton at the bale level. Interestingly, this was the first time RFID was being used anywhere in the world for managing cotton at gins.

Spinning: In the spinning process, lint cotton is spun into yarn. Esquel decided to bring yarn production in-house due to difficulties the company experienced in purchasing high quality yarn from the outside. Some of these challenges included limited quantity and high prices of yarn from Egypt, and unreliable local sources in Xinjiang. To ensure high yarn quality, and due to the high impurity of cotton from China and low labor cost in Xinjiang, before the cotton entered the spinning process it was manually cleaned, to remove foreign fibers and other impurities. In addition, HVI tools were used for checking the cotton that was not processed at Esquel's ginning mills. The cotton used for each batch of yarn was selected based on the characteristics of each bale, so as to ensure consistent quality.

Spinning of all gray yarn, which was the majority used by Esquel, took place in two spinning mills in Xinjiang. The first mill was established in 1995 and the second in 1998, and together they produced 9,000 tons per year. The spinning mills used the ELS cotton to make high-count yarn up to 200s (the higher the yarn count, the thinner the thread and the higher its quality). The newer of the two mills used advanced technology to make compact yarn, which is characterized by improved strength and reduced yarn hairiness compared to regular yarn. Despite the huge capital investment required, Esquel decided to invest in compact spindles to produce better quality yarn, which was crucial to achieving the fabric hand-feel that was previously characteristic only of yarn from European mills. Furthermore, the use of compact yarn resulted in improved fabric strength and lowered costs downstream by allowing for faster weaving and knitting.

In addition to the spinning mills in Xinjiang, Esquel operated a smaller spinning mill in Gaoming, which is located in the province of Guangdong in the southern part of China. This mill specialized in heather and fancy yarn. Unlike gray yarn, heather yarn is made from dyed cotton, and is used for making fabrics with unique color patterns. Given the wide selection of colors and types of yarn, most heather yarn was produced based on actual orders, with an average lead-time of 2 to 3 weeks. Since this top dyeing process originated from the wool industry, heather yarn was typically used for winter clothes, which made demand very seasonal. To make the best use of the available capacity, heather yarn was also sold to sweater manufacturers, with whom Esquel had no direct competition.

During the spinning process, short fibers were removed from the cotton to ensure high yarn quality. Rather than throw away the leftover cotton fibers, Esquel used it through the "open end" spinning process to produce coarse yarn, which was later sold to other OEMs that used it to make denim and other coarse fabrics.

At the end of the spinning process, the quality (strength, evenness, etc.) of a sample yarn was tested at a QA lab located in each of the spinning mills. After spinning, the yarn was transported to Gaoming for weaving and knitting into fabric.

Fabric: As Esquel moved up-market and started to supply garments to customers such as Polo Ralph Lauren, these customers demanded better fabrics. However in the late 1980s textile from Japan, which used to be Esquel's main source for high-quality woven fabrics, became too expensive to be competitive and its quality was deteriorating. Esquel was unable to find other sources that would be able to consistently provide high-quality woven and knitted fabrics at a reasonable lead-time and price, and so eventually the company decided to set up its own fabric mills.

The production process includes dyeing, knitting or weaving, finishing, and quality inspection. All these steps took place in Esquel's production facilities in Gaoming, the company's largest operation center, with 21,000 employees as of 2004.

- **Dyeing**: As a first step, the dye lab determined the exact color formula for each fabric based on customer specifications. Once the formula was determined and approved by the customer, it was saved in a computer, which then controlled the dyeing process of the yarn or the fabric. This automatic control ensured consistency of the dyeing process across different batches.
- **Weaving**: The yarn-dyed weaving factory was built in 1989, and its capacity grew over the years to meet the growing needs of the company. By 2004 the factory had 560 weaving machines, providing capacity to satisfy all internal demand for yarn-dyed woven fabric. In addition, a very small portion of the output was sold to external domestic customers, to fill capacity as needed. The weaving process is highly automated, with minimal human intervention. To minimize defects and ensure high quality of the woven fabric, Esquel equipped the factory with the best machines available.

  In late 2004, Esquel decided to further expand its woven fabric facilities. With an investment of US$150 million, the company planned a new factory in Gaoming, adjacent to the existing weaving mill and installed with the latest European machineries and equipment. This new facility was scheduled to be completed in April 2006 and was expected to produce over 40 million yards of woven fabrics of the highest quality, equivalent to Italian fabrics which are regarded as the best in the world. Esquel expected this move to further consolidate its leadership in the woven shirt business.
- **Knitting**: The knitting factory was founded in Malaysia in 1988, and was relocated to Gaoming in 1996. As of 2004, it had 179 knitting machines

that provided capacity to satisfy the majority of Esquel's internal demand for knitted fabrics. Dyeing could take place either prior to the knitting process (yarn dye) or after its completion (piece dye), depending on the color pattern of the knitted fabric.

- **Finishing**: In this step the fabric was treated to provide it with certain desired characteristics such as softness, dimensional stability and durability, as well as special qualities such as hand-feel, certain surface characteristics, stain resistance, or wrinkle-free. This process was customized based on customer requirements.
- **Quality inspection**: After production was completed, the fabric went through 100 percent inspection. All quality problems were recorded. Designated quality control groups then worked with the production people to identify the source of each problem and determine ways to eliminate the defects. This method had proven to be very successful, as it resulted in substantial and sustainable improvement in the quality of fabric.

Garment Manufacturing: As of 2005, garment production took place in multiple places, including four different locations in China, plus factories in Malaysia, Vietnam, Sri Lanka, the Philippines, Mauritius, and Hong Kong. Work orders were distributed among the factories based on the required customer lead-time, individual factory capability, buyer preferences, as well as various trade restrictions and quotas (for example, the U.S. set quotas on the quantities of garments that could be exported from China, while other countries, such as Japan, had no such trade quota restrictions). In fact, Esquel chose to have geographically dispersed operations, even though the multiple locations complicated material flow and reduced economies of scale, because of quota restriction and availability of labor. As an example, special import privileges given by the U.S. in favor of African countries was the main reason for Esquel to continue successfully operating its factory in Mauritius. Another benefit of the multiple locations was that they provided Esquel with relatively high levels of flexibility during production. However, it also required the company to standardize quality across the factories to make sure that, regardless of the production location, customers always received the same quality of end products.

Outward Processing Arrangement (OPA) was another way for Esquel to balance labor costs and quota restrictions. Under OPA – the process approved by relevant government authorities – the majority of Esquel's production took place in China, with the semi-finished garment panels then being transferred to other countries for the final steps remaining in the production process. While such practices resulted in efficiency losses, the financial benefits were significant and more than compensated for the lower efficiency.

Garment manufacturing included the following main steps:

- **Fabric cutting**: Took place either automatically or manually. While automatic cutting might have been more efficient, manual cutting was required for fabrics with special patterns or for small orders. Cut fabric pieces that were intended for high-end products then went through 100 percent inspection.
- **Bundling**: In this step 5 to 15 pieces of the same part were bundled together, and a working ticket was attached to each bundle, specifying the garment order to which the bundle belonged. Bundles were then sent to the sewing floor, or if needed, to panel embroidery and then to the sewing floor.
- **Sewing**: This part of the process was manual-intensive. Each job could have dozens of sewing operations, with each worker specializing in a few of them. Sewing instructions were hung above the workstations, to ensure that orders were processed correctly. Typically, sewing of knitted and woven fabrics took place at different workshops, with the layout of the sewing floor following the steps of the production process.
- **Washing**: When needed, garments were washed for special effects.
- **Final inspection**: At the end of the process, the garments went through 100 percent inspection. In addition, process measurement control was done for a sample of the products.
- **Finishing & Packaging**: Included ironing, folding or hanging, attachment of various accessories such as hand tags, and packaging of the garments. Garments to Japan and those intended for children went through an extra step of needle inspection, to ensure there were no needles in the garments.

Accessories: Esquel entered this business mainly because it was unhappy about poor trade practices and quality issues, and the high margins associated with outsourcing this part of the business. In 1985 the company formed a joint venture with Rochester Button Corporation, and five years later it bought Rochester's share of the company. Over the years Esquel expanded this business, and by 2004 the EAP (Esquel Accessories & Packaging) division was in charge of making about half of all the accessories required for Esquel garments, including buttons, hand-tags, care labels, wrapping tissue, collar bands, hangers, and poly-bags. In addition to meeting internal demand, some of the accessories produced by the EAP factories were also sold externally. At times when customers chose Esquel as the sole provider of accessories while dividing their garment orders among multiple manufacturers to spread the risk, Esquel might ended up selling its accessories even to competing garment manufacturers. To increase capacity utilization at the different EAP factories, idle capacity was used to make other products that were not directly related to the garment business, such as shopping bags and gift and other boxes. To further improve profitability, while at the same time preserving the environment, some of the waste, such as plastic, was recycled to make goods that were later sold to external customers.

Over time, the EAP division further expanded its scope of service by doing more product development and by acting as traders for its customers for those accessories that could be made by Esquel, such as patches, lace, and feathers. Overall, the accessories business was quite profitable, accounting in 2004 for about 7 percent of the company's total sales, with about 40 percent of these sales made to internal customers within Esquel.

Retail Stores – PYE: In the early 1990s, Esquel decided to start selling its apparel within China, and to do so it created its own retail network. At the peak PYE products were available for sale at more than 100 stores, but due to multiple hurdles, such as immature transportation, inconsistent quality of the franchisees, and inability to exercise management control at the time due to infrastructure limitation (e.g., no telephone connection to update the POS), Esquel was unable to provide end consumers with the desired shopping experience. Consequently, the company decided to close all stores and re-position itself. While such a move had no doubt negative financial consequences for Esquel, it clearly demonstrated the company's determination not to compromise its stated goal of providing high-quality goods and services to its customers, a determination that brought the company to its current market position as the best-quality garment manufacturer in China.

Esquel re-launched PYE in 2001 at a higher price point with a range of shirts and tops that retailed at RMB 500 (US$ 65) and higher. At that time, two PYE boutiques were opened in Beijing. The main purpose of these boutiques was not to generate additional profits immediately, but rather to build up brand equity for the future, offer opportunities to try out new products and demonstrate Esquel's capabilities in terms of product quality and design. Furthermore, the boutiques provided Esquel with better understanding of the consumers. In 2004, a PYE outlet was opened in Gaoming, mainly for the use of Esquel's own staff. With such a limited number of stores in China, they were not perceived as competition to Esquel's current customer base.

13.3.2   Research & Development Activities

Esquel's research and development (R&D) center assisted the company in improving the quality of its garments, expanding its product offering, and addressing requests made by its customers for special product characteristics. R&D activities provided a significant competitive advantage for Esquel and helped the company to distinguish its products from its competitors, to maintain current customers, and to acquire new ones.

The R&D group was in charge of conducting research activities for the entire company, starting from cotton growth all the way to making the final garments:

Cotton: A research team in Xinjiang looked for ways to modify the cottonseeds so as to achieve a higher quality of cotton, with better strength and fiber length. A higher cotton quality led in turn to better yarn and fabric, which allowed the company to apply special finishing to the fabric while still maintaining its quality. Research projects included those involving various breeding techniques. In addition, the research team studied various irrigation methods in order to conserve water, a scarce resource in Xinjiang. A dedicated team worked with local farmers in Xinjiang on sustainable farming techniques, and advised them on ways to grow and collect the cotton so as to improve cotton quality while at the same time increase the farmers' income.

Fabric: Research activities focused on special processing, dyeing, and finishing methods, which aimed to improve either the process or the end product, or to add certain performance features. Process improvements included increased efficiency, lower cost, environmental issues, and lower energy / water / chemicals consumption. In product development the focus was often on fabric, but may have also included modifications to the cotton, yarn, or garment production.

Some of the special products developed by Esquel included:

- **Wrinkle-free**: To achieve this quality, the fabrics went through a special chemical process. To prevent the chemical process from weakening the fabric, Esquel used a specific type of cotton, and customized spinning and finishing processes. Esquel also developed a patented technology to make pucker-free shirt seams, gaining an important share of the market in a relatively short time.
- **Nano-technology**: Esquel used nano-scale polymers to treat the fabric to make it water, oil, and stain-repellant.
- **Performance care**: Maintained newness in the look of the fabric even after 20 to 30 washes.
- **Anti-bacterial finishing**: Prevented growth of bacteria in garments, which could create bad odors.
- **Dry-fit**: Provided sportswear with the ability to transform moisture from inside to the outside, and maintained a dry feeling to the one wearing the shirt. The knitting process was modified to achieve this quality.
- **Bamboo fabric**: A special fabric made of treated bamboo, characterized by the ability to absorb moisture and discharge it quickly.

- **Cashmere-like yarn**: Esquel used a special process to produce this type of yarn, which when used, gave fabrics and garments the look and feel of cashmere.

Esquel's R&D center was also involved in other, longer range and strategic research projects, many of them conducted in collaboration with academic institutions. They focused on such areas as conservation and functional luxury. In particular, some of the research activities that aimed to reduce the impact of the company's operations on the environment included:

- **Reduced dyeing cycle**: Esquel developed and implemented a new technology to shorten the dyeing cycle, which helped to save water and energy and to reduce the amount of dye used for the process.
- **Use of natural resources**: Rather than chemicals, Esquel started using natural materials such as clay and enzymes.
- **Use of organic cotton**: Since 2000 Esquel grew organic cotton in a farm and ginning mill that were certified by OCIA (Organic Crop Improvement Association). Even though by 2005 market demand was relatively low, it was expected to grow in the following years.
- **Recycled heat**: Extra steam generated from the power plant for heating purposes was used during the dyeing and finishing process.
- **Recycled water**: Esquel searched for ways to improve the quality of treated wastewater, so as to make it reusable.

Some of the new innovations were initiated by the R&D group, and required the sales and marketing people to analyze their profitability. Other research initiatives were triggered by requests from Esquel customers. One such example was Nike's request for Esquel to re-engineer the collar of its golf shirts. In April 2002, as Tiger Woods was winning his second consecutive Masters Tournament in Augusta, Georgia, executives at Nike were pleased with the victory but horrified by the looks of his shirt – by the 18th hole, the heat and humidity were laying waste to the collar of his signature Nike golf shirt. Nike was very concerned with the potential bad publicity, and so the next morning they contacted Esquel, asking the company to re-engineer the golf-shirt collar from scratch and create one that would not buckle in sweat and heat. Esquel tried different technologies, and within weeks already had multiple prototypes ready to be tested. By October 2002 Esquel was already mass-producing the new line of shirts – which was a big win for the company since the previous version of the Nike shirts were made by one of its rivals (Kahn, 2004).

An annual plan determined the group's development activities for the following year based on market needs as specified by the sales and marketing people. In addition, the R&D people were encouraged, and provided with the time and tools to generate and test new ideas for product and process innovations. Each of the new prototypes was first developed in the lab. Then, the R&D group started working with the Technical Development Center to determine the best way to work with the new fabric when making garments. In parallel, the R&D group worked jointly with the relevant factories, to help stabilize the process and bring it to mass production. Once the process was stabilized, marketing teams started promoting the new products.

13.3.3   Pros and Cons of Vertical Integration

Esquel's vertical integration was quite a unique strategy, at a time when a growing number of manufacturers in a variety of industries chose to outsource some or all of their non-core operations. The most common reasons for outsourcing are expected cost reductions, improved production, growth of capacity, expanded skills / capabilities, and enhanced service quality (A.T. Kearney, 2005). It is clear then that Esquel's strategy of having a vertically integrated and geographically dispersed supply chain might have had some negative implications, such as higher financial burden, increased operating costs, and inflexible capacity that might be hard to fully utilize at all times and that required high capital investment to maintain and upgrade as needed. Furthermore, production planning and coordination of internal operations was also a challenge, requiring the company to make significant investments in IT capabilities in order to improve the efficiency of these tasks.

At the same time, this unique supply chain supported Esquel well in their commitment to provide their customers with high-quality products and services. As described in the previous pages, each of the steps in the production process was brought in-house mainly to maintain its high quality, and so eventually Esquel's vertically integrated operations ensured the highest quality in every step of the manufacturing process that creates each shirt, a quality that could not be easily imitated by other garment manufacturers that were not vertically integrated. Other advantages Esquel gained from being vertically integrated include shorter order cycle time and the ability to maintain better control of the operations, as well as improved response time to new trends in the market. Furthermore, the company took advantage of the fact that the earlier in the supply chain quality was assured, the cheaper it was to reach a certain level of quality. Vertical integration also enhanced the company's R&D capabilities, and allowed for new ideas to be explored throughout the supply chain in a relatively short time. In addition, it was easier for Esquel to devote the required resources throughout the supply chain to develop samples for customers, which resulted in fast response to customer requests. All these benefits helped Esquel to provide its customers with the desired high quality of products and services, and strengthen customers' confidence in continuously receiving the promised high quality of goods and services. Another benefit of vertical integration was Esquel's ability to better absorb fluctuations in cotton prices compared to smaller companies and yarn manufacturers that might have found it hard to transfer all cotton price increases to the yarn. Esquel, on the other hand, could improve efficiencies throughout the supply chain to absorb part of the price shock.

Overall, it seems that vertical integration was a good strategy for Esquel, supporting its market position as a producer of high-quality garments. At the same time, even though Esquel moved away from focusing on cost, to ensure profitability the company had to still pay close attention to monitoring and controlling its costs and to increasing the efficiency of its operations. After all, even those customers that looked for high-quality garments were still cost sensitive, and were seeking to find the best deal for their apparel goods.

13.4. Value Proposition to Customers

Due to the over-capacity in the apparel world, Esquel chose not to target the commodity market, where competition is based mainly on price, but rather to position itself as a quality vendor. As such, Esquel focused on the more demanding customers, developed with them tight relationship, and provided them with quality products and value-added services. The company's philosophy was to do whatever it took to meet its commitments and maintain good relationships with its customers. For example, if production was delayed, the company sometimes shipped products by air to customers to meet the promised delivery date, despite the higher costs associated with air shipment.

At the same time, Esquel was cautious not to offer its customers services that would expose the company to too much risk. For example, Esquel was not in favor of offering customers such services as maintaining inventory in customers' retail territory, which would change the company's risk profile. Instead, the company worked with customers to use proper supply chain management techniques to minimize overall inventory in the pipeline. This way, customers carried fewer inventories while maintaining or expanding their business. At the same time, Esquel was not put in a position to overstretch its capital taking on retail risks without charging the relevant risk premium.

Below are examples of some of the services Esquel offered its customers, to strengthen its value proposition.

Product Merchandizing and Design. Several months prior to the beginning of the season, the merchandising group started working with their customers' chief designers to decide on yarn finish, color collection, and patterns for each of the product lines for that season. When needed, Esquel also provided design services, and assisted customers during the design process by providing information regarding the latest fashion and material trends, access to show rooms with a selection of garments made by Esquel, and a fabric library that included samples of all fabrics that Esquel made in the previous few years. Since trends and fashions in the apparel industry tend to be cyclical, Esquel customers found such services very helpful. If a customer was interested in a new type of fabric or finish, the query was passed on to the R&D group to determine whether it would be possible for Esquel to make such a fabric. Once garment specifications were defined, Esquel's Technical Development Center (TDC) developed a proto-sample and sent it for approval to the customer. The TDC then prepared a fit-sample, which was more detailed and included all the final specifications of the garment. In parallel, the TDC determined the set of accessories required for the garment. Once the customer approved these, their sources were determined. In addition, the TDC prepared a complete set of detailed manufacturing instructions, including machinery adjustment and sewing instructions, and transferred it to production. A similar process took place to determine the washing effects for each garment. Following this process eliminated almost completely any problems that might occur once mass production began. Any changes made by the customer to the product requirements were followed by an update to the manufacturing instructions. Total duration of the development cycle-time varied among customers, ranging from a few weeks to a few months.

To maintain close direct contact with customers, Esquel opened offices in the U.S., U.K. and Japan, which supported customers through merchandising and liaisons. As an added service, customers could choose to work directly with TDC when preparing samples, to avoid information distortions that could occur when the information was transferred through the sales people. In those instances where customers preferred to be highly involved in the product design and selection of accessories, Esquel sometimes assigned an employee to the customer site. Overall, customers were very pleased with Esquel's product design services, and with the resulting consistent quality that was maintained when moving from sample to bulk production. Consequently, the company had seen a significant reduction in the number of customer complaints.

Product Innovations.   In addition to responding to customer requests, Esquel also took the initiative to develop and provide customers with samples of new innovations. The merchandising group developed samples of new products based on inputs received from multiple resources such as trade shows, combined with the company's own capabilities. Technical know-how was used to improve the yarn and fabric so as to achieve the desired product characteristics. In addition, the R&D group sought for ways to improve the product characteristics. Esquel customers were happy to see samples of new innovations, which they could then decide to apply to their product lines.

Management of Customers' Inventory. Starting from 2000, Esquel had been engaged with Lands' End in the Net Position Management (NPM) program, which focused on the management of Lands' End inventory and was similar in nature to the concept of Vendor-Managed-Inventory (VMI). Under this arrangement, Lands' End expected Esquel to manage the replenishment of its inventory, so as to ensure that the net inventory maintained its target level. The process started with a demand forecast that was provided by Lands' End 12 weeks in advance. Esquel used this forecast, which was usually 60–70 percent accurate, to prepare material and space. After that, Lands' End provided Esquel exact sales figures on a weekly basis, and Esquel was committed to an actual response time of four weeks. To achieve this response time Esquel relied on a control management system, which allowed the company to continuously monitor inventory and anticipate demand. Esquel expected more customers to move in this direction and engage in a similar type of service arrangement, which transferred some of the risk from the customer to Esquel. In fact, by 2005 Esquel had already started collaborating with other customers such as Brooks Brothers, not only actively managing the customer's inventory but also acting as a key participant of the business. In particular, Esquel worked with Brooks Brothers closely on Demand Chain Management with full collaboration from the set-up of the program, understanding the forecast profile and assumptions, actual retail performance, and replenishment expectation. Here, again, Esquel viewed its size as a competitive advantage, since only large garment manufacturers could afford the risk burden of providing a VMI type of service.

13.5. Internal Operations Management

Over the years, Esquel has taken multiple major initiatives in an effort to improve the efficiency of its internal operations. One of these initiatives was the development and adoption of a central production planning process that aimed to provide the best response time to customer orders while simultaneously optimizing production capacity loading at the different production facilities. In addition, the company put in place dedicated teams that looked for ways to improve operational efficiencies. Esquel also adopted a set of global and local metrics to monitor actual performance and identify opportunities for improvement. Furthermore, significant investments were made in IT applications to support internal operations, enhance supply chain management capabilities, and further improve operational efficiency. The following paragraphs describe these initiatives in more detail.

Production Planning.   In the past, production planning was conducted in a decentralized manner: sales people received orders from the customers, and decided how to distribute them among the factories. Their decisions were based on multiple criteria such as customer preferences, quality, price, and lead-time, and resulted in sub-optimized operations.

Starting from 2004 the process was revised, and a Central Production Planning Control (PPC) group, based in Hong Kong, was put in place to be in charge of decision making. The process started when sales bookings and orders were placed in a centralized production planning system at the Esquel head office. The Central PPC Group reviewed and allocated the bookings and orders to the most appropriate garment factories so as to balance capacity utilization and production time. The orders were transferred automatically, on a real time basis, from the PPC group to a Factory Production Planning Group in each plant. In addition, the Central PPC Group notified the spinning and fabric factories of the material requirements. A dedicated PPC Accessories team determined the list of accessories required for each order, and notified the different EAP factories accordingly. A central procurement team was in charge of ordering all the accessories that were purchased from external suppliers. The orders for accessories were placed based on their expected lead-time, which could be much longer than the one-week average lead-time provided by the EAP factories. Any changes customers made to their orders, as well as orders for samples, followed the same process. In addition to actual production planning based on customer orders, PPC used sales forecasts to generate on a monthly basis a rough plan for each factory, which was used for planning capacity utilization, estimating material needs, and determining expected lead-times. Esquel made an effort to maintain a sufficiently flexible production process to accommodate last-minute changes in customer orders, but nevertheless, they could still be problematic and affect capacity planning.

Based on the bookings and orders received from the Central PPC Group, the local planning system determined local production plans and alerted a local production planning team to any scheduling conflicts, such as inability to complete an order by the specified due date. The local production planning team was in charge of this process, and of resolving any potential scheduling issues. All factories used the same production planning tool, which was integrated to the central production planning and ordering system and provided production planning and tracking capabilities. Furthermore, the Central planning system linked not only garment but also the fabric production planning processes to ensure fabric capacity, and that lead-time was matched to garment demand.

On average, total order-to-delivery cycle time took about 60 days (excluding shipping), including 30 days for fabric production and 30 days for garment production. However, actual cycle time could be as low as 45 days, usually if all production took place in China, or as high as 90 days – when production spanned multiple geographic locations.

As for purchase orders placed by customers for non-garment goods, such as heather yarn and accessories, those orders were placed directly with the local plants rather than with the central sales organization. Production plans for these orders also took place locally, with the help of the local planning system.

Some of the benefits that Esquel achieved from the centralized, automated production planning system included improved capacity utilization and balanced workload, and increased profitability through better selection of the production location for each order. In some instances the improved processes also resulted in reduced lead-times to customers. Still, Esquel saw further opportunities for improvement, for example through automation at the line-planning level.

Improving Operational Efficiency. A common practice among many garment manufacturers is to solve operational inefficiencies by hiring more people. Esquel, on the other hand, used organizational approach, advanced technology, and management discipline as means to improve productivity and operational efficiency. A central Technology Engineering Department (TED) determined the standard time for each task in the garment manufacturing process, information that was used for production planning and line balancing. In addition, TED worked with the factories on productivity improvements and standardization to ensure consistent quality across the different factories. Within each factory, local TED groups looked for various ways to improve efficiency. For example, the company trained different groups of employees, each with a different set of skills and associated compensation level. The local TED group determined how to balance the number of employees in each level, so as to maintain the high quality of end products while not incurring excessive wage expenses.

In addition to on-going tasks conducted by TED, the company also tested from time to time various ways to revise and improve its production processes. For example, in 2004 the garment factory in Gaoming tested a transition away from a batch mode of operation; that is, rather than transferring WIP from station to station in bundles, the batch size was reduced to one unit. This test showed promising results: cycle times were significantly reduced, allowing for faster response to customers. In addition, not piling up the garments during production made them less prone to wrinkles, which in turn made the finishing work at the end of the process easier and faster to complete.

Performance Measurement.   The actual performance of the organization was measured through a combination of global metrics, which measured the performance of the entire organization, and local metrics for each operation. As of 2004, metrics at the corporate level included financial Key Performance Indicators (KPIs) such as return on operating assets, profit, revenue, overhead expenses, and inventory levels. In addition, Esquel measured garment sales performance at the corporate level by using such KPIs as garment sales revenue, claims and airfreight expenses, sales process performance, and customer satisfaction. A number of local KPIs were in use to measure the performance of each of Esquel's spinning, fabric, and garment facilities. They included such KPIs as inventory holding days, capacity utilization, first-pass yield (FPY), cycle time, on-time arrival (to an internal customer downstream)/on-time delivery (to external customers), productivity, unit cost, and quality. All the performance indicators were reviewed and analyzed on a weekly or monthly basis as appropriate by cross-functional teams, with corrective actions taking place to improve performance.

Supporting IT Applications.   In the past, Esquel used a decentralized IT system that focused on order management and fulfillment. Information was captured only during order processing, and was used mainly for initiating shipments and payments. With time, however, the company realized the potential of a more sophisticated system as a means to increase productivity and improve efficiency of internal operations. Consequently, in 2003 the company started implementing a new system that had several major advantages compared to the old one. First, information was captured much earlier, starting with information collected or generated by the merchandizing and product development teams during the design process. All the information regarding each garment, such as manufacturing instructions, paper pattern, fabric and accessory information, and historical customer comments, was stored in three libraries (style, fabric, trims), and was used for production planning, procurement, and other operational activities. Since the system used a single input point and the information was stored in a central IT system, the same database was accessible by all parts of the organization, which left less space for errors and improved operational efficiency.

Furthermore, style information captured during product design could be reused for future products, which improved knowledge management. In addition, the switch to automatic processing resulted in multiple productivity gains – for example, with the new system the time required for generating a purchase order was reduced from one week to 15 minutes, and the cycle time to transform a designer's drawing into an actual garment was reduced from five months to as little as 45 days. The system also improved production and inventory management by providing Esquel with the means to plan and monitor production and by supporting execution of VMI programs.

In addition to internal benefits realized within Esquel, the new IT system improved customer service in multiple ways:

- It allowed customers to place orders electronically, and later to check online the status of their orders.
- It provided inventory information to those customers with whom Esquel had a VMI agreement.
- It significantly shortened the duration of the development cycle if customers used the system to electronically connect with Esquel during product development.
- It allowed customers to see how a fabric would look like even before an actual sample was made.
- It allowed customers to connect with Esquel as early as at the beginning of the product development stage.

To realize the most benefits from this capability, Esquel worked with customers to integrate the process and system at this early stage.

Some of the characteristics of Esquel operations where common in the apparel industry, such as three-dimension SKUs (for color, size, and style), and a built-to-order environment that resulted in a unique process flow with almost no inventories. However, Esquel was unique in the industry in the extent of its vertical integration from cotton to garment manufacturing. It also had strategic value proposition in its strong merchandising and product development cycles, which fitted the needs of each customer and required communication through the whole order management process. Due to these special characteristics, many of the IT solutions available in the market did not meet the needs of the company, and so eventually Esquel decided to develop the new IT system internally. The system was based on Web technology, and was connected to an Oracle system that provided such modules as finance and procurement.

Another major initiative was the implementation of RFID technology in parts of its operations. As of 2005, the company used RFID to track cotton quality at the bale level, which was a major improvement compared to the past, when cotton quality could only be recorded at the batch level (each batch includes 250 bales). The information was used for cotton blending and during production planning, so as to ensure consistent yarn quality in the spinning process. In addition, RFID was used for cotton warehouse management and to control the delivery of cotton orders. Some of the benefits Esquel realized after implementing the Bale Cotton Management System included improvements in warehouse efficiency, logistics, and the efficiency of the blending process, as well as better information and processing control at the cotton level. In addition, it improved the quality and consistency of the spinning mills, blending methodologies, and cotton selection and mixing operations.

Replacing all legacy systems in the entire organization was a daunting task, which was expected to take years to complete. By 2005, Esquel's ambition was to at least achieve system integration within its core businesses, which was already a big improvement compared to the past. In addition, Esquel planned to further expand the use of its centralized IT system and to enhance its capabilities. One of the major areas Esquel intended to focus on was optimized capacity utilization. The plan was to use the information already available in the system to optimize the planning process, while considering local vs. global optimization. In addition, the company could benefit from automating the pricing process for new products – information that could later be used during negotiations with customers. As for RFID adoption, Esquel planned to use it to capture information throughout its entire supply chain, and to utilize it in local production monitoring and planning, inventory control, and overall optimization planning. Esquel also planned to adopt the EPC Global standard for RFID coding.

13.6. E-Culture

13.6.1   What is Esquel's E-Culture?

Esquel's business philosophy was rooted in quality control, operational efficiency and corporate responsibility. For Esquel to develop into a successful business, the company recognized that it had to meet the needs of its employees and the expectations of its customers and other stakeholders. This meant providing an equitable, just and fulfilling work environment for its employees, minimizing the environmental impacts of the business and contributing positively to the communities in which the business operated.

With operations spread across multiple countries, Esquel naturally faced challenges associated with culture and language differences. In an effort to manage these challenges, a strong corporate culture had been established. It was a corporate culture by which Esquel could manage, develop and also share the company's goals, beliefs and expectations across the entire organization. One step the company took to communicate these values was by establishing what was called the E-Culture. Initially, the E-culture included four areas, namely: Education, Exploration, Environment, and Electronics. In 2003 its structure was revised to include the following five areas:

- **Ethics**: Be a good citizen and a good employer.
- **Environment**: Cherish the environment.
- **Exploration**: Explore and embrace innovative solutions and enhancements.
- **Excellence**: Reduce wastage through functional excellence.
- **Education**: Promote a culture of learning and support education.

The E-Culture reflected the business philosophy of Esquel Chairperson Marjorie Yang. On the one hand, to be successful in a highly competitive landscape it was important to strive for continuous improvement and increased efficiencies. At the same time, people must not abandon such fundamental values as ethics and integrity while rushing for success. Below is a more detailed description of each part of the company's E-culture.

Ethics.   Different cultures have varying viewpoints on what is right and wrong in business ethics, especially in places that do not have much exposure to business with the Western world. For Esquel, however, it was very important to make profits in an ethical way, so the company put in place control systems and a continuous education process to achieve this goal. As a by-product of the company's emphasis on business ethics, by 2005 Esquel had already been awarded for 10 years by the local customs the title of Credible Enterprise and AA Level Enterprise. In addition, Esquel was working to make sure all garment factories received the WRAP (Worldwide Responsible Apparel Production) certificate of compliance.

Environment.   The company emphasized the importance of minimizing its effect on the environment. In Xinjiang, where water supplies were scarce and relied mostly on underground resources, Esquel advocated against investments in fabric production, a segment of the textile supply chain where large amounts of water were used both in the production process and to treat dyestuffs and chemicals before discharging them. Rather, Esquel had in Xinjiang, a major cotton producing region for China, only cotton farming, ginning and spinning facilities. These operations did not create excessive output of polluted water. Furthermore, in the company's spinning mill in Turpan, Xinjiang, water was reused to improve the natural environment of the campus. Over a period of seven years, over 3,800 fruit trees were planted throughout the campus. Treated facility wastewater was redirected to irrigate these fields resulting in a closed loop system in the facility's management of their water resources.

Esquel also sponsored pilot studies at two cotton farms in Xinjiang to test the effectiveness of drip irrigation versus conventional open field watering. Based on crop yield results, the yield from the drip irrigation field increased from 170kg/mu in 2003 to 330kg/mu in 2004, while the yield in the non-drip irrigation field was 200–230kg/mu in 2004. This success was recognized by the local governments and farmers alike, with the latter looking to apply this technology in their fields.

Esquel's fabric production was centralized in a single location, at Gaoming, where the company invested in two major initiatives to minimize damage to the environment – a water treatment facility and a power plant:

- **Water Treatment Facility**: The dyeing and finishing processes generated large amount of polluted water – about 20,000 tons per day as of 2005. Unlike some other fabric manufacturers who treated only a portion of the polluted water due to high costs, Esquel treated all its polluted water to a standard that conformed with environmental requirements, to minimize damage to the environment. Two treatment facilities were in use, one that was built in 1995 and a second that was built in 2000. With a new phase of the company's fabric mill coming into operation at the end of 2005, a third water treatment facility was also being constructed. While as of 2005 the treated water was not reused, Esquel's R&D group was studying how the company might reuse the treated water in the fabric finishing operations.
- **Power Plant:** In 2002 Esquel started building its thermo-power plant in Gaoming. Completed in late 2004, the plant satisfied the electricity needs of all of the company's facilities in Gaoming. One of the advantages of the power plant, which made it environmentally friendly, was its high utilization of natural resources. In addition, 99.5 percent of static dust was collected to prevent air pollution, and 100 percent of the dust and residues were recycled. In addition to being environmentally friendly, another major benefit of the power plant was that it ensured consistent power supply to Esquel production facilities (the local electricity supply grid tended to be unreliable). This, in turn, contributed to more consistent production time, which resulted in more reliable lead times and fewer instances of production stoppages and late deliveries to customers. While building and operating the power plant was quite a challenge for Esquel, given that its core competencies lay in textile and garment manufacturing, Esquel nevertheless had been successful in managing this project. Training was provided to all the people working in the power plant, to ensure safety and smooth operations.

Another initiative the company had taken was the update of its product offering, to include garments made of organic cotton. Esquel had started growing organic cotton in 2000, and the farm and ginning mill were certificated by OCIA (Organic Crop Improvement Association International). Esquel was also closely involved with the Organic Exchange, a nonprofit organization dedicated to moving organic fiber into mainstream markets. Members of the Organic Exchange included individual farmers from all over the world as well as household brands. They were committed to creating fair, transparent and sustainable supply chains to make products like clothing, home furnishings and personal care products that met basic human needs.

Exploration.   The company put a lot of emphasis on continuous improvement. It encouraged its employees to be creative, explore and try new things, go to far places to find opportunities, and always push the limits. Within each factory, internal teams were in charge of exploring ways for process improvements, such as reduced machine idle time and increased capacity utilization. Cross-functional teams further enhanced creativity. In addition, the technical development center focused on apparel manufacturing research, while the R&D group focused on developing new products, improving product quality, and exploring new processes, techniques and materials. These initiatives were viewed as one of the key contributors to the company's success.

Excellence.   The stated goal of the company was to be the best cotton shirt partner. To achieve that, the company encouraged its employees to innovate to deliver value through leadership, speed and quality, and to be the best in everything they did. At Esquel, excellence was a continuous quest in every sphere of activity. The company emphasized the cycle of learning, and the importance of learning internally as well as from others – not only from companies within the apparel industry but from leading companies in other industries as well.

Education.   Esquel was a learning organization. People were encouraged to try new ideas and not to be over-concerned about possible mistakes. At the same time, people should also be quick to learn from mistakes and to improve practices. Early on, Esquel realized the importance of positive reinforcement as a means to encourage its employees to take chances and try new things. To achieve that, Esquel gave credit to employees when they were doing something good. For example, the company posted "Star worker" placards on the walls of the factories, which acknowledged those employees who had performed exceptionally well.

In addition, the company organized educational events for its employees, and provided them with such means as computer rooms to encourage learning. Esquel educated its employees about such issues as health and safety, and encouraged them to broaden their horizons (go to museums, read more books, etc.). While such activities were good for the employees' well-being, they also helped to increase their creativity.

This attitude was reflected in the profile of Esquel employees. Unlike many other garment manufacturers, Esquel had many well-educated employees, with advanced degrees from the U.S. and Europe. For example, at the Gaoming campus, which by 2005 had over 20,000 employees, close to 7 percent had a college or higher degree.

13.6.2   Internal Adoption of Esquel Vision and Culture

Given the size of the company, the geographic span of its operations, and the cultural differences among its employees, ensuring that all employees shared the same values and goals was clearly a challenge. Esquel continued to work on multiple initiatives in an effort to spread its vision and core values throughout the organization.

At the end of 2003 a new human resources director joined the company, to lead the transition of the organization from an administrative-intensive to a modern one. The mission of the new HR organization was to support the company's business strategies by constantly improving the effectiveness of the company through people practices. An organizational infrastructure was put in place, including such tools as metrics, reward mechanisms, and new management methodologies, with a focus on building trust across the organization. Training sessions were organized across all levels of the organization to spread the company's culture and goals. Since communication is always an issue in a large and diverse organization, communication champions were put in place in each location to lead the channels of communication between the local people and the company's management. Emphasis was put on interpreting the culture dynamically in each part of the organization, to ensure that the message was relevant to each location.

Overall, these initiatives had proven to be very successful, and it became evident that with time the number of employees that internalized the culture had increased. Still, achieving a cultural change is a hard task – much harder than developing the right technical skills. By 2005, there were still locations in which the level of adoption of the company's culture was relatively low. In those places, the HR team worked closely with top local management to determine the approach that would be most effective in ensuring that the Esquel culture was absorbed.

Creativity 2000.   In an effort to strengthen the adoption of the company's E-culture within the organization, the Creativity 2000 competition was initiated in the year 2000. Every division across the entire organization was invited to join this annual competition, and work on projects that aimed to improve one or more of the "E"s in the company's E-culture. Every year, the executive board of the company chose the winning team from a shortlist of entrants who were invited to the headquarters in Hong Kong to present their projects. Overall this initiative had proven to be very successful, yielding both quantitative and qualitative improvements. Some of the ideas developed by the participating teams included the use of recycled water in Xinjiang for growing fruit and flowers, and the modification of sewing work flow at the Sri Lanka factory, which resulted in improved efficiencies.

The Creativity 2000 competition was also instrumental in encouraging many bottom-up initiatives, projects that were not driven by corporate head office in Hong Kong.

13.7. Corporate Social Responsibility

In addition to its business goals of growth and financial success, Esquel has always strived to impact the community and bring about a positive change in society. Some of the activities the company focused on include environmental protection, donation to unprivileged areas, donation to local schools, and agricultural cooperation.

Environmental Protection. As the largest foreign investor in Xinjiang, Esquel always kept an eye on environmental protection and sustainable development. Some of the steps Esquel took to increase awareness to this issue include the hosting of conferences to educate the people on the importance of protecting the environment. First in Gaoming in 2001 and subsequently in Urumqi, Xinjiang in 2002, these conferences brought together stakeholders and experts from within and outside China, and were instrumental in addressing various sustainability issues pertinent to China's southern and western regions. In addition, Esquel operated in Gaoming a wastewater treatment center and an environmentally friendly power plant, for internal use.

Donation to Unprivileged Areas. Esquel donated RMB 1.5 million (US$181,000) for projects in drinking water, electricity and orphanage in Hetian and Kashi districts of Xinjiang during 1995 and 1996. In 2003, Esquel donated 31,000 garments, worth over RMB 1.5 million, to earthquake-stricken areas of Xinjiang.

Donation to Local Schools. Through the Esquel-Y.L. Yang Education Foundation, the company worked to improve local education in Xinjiang in multiple ways. First, the company financed the rebuilding of decrepit schools – by the end of 2005 Esquel had rebuilt 12 schools in various rural locations in Xinjiang. Another activity conducted by Esquel to improve local education was the donation of mini-libraries, which were often difficult to afford for rural communities. By 2005, around 800 such mini-libraries had been set up throughout Xinjiang. Employees at Esquel also tried to help in less fortunate communities. With employee and company contribution, Esquel provided thousands of needy children with financial support, which was used for basic education expenses such as tutorial and exercise books. In addition, Esquel sponsored college students for science works, provided scholarships to out-standing high school graduates to attend university, and supported the MIT IMBA program at Tsinghua, Fudan and Lingnan universities.

As part of the Tsunami relief, staff and company also contributed to rebuild-ing homes and to the education of the affected workers' children in Sri Lanka and Malaysia.

Agricultural Cooperation.   Esquel is devoted to developing the local agri-cultural economy in Xinjiang, and to protecting the benefits of farmers. In fact, in 2002 the company won the title "Pivot Dragon Enterprise" from the Xin-jiang government for its activities in this area. In addition, to improve the qual-ity of the cotton and minimize impurities, Esquel provided farmers – free of charge – such items as pure cotton garments to be used when collecting the cotton. Esquel also offered farmers workshops on cotton farming, and invited them to visit its spinning mills to demonstrate the impact of cotton quality on the quality of the yarn, and eventually on the quality of the garments. While these activities helped Esquel to maintain high cotton quality, they also helped the local farmers improve their practices and become more profitable. In ad-dition Esquel placed orders with the farmers in advance, when the cotton was planted, and guaranteed to farmers a minimum price for the cotton, with the maximum set by the market. Both Esquel and the farmers benefited from this arrangement: farmers were not wiped out if cotton prices went extremely low, and at the same time Esquel secured its cotton supply, and also had better rap-port with the farmers, who were wary of dealing with foreign capitalists after decades of selling only to the government (Murphy, 2003).

All these activities, which aimed to protect the environment and make a positive impact on the communities in which Esquel operated, no doubt added a financial burden on the company and resulted in higher costs. However, those activities were in line with the values of Esquel's management team and with the company's culture, which is why Ms. Yang and her management team initi-ated and supported such activities. It is important to note that being a privately held company clearly provided the company with more freedom to pursue these initiatives, which did not necessarily make financial sense. In contrast, public companies, which must usually focus on performing activities that aim to improve financial strength and growth, may find it harder to justify such initiatives which do not directly contribute to the company's bottom line.

13.8. Taking Care of Employees

While in many Chinese companies the owners did not invest in their employees adequately, at Esquel the attitude was different. Esquel took multiple steps to ensure the well-being of its employees, based on the belief that if the company took good care of them, they in turn would also care about the company. Even when employees had to be downsized (e.g., when the Mauritius pants factory was closed), Esquel took steps to ease the process and help the laid-off employees find new jobs.

In contrast to some manufacturers in less developed countries, which might provide employees with inadequate working conditions in an effort to reduce operating expenses, Esquel always made sure that its employees had good working conditions, did not work excessive hours per week, took lunch breaks, etc. In addition, Esquel taught its employees how to calculate their wages, so that they understood how their salary was determined. These efforts started even before labor compliance became mandatory. Ironically, the issue of limited working hours was not always welcomed by employees: since many of them joined the work force for a limited number of years, their goal was to earn as much money as possible during this limited period of time. Therefore, some of them viewed limitations on the length of the working day as a disadvantage, since it reduced their potential compensation. Esquel organized activities for employees to improve their quality of life, and educated them to appreciate the free time they had, rather than focus only on net wages. In addition, in an effort to help its employees to reach a sufficiently high compensation level without working an excessive number of hours per day, Esquel invested in improved productivity and efficiencies, to increase first-pass-yield and ensure high output rates. While these initiatives were beneficial for the workers, the improved FPY also helped Esquel to more accurately forecast working-hours requirements, and not to count on overtime as a cushion for unexpected low yield.

Finally, training sessions helped Esquel employees to develop their skills. Esquel believed that its people represented its single most important asset, and that training and development were key to having the most skilled, dynamic, creative and effective workforce. Esquel's training and development efforts focused on helping its employees to improve their technical skills as well while becoming better leaders and people managers throughout their stay with the company.

The people improvement plan that was in place as of 2005 consisted of developmental programs and initiatives for all levels of employees, in both technical skills and management development areas. These initiatives included:

(1) A Core Curriculum that aimed to promote employee development in the corporate cultural identity at all levels; appreciation of the company's business, its customers, competitors and the best practices of others; and a set of specific workplace competencies. Examples of programs include: Orientation Program; Lunch and Learn Series; Language Skills; Problem Solving Skills; and Total Cycle Time.

(2) Specific product knowledge / technical training such as the "Core Factory Technical & Product Training" program.

(3) Sales Training Program for the sales organization.

(4) Supervisory Skills Program, which consisted of a series of both technical and management skills for supervisors on the factory floors.

(5) Skill Certification Program aimed at helping workers to improve their sewing skills.

Esquel also implemented a Management Training Across Esquel program (MTAE) where a group of selected young talents were assigned to a rotational program in locations and functions across the company for a period of one year. The aim was to develop these individuals by giving them exposure to multiple functions and major activities across the company to build both depth and breath of experience as well as contacts in the matrix organization.

13.9. Future Perspective

Esquel prided itself on being the best-quality fabric producer in China. However, given the evolving marketplace and increasing customer expectations, Esquel could not stand still, and had to continue exploring new ways to further improve quality and maintain its leading position in the market.

The goal of Esquel was to double revenues by 2010, and to become a US$1 billion company. Achieving such growth was a big challenge for the company, and would require aggressive sales and marketing efforts, both in existing markets and in new geographic regions, in addition to product development initiatives and new innovations. In the U.S., consumers were looking for quality and were willing to pay for it, which was a good opportunity for Esquel. On the other hand, sales to Europe were more difficult, since the European market was much more fragmented, with smaller retailers. In addition, the European market had access to closer textile and garment supply bases in Eastern Europe and North Africa, whereas the longer distance from China to Europe increased order lead times and made the logistics operations more complicated and costly.

Achieving this growth target would also require Esquel to expand capacity (or alternatively to subcontract parts of its operations), take steps to match the marketing strategy with the execution, seek ways to improve cost control while maintaining high quality, acquire the right management capabilities, develop the appropriate supporting IT applications, and take aggressive steps to recruit additional people and to maintain synchronization between the people and the organization. It was expected though that the steps the company had been taking to improve productivity would make expansion plans somewhat easier to implement, since they would allow for a more gradual increase in the number of employees. In fact, a 30 percent increase in the number of employees was estimated to be sufficient to meet the goal of double sales. Similarly, the company explored ways to improve the efficiency of its equipment whenever possible, to allow for a more gradual capacity expansion.

While it was clear that Esquel would need to expand its product and service offerings to support growth, the company did not plan to enter new markets outside its established area of expertise in cotton shirts. Rather, Esquel expected 80 percent of the growth to come from existing core businesses. The remainder would come from non-core businesses, such as accessories and novelty yarn. To achieve that, Esquel might need to grow some parts of its product offering, such as shirts for ladies and children as well as accessories, and also to offer more merchandise services and further enhance customer relationship management. Esquel also planned to look for new innovations as a means for expanding its business with current and new customers. Esquel realized, however, that this would not be an easy task. For example, the market for ladies' shirts was less stable and predictable compared to men's shirts, due to a larger variety of styles and materials, frequent changes in fashion, and no consistent demand for basic items over the years.

The elimination of all quota restrictions on January 1, 2005 was expected to help the company to achieve this growth target, since it was likely to give Esquel more flexibility when deciding on the location of its production facilities, and might lead the company to expand production in China, which had so many quality and cost advantages. Still, executives at Esquel anticipated that other trade barriers might be in place to prevent China from dominating the market, such as safeguard measures and anti-dumping rules, which would prevent Chinese companies from selling products to the U.S. and Europe in conditions resembling a totally free market environment. In fact, this prediction started to materialize in 2005 when in May the Bush administration, reacting to a flood of Chinese clothing imports to the U.S. since the beginning of the year, announced that it would impose new quotas on cotton shirts, trousers and underwear from China (Becker, 2005). In parallel, the European Union had also been weighing new restrictions on Chinese textile and apparel goods (Barboza and Bradsher, 2005). In response, a few days later China declared it would raise its tariffs on many textile and apparel exports, in the hope of easing trade frictions with the U.S. and Europe (Barboza and Bradsher, 2005). Therefore, when making its expansion plans, it was critical for Esquel to weigh very carefully all the different options, considering these and other potential safeguard barriers and tariffs, as well as such issues as bilateral trade agreements, duty differentials, and quotas on locations outside China, such as Vietnam. Furthermore, Esquel would have to maintain flexibility to allow it to respond to changes in the environment as they took place. The company also had to keep in mind that cancellation of the quota might encourage more competition to enter China, which would require Esquel to further strengthen its capabilities. In summary, while the opportunity was clearly big, this was also a very challenging and risky time for Esquel, since it was not clear how the future marketplace and the relative positions of different supply bases would look, and making the wrong decision could have severe consequences.

Above all, Esquel and its chairperson, Marjorie Yang, emphasized the importance of achieving the growth goals without giving up their culture and values. Monetary return was not enough; their mission was to create a company where everybody was happy, and in parallel to bring about a positive change in society. Only that would bring them a real sense of satisfaction.

References

A.T. Kearney Technology Innovation Study (2004–2005). http://www.atkearney.com/shared_res/pdf/9029_ATK_Reportv15-pdf.pdf

Barboza David and Keith Bradsher (2005). "China to Raise Tariffs on Textile and Apparel Exports." *The New York Times*, May 21, 2005.

Becker Elizabeth (2005). "U.S. Moves to Limit Imports From China." *The New York Times*, May 14, 2005.

Blustein Paul (2004). "China Could Rule Textile Market After 2005." *The Washington Post*, August 12, 2004. http://www.washingtonpost.com/wp-dyn/articles/A58429-2004Aug11.html

Economy, Elisabeth C. (2004). "Congressional Testimony: China's Environmental Challenges." *Council on Foreign Relations*, 9/22/2004. http://www.cfr.org/pub7391/elizabeth_c_economy/congressional_testimony_chinas_environmental_challenges.php

Kahn, Gabriel (2004). "Esquel Weaves Tight Web." *The Wall Street Journal* (Europe), March 26, 2004.

Murphy, Cait (2003). "The Daughter Also Rises." *Fortune*, October 13, 2003.

Peleg-Gillai, Barchi (2005). "Esquel Group: Transforming into a Vertically-Integrated, Service-Oriented, Leading Manufacturer of Quality Cotton Apparel." *Stanford Global Supply Chain Management Forum Case*, SGSCMF-003-2005, Stanford University, 2005.

U.S. Customs and Border Protection (2004). *World Trade Organization (WTO) Quota Elimination – Overview*. http://www.customs.gov/xp/cgov/import/textiles_and_quotas/wto_quota_climination/wto_quota.xml

U.S.-China Business Council (2004). *US-China Trade Statistics and China's World Trade Statistics*, February 23, 2004. http://www.uschina.org/statistics/tradetable.html

Hau L. Lee and Chung-Yee Lee (Eds.)
*Building Supply Chain Excellence
in Emerging Economies*
© 2007 Springer Science + Business Media, LLC

Chapter 14

END-TO-END TRANSFORMATION IN THE CEMEX SUPPLY CHAIN[1]

David Hoyt and Hau L. Lee
*Graduate School of Business
Stanford University, USA*

Abstract: Operating a supply chain in an emerging economy is especially challenging, given the diverse needs of the customer bases, the evolving infrastructure, and the operational uncertainties in the field. But CEMEX, a company in a very traditional industry – cement – shows that it is possible to use information technologies, innovative processes, human resource and cultural developments to create a highly efficient supply chain. CEMEX was able to test their new concepts and practices in a very challenging supply and demand environment of Mexico, and eventually applied them worldwide. This chapter describes how CEMEX transformed its cement supply chain from end-to-end, leading to world class business performance.

14.1.   Introduction

Cement was first made in the early 1800s in England, and today is still in great demand in every nation as a construction material. Cement manufacturing is therefore an industry that is over 200 years old. It is an industry with many established, mature players, and is highly competitive. In this "old," traditional industry, CEMEX is a shining star. The Mexico-based company has enjoyed consistent growth over the years (see Exhibit 14.1). Although only the world's third largest cement manufacturer in volume in 2005, it was the most

---

[1] The authors gratefully acknowledge CEMEX's support in writing this chapter.

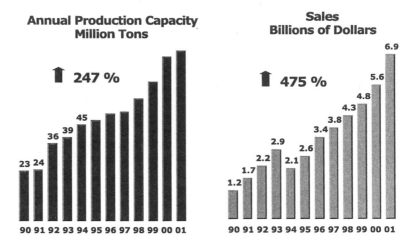

Exhibit 14.1. Growth of CEMEX Capacity and Sales.

profitable one. Its revenue in 2004 was only 40 percent that of the world's leading manufacturer, Lafarge; but its profit was more than 10 percent higher than that of Lafarge.

CEMEX is based in Mexico, and from humble beginnings, it became a global cement manufacturer, competing head to head with others that are based in countries with advanced, developed economies. Competitors pursued aggressive acquisition strategies to increase scale capacity, diversify geographical operations, and expand vertical integration. In 2005, the largest manufacturer, Lafarge S.A., was a French building materials firm with 40 percent of its sales in Western Europe. Lagarge had sales in 2004 of $19.6 billion, of which about half was from cement. Its net income from 2000 to 2004 ranged from 3.1 percent to 6.0 percent of sales.[2] Unlike CEMEX, which was built around a core business in cement, Lafarge had always been vertically integrated, incorporating other building materials such as gypsum and roofing. The second largest cement maker in 2004 was Holcim Ltd., a Swiss company with revenues of about $10.4 billion. Holcim's core business was cement, with cement and clinker accounted for about 70 percent of the company's sales. Like Lafarge, Holcim's margins were considerably below those of CEMEX, with net income in 2000 through 2004 ranging from 3.9 percent to 6.9 percent.[3] Exhibit 14.2 compares the world's leading cement manufacturers.

---

[2] Hoover's Online.

[3] Ibid.

Exhibit 14.2. Cement Company Competitors (2004 Data).

	CEMEX	CEMEX +RMC	Lafarge	Holcim
Revenues ($U.S. million)	8,149	>15,000	19,555	10,657
Net Income ($U.S. million)	1,307		1,176	737
Net Income (percent)	16.0%		6.0%	6.9%
Cement Production Capacity (million metric tons)	81.7	97	138	124
Cement Capacity Ranking	3	3	1	2
Compounded annual growth in EBITDA, 1990–2004	16%		12%	7%
Geographic concentration	Mexico: 36% of sales, 50% of EBITA; U.S. 25% of sales	Mexico: 17% of sales; U.S. 25 % of sales	About 50% of sales in Western Europe and North America	30% Europe, 24% Asia/Pacific, 22% Latin America
Cement as % of total revenues	73%	54%	47%	

[a] Sources: Hoover's Online, company websites, CEMEX, Imran Akram, Paul Roger, Daniel McGoey, "Global Cement Update," Deutche Bank, November 26, 2004.

CEMEX's profitability advantage could also be seen in its operating profit margins, at 20–30 percent of sales in 1999–2004, compared to about 20 percent for Lafarge and 15 percent for Holcim for the same period. Among many factors contributing to the success of CEMEX was its efficient supply chain. It operated in an industry not known for aggressiveness in technology investments, yet it invested heavily in information technology and people, and transformed its supply chain from end to end to support its cement business. The success story of CEMEX can provide us with many useful lessons. In this chapter, we describe the journey that CEMEX has gone through, and draw out these lessons.

14.2. The Cement Supply Chain

Cement is a key ingredient in concrete, one of the world's most important and versatile building materials. It is a mixture of limestone (the primary ingredient), sand, clay, and iron. About 1.6 tons of raw materials are required to produce one ton of cement, so factories are generally built near limestone mountains in order to minimize transportation costs.

Cement manufacturing requires heavy machinery and equipment, a large amount of heat and energy, and about 80 production steps. The raw materials, in particular limestone, are often quarried as rock, and are put through a series of crushing stages until the pieces are about 3 inches or less. They are then mixed in the appropriate proportions and heated to 2,700° F in huge cylindrical kilns – often as big as 12 feet in diameter and as long as a 40-story building is high. The kilns rotate, slightly inclined from the horizontal, and raw material is fed into the higher end. The product, marble-sized pieces called "clinker," is removed from the bottom.

Clinker is the common starting point for producing different types of cement, and is sometimes sold to cement companies for processing into the final product. To process clinker into cement, it is ground to a fine powder, and additives are introduced. The precise mixture of additives determines the performance specifications of the resulting cement. Gypsum is added to control the rate at which the cement hardens when mixed with water. The finished cement particles are so fine that one pound contains about 150 billion grains, and can pass through a sieve that is capable of holding water.[4]

The cement industry is capital intensive. A cement factory might cost $300–400 million due to the heavy machinery and equipment that are required.

Cement makes up 10–15 percent of concrete, which also contains aggregate particles. Water and cement form a paste that coats the surface of the aggregate particles, then hardens to form the rock-like mass of finished concrete. Concrete must be poured within about 45 minutes after the components are mixed. Concrete can be made directly at the construction site by mixing bagged or bulk cement with water and aggregates (sometimes with the help of a semiautomatic mixer), or it can be mixed at a factory and delivered to the construction site in rotating drum trucks (this is called "ready-mix").

The production and distribution of ready-mix concrete is an industry in itself. There are many independent ready-mix concrete companies who buy cement and aggregates from third parties, produce the concrete, and deliver it to construction sites. Some of these companies are vertically integrated, producing their own cement and aggregate.

Cement is sold in bulk and in bags. In Mexico, bagged cement was generally purchased by "self-constructors," people who did their own small construction projects – often building their own houses or room additions. Thus, cement usage in Mexico in 2005 was closely tied to housing construction. CEMEX also sold bulk cement to independent ready-mix cement companies as well as to its own ready-mix operation. CEMEX's ready-mix cement business accounted for more than 20 percent of the company's Mexican sales in 2003.

---

[4] Portland Cement Association, http://www.cement.org (6/2/04).

14.3. CEMEX's Corporate History

Founded in Mexico in 1906, CEMEX grew from a small local player to become one of the world's leading cement companies. With headquarters in Monterrey in the north of the country, CEMEX expanded throughout Mexico in the 1960s and 1970s through both acquisition and construction of new plants. The company went public in 1976.

In 1968, the grandson of one of the founders, Lorenzo Zambrano, joined the company. Also named Lorenzo Zambrano, he worked in the engineering organization, then as head of operations, before being appointed chief executive officer in 1985. In 1989, CEMEX acquired the country's second largest cement maker, thus becoming one of the ten largest cement producer in the world. [Note: CEMEX became the leader in Mexico in 1976 after acquiring Cementos Gudalajara]

In 1992, CEMEX expanded outside of Mexico, buying the two largest Spanish cement companies. In the next few years, CEMEX bought the largest cement company in Venezuela, cement companies in Panama, the Dominican Republic, Colombia, and a Texas plant – making it the world's third-largest cement maker in 1996. The company continued its geographic expansion, buying the second largest U.S. cement producer (Southdown) in 2000 as well as companies in Asia and elsewhere throughout the world. Overall, CEMEX acquired 16 companies (either wholly or in part) from the time Zambrano became CEO until the end of 2002, as well as purchasing additional production and distribution facilities.

By the end of 2003, CEMEX was the third largest cement producer in the world. Its 54 cement plants around the world had an annual capacity of about 81.5 million tons. It had 466 ready-mix plants, 191 distribution centers, and 60 maritime terminals.[5] The company employed over 25,900 people.[6]

In September 2004, the company announced its largest acquisition ever, the purchase of the British ready-mix company RMC Group p.l.c. for $5.8 billion. RMC was the world's largest ready-mix company, with CEMEX the sixth largest before the acquisition. The acquisition increased the company's geographic diversification, adding substantial business in Northern and Eastern Europe, and increasing its presence in the United States.[7] The acquisition of RMC was completed in March 2005.

---

[5] "Logistics in CEMEX," presentation to the Stanford Graduate School of Business, December 2003.

[6] http://www.cemex.com/oe/oe_lp.asp (June 9, 2004).

[7] Presentation by Lorenzo Sambrano, September 27, 2004. Available online at http://www.cemex.com/mc/mc_pr2004.asp (January 24, 2005).

Exhibit 14.3.   Selected CEMEX Financial Results (in $US million, except percentages).

	1999	2000	2001	2002	2003	2004
Net Sales	4,828	5,621	6,923	6,543	7,143	8,149
Operating Income	1,436	1,654	1,653	1,310	1,455	1,852
Operating Margin	29.7%	29.4%	23.9%	20.0%	20.3%	22.7%
EBITA	1,791	2,030	2,256	1,917	2,108	2,538
EBITA Margin	37.1%	36.1%	32.6%	29.3%	29.4%	31.1%
Net Income	973	999	1,178	520	629	1,307
Net Income %	20.2%	17.8%	17.0%	7.9%	8.8%	16.0%
Debt/Equity	68%	74%	66%	83%	87%	68%
Free Cash Flow	860	886	1,145	948	1,143	1,478

Sources: 2003 Annual Report, 2004 Q4 Report.

CEMEX was extremely successful in integrating its acquired companies by using technological and managerial processes that it developed and proved in Mexico to drive down costs, improve operational efficiency, improve customer service, and establish the CEMEX brands in a commodity market.

Financially, CEMEX posted impressive profits, even through difficult financial times. Through the second half of the 1990s, net profits were between 17.9 percent and 29.6 percent of revenues, before a cyclical decline in the construction sector led to decreased margins of 8.8 percent in 2002 and 8.9 percent in 2003, before increasing to 16 percent in 2004.[8] Over the 15-year period of 1987–2002, CEMEX grew sales by 23 percent compounded annually, and grew EBITDA by 21 percent compounded annually.[9] (See Exhibit 14.3 for financial information)

14.4.   The Roots of Culture and Technology Innovation at CEMEX

The use of information technology was central to CEMEX's growth through acquisition, and to its high profitability. When Zambrano became CEO in 1985, the company operated in a very traditional manner, with inefficient procedures designed to take advantage of low-cost labor. Information was not readily available – certainly not in a computerized manner that facilitated real-time understanding and analysis.

[8]   Hoover's Online. 2004 results from Q4 2004 Quarterly Report.
[9]   Comments by Lorenzo Zambrano at analyst's meeting, July 2–3, 2003. Online at: http://www.cemex.com/pdf/ir/03_LHZ_d.pdf (June 9, 2004).

At that time, CEMEX was still a regional company focused on northeastern Mexico. However, Mexican trade barriers were lowering, and CEMEX began to feel the pressure of large foreign companies. In order to compete over the long term, CEMEX had to grow, and to become extremely efficient. Zambrano also believed that improving and standardizing operations was necessary for the company's competitiveness – and that it was essential to integrating acquired companies and to conducting a profitable acquisition strategy.

In 1987, Zambrano hired CEMEX's first chief information officer. With new IT capabilities, Zambrano created a revolution within CEMEX. In the late 1980s, the company made an important step forward by improving their communications capability. The Mexican phone system was unreliable, and insufficient to network the company's operations. CEMEX invested in a satellite communications system that it called CEMEXNET. This enabled all the CEMEX facilities (then 11 cement factories) to communicate and be managed in a coordinated fashion instead of as a set of independent operations with no knowledge of what was happening at the other locations. Demand requirements could be balanced against the company's overall production capabilities, and fulfilled by the appropriate operation.

In 1990, CEMEX implemented an Executive Information System, requiring managers to input manufacturing information that could be viewed by other managers and used for decision-making. This system allowed Zambrano to make "virtual inspections" of all aspects of the operation, down to the operating performance of individual cement plants. When he had questions or found problems, he would call the responsible manager directly to discuss the issues.[10] The use of this system greatly improved operational transparency, as executives were immediately informed of activities throughout the organization, and increased responsiveness and accountability throughout the organization.

The company culture changed dramatically. Zambrano set an example for the company's management. He was among the rare CEOs in Mexico who used emails long before they were popular. He took a hands-on interest in every aspect of the business, using the database and network to gather information and contacting people throughout the organization directly with questions, rather than going through the chain of command. As the network developed, he could observe the operating conditions and performance of plants around the world from anywhere on his laptop computer. Throughout the organization, managers either changed their approach, or were replaced by those who would embrace the new initiatives.

---

[10] Chung, op. cit., p. 4.

14.5.   The CEMEX Supply Chain in Mexico

CEMEX was a highly international company by 2005. However, Mexican operations remained critical to the company's success. Mexico accounted for 36 percent of corporate sales, and 50 percent of EBITA in 2004.[11] Distributors sold between 60 and 80 percent of CEMEX's cement, but they were a highly fragmented group. In 2003, about 73 percent of CEMEX's Mexican cement sales were made through 5,900 distributors, the five largest of which combined to account for just 4 percent of sales.[12]

CEMEX had 15 cement plants in Mexico, 220 ready-mix plants, 79 distribution centers, 8 marine terminals, and 12 bases for CEMEX-owned trucks in 2005.[13] CEMEX managed an extensive logistics operation to move the large amounts of product from its cement plants to its distribution centers and other customers, as well as to move raw materials to its plants. The fleet of 5,000 trucks managed by CEMEX was one of the largest in the country. Most of the CEMEX trucks were contracted by CEMEX from independent operators and trucking companies, with just 250 owned by CEMEX. These trucks delivered about 4 million metric tons of material (products and raw materials) each month. CEMEX also managed the largest railroad cargo volume in the country (5,000 railcars, moving about 1 million metric tons per month), as well as about 750,000 metric tons of material by ship every month. Exhibit 14.4 shows the supply chain of CEMEX.

Supply chain management at CEMEX became increasingly challenging due: to the proliferation of product variety, and the increasing demand on the reliability and responsiveness of deliveries. Before an initiative to expand product offerings began in 1999, there were about 25 SKUs. In 1999 the number of items grew to about 800. By 2005, there were about 5,000, as more construction industry finishing products (such as floor tiles) were added.

A substantial amount of the bagged cement was picked up by customers at CEMEX distribution centers. These customers were small construction material distributors with their own trucks. CEMEX charged to deliver cement, so customers with their own trucks often preferred to pick up cement. These distributors might also be delivering other construction materials to their clients, so they could pick up cement at the CEMEX distribution center and then deliver it (together with other products the builder had purchased) to the construc-

---

[11] CEMEX 2004 Fourth Quarter Results, http://www.cemex.com/ic/PDF/2004/CXING404.pdf, (January 26, 3005).

[12] 2003 Form 20-F, pg. 21-22.

[13] The marine terminals centers were used for exports to Asia, the Caribbean, Central and South America, and the U.S.

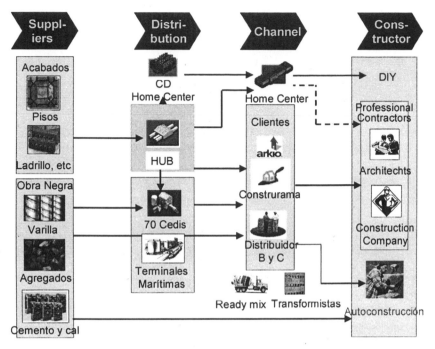

Exhibit 14.4. CEMEX Supply Chain.

tion site. The portion not picked up by customers was delivered to retailers by CEMEX.

Bulk cement was also shipped from the factory to the CEMEX distribution centers. Unlike bagged cement, almost all bulk cement was delivered to customers by CEMEX. Most bulk cement customers were ready-mix companies that had cement mixer trucks, but did not have the special trucks needed to transport bulk cement.

The logistics challenge was not limited to product shipments. Raw material and fuel shipment was also an important logistics issue. In fact, the transport cost of fuel for kilns was one of the factors that focused the company on logistics as an opportunity for increasing efficiency. In the late 1990s, CEMEX converted many of their kilns from fuel oil or natural gas power to pet coke due to fluctuations in oil prices. They then discovered that the cost of transporting pet coke (a solid) resulted in very different raw materials costs at each plant, depending on the distance it had to be transported – in some cases making the cost higher than it was using fuel oil or natural gas. Since a kiln used about eight tons of pet coke per hour, this represented a substantial cost.

14.6.   End-To-End Supply Chain Transformation

The culture of innovation and technology utilization established by Zambrano set the stage for a dramatic transformation of CEMEX's Mexican business in 2000–2005. The company had previously focused largely on its internal operations, with key indicators that reflected production efficiency. However, management realized that this was no longer sufficient, and that there were important opportunities to be gained by a shift of focus.

In the 1990s, while plants were becoming more efficient, there was often a line of trucks waiting at plants for several days to pick up cement (a situation that was common for all cement producers in Mexico). In times of high demand, order-takers often overcommitted by promising product that could not be delivered. Clients were angry, resulting in a loss of business. In 1997, the newly-appointed president of CEMEX Mexico, Francisco Garza, conducted an overall review of the company, along the lines used to evaluate new acquisitions. The result was a new focus on meeting the needs of customers, in addition to production efficiencies. Efforts at improving operations were continued, but many new initiatives had a customer focus.

14.6.1   Supplier Integration

Prior to the product line expansion, the inputs to CEMEX consisted primarily of fuel to fire the furnaces, and maintenance parts to support the manufacturing equipment. CEMEX had tight information linkages with a small number of suppliers of these parts and services.

However, CEMEX realized that it could use its national scope to negotiate favorable prices for large quantities of construction materials, and then sell these materials to its distributors at better prices than the distributors could negotiate directly. It therefore aggressively expanded the products that they offered to their customers, and by 2002, CEMEX was one of the largest rebar distributors in Mexico.

This change proved beneficial to other producers of building supplies, as CEMEX provided an efficient distribution system for their products. Steel producers were originally wary of CEMEX entering the rebar business, fearing that prices and margins would be depressed. However, they found that their businesses became more efficient, as they could deal with one large customer rather than a large number of very small customers, and the increased efficiency more than offset the discounts they gave to CEMEX.

Given the long lead times of some of these products from outside suppliers, CEMEX practiced the principle of postponement. It ordered in advance from the suppliers, but postponed the decision of where the products are to be shipped as late as possible, until concrete order information from customers was received.

14.6.2 Inventory Management for Builders

CEMEX realized that one problem facing building contractors was managing on-site inventory. A work stoppage due to a material shortage was expensive and delayed project completion. CEMEX began offering to manage inventory for builders, with an on-site warehouse for projects such as housing complexes. The material was owned by CEMEX until it taken by the builder. CEMEX did the materials planning, and updated plans according to the construction progress. This arrangement saved storage, working capital, and provided security for the constructor. It allowed builders to work with one supplier to plan their entire materials needs, and eliminated shortages. CEMEX did all the back office tracking and scheduling. In return for providing this added value, CEMEX received all the builders' business, and turned the purchasing decision away from price towards other value added services, including supply chain management.

14.6.3 Construction Solutions

One challenge faced by construction companies was a shortage of qualified labor. Some customers told CEMEX that they had to recruit qualified workers from other parts of Mexico and to fly them to their job sites. Builders also needed to accelerate their construction processes, so that they could collapse the time between making investments and collecting payments. CEMEX worked on systems to simplify construction so that labor was less critical, and so that construction projects could be completed more quickly. For instance, CEMEX developed a method that accelerated construction for companies building low-income housing developments. Using this method, construction firms cut their costs by about 18 percent, and reduced construction time by about half.

Innovative construction solutions also presented an important marketing opportunity. Consider, for instance, the construction of a new hotel. There were several levels of potential customers. First was the end-user, the person who would eventually own and operate the hotel. That person hired an architect (a second customer level) to develop the plans and hire contractors. One of the contractors would purchase cement (a third level). CEMEX could wait for a bid request from the contractor, at which point the purchase would be strictly a price competition. Alternatively, CEMEX could work with the owner and architect to help them find efficient ways to design and build the hotel. If that were done, then the eventual cement purchasing decision would not be based strictly on price, but would incorporate other value added components.

Efficient construction·was also a problem for self-constructors. Owners would buy concrete blocks and cement as they earned money, gradually building a wall, then a small room.[14] Eventually, they could build an entire small house. CEMEX-developed casts could be used by self-constructors to speed up their projects and to reduce costs.

14.6.4   Transforming Planning and Logistics Management

In the 1990s, customers (generally CEMEX distributors, or CEMEX-contracted trucks which would take cement to distributors) had to wait in line for days at a cement plant to pick up cement. While this was extremely inefficient, it was standard in the cement industry in Mexico at the time.

Recognizing the importance of logistics, in 1994 CEMEX created a logistics department that reported directly to the president of CEMEX Mexico. In late 1999, the newly-appointed vice president of logistics for CEMEX Mexico, Jesus Lopez, realized that these initiatives had to be supported by structural changes to the supply chain, so that customers could have the products they wanted at the time and location they needed them. In 2000, the company implemented the first stage of a sophisticated, computerized, planning process to optimize the output from each plant based on where the product was needed. By 2005, CEMEX Mexico had implemented powerful tools that provided visibility of product shortages and potential shortages, and was developing additional tools to provide control and visibility throughout the supply chain. There were no long lines at cement plants anymore, and CEMEX had successfully implemented a policy of delivery within 24 hours of receiving an order. Details of these tools are described in the next section.

Sophisticated tools were only useful if employees knew how to use them properly to improve business performance. To ensure that employees were thoroughly trained, CEMEX established an intranet-based "Logistics University." The university was started in late 2002. Its mission was to identify knowledge that resided within the company, transmit it to others that needed it, and make it possible for people to use this knowledge in their jobs. As a result, they developed all training courses in-house. About 75 percent of courses were developed by people in the logistics group, with the balance developed by the staff of Logistics University (which consisted of just two people). Course modules were presented in a consistent format, and designed to be completed in

---

[14] Low income housing was traditionally done using block-to-block construction, in which concrete blocks were assembled into walls, and concrete (made using bagged cement) was used to join and bind the blocks.

an hour or less – the basic idea being "in one hour we will teach something that you can use in your job." Courses dealt with policies, logistics processes, management and administration, inventory, transportation, systems, and organization.

By 2005, all 700 people in the logistics organization were registered in Logistics University, along with 1,300 people from other parts of the CEMEX organization. When a person registered, he/she was assigned a curriculum based on their job. When the person logged into the Logistics University website, their courses were color-coded based on whether they had been completed, were in process, or had not been started. Since new courses were continually being developed, there were always additional courses to be taken. As part of their annual performance and appraisal program, each employee was required to take a certain number of courses, and many employees were required to develop training programs.

14.7. Supply Chain Integration with Information Technology

The transformation initiatives described above require extensive data integration and application of advanced planning application software. In this section, we describe some of information technologies deployed by CEMEX. A summary of major programs is given in Exhibit 14.5.

14.7.1 SCS Tactician

SCS Tactician was a product from i2 that provided long-term visibility of the cement production process and delivery requirements throughout Mexico. Input came from a statistical analysis of sales from an i2 demand planning software system, with additional management input, such as plant maintenance schedules, freight costs, and planned product promotions. The output provided information on how much production was needed from each plant (and kiln) at what time, and where it needed to go. The output was also used to plan logistics requirements, such as the number of trucks or rail cars needed at a particular time in a specific place.

SCS Tactician estimated the costs of closing a plant or a distribution center, and could optimize production and distribution. The tool provided "what-if" scenario planning, allowing evaluation of when kilns needed to be brought online or could be taken offline. If a kiln broke down, this tool could be used to evaluate alternative places to produce the needed product, and reallocate production and distribution from plants to distribution centers. This capability

Exhibit 14.5.   Summary of Major Logistics IT Programs.

Program	Status in mid-2005	Description	Benefits
SCS Tactician (from i2)	Operating	Model needs, demand, resources, and constraints, optimizing supply chain by week.	Provided short, medium, and long-term visibility of product and resource needs. Allowed scenario testing. Reduced distribution costs.
Daily Load System (Sistema de Cargas Diarias, "SICADI")	Operating	Developed an hourly production and delivery plan for each plant, using input from SCS Tactician, with real-time customer orders and commitments.	Ensured commitments to customers were based on accurate availability data. Optimized resource utilization. Assured timely replenishment of inventories based on actual need.
Exception Handling Administrative System ("SAPE")	Operating	Provided alarms for stockouts and potential stockouts.	Filtered data so that items needing attention were highlighted. Allowed proactive steps to be taken to reduce inventory shortages. Enabled rapid response to stockouts.
Intelligent Multi-Agent Transportation Organizer	Rolling Out	Real-time transportation organizer, optimizing and monitoring status of all delivery vehicles.	Facilitated efficient fleet management. Improved customer service. Saved transportation costs.
Vehicle Visibility System	Pilot Testing	Enabled real-time monitoring of vehicle location, and communication with driver.	Improved fleet administration.
Delivery Visibility System	Testing	Loaded delivery documentation on drivers' PDAs, for acceptance by customer and transmission to CEMEX.	Sped invoice process. Improved communication between driver and base.
Delivery Metrics	Rolling Out	Automatically identified late deliveries. These were categorized by administrators, and performance indicators calculated.	Provided measurement of overall performance in meeting delivery commitments. Provided information on reasons for late deliveries.

(Note: all of these programs were Intranet based.)

provided about $20 million per year in savings. Each week management used the system to make decisions that could save $500,000 to $1 million.

In 2005, CEMEX was adding the capability to include raw materials into the analysis – how much of each is needed, where is it needed, and when. By 2005 this tool had been implemented in Colombia, Venezuela, and was planned for introduction in Europe.

14.7.2   SICADI (Demand Fulfillment Tool)

SICADI (Sistema de Cargas Diarias, or Daily Load System) was a tool developed in-house. Its input was the weekly plan from SCS Tactician, from which SICADI developed a daily plan. The daily plan was further broken down by SICADI into an hourly plan for each plant for both production and delivery. All new orders in Mexico were sent to Monterrey to be processed through SICADI, which enabled order processors to make commitments based on actual availability, and for production and delivery requirements to be updated in real time.

SICADI showed the available capacity of each type of transportation method (truck and rail) by hour, and the committed or planned shipments. The system also maintained a real-time status of the inventory in each distribution center. When a customer called to place an order, the call center operator knew what material was available, as well as the transport capability, hour-by-hour. A commitment could then be made to the customer, and fed back into the plan.

CEMEX started using SICADI in one plant in 2003. At that time there were 100–120 trucks waiting at the plant, and plant output was highly variable. By 2005, there were at most 10 trucks waiting. Trucks were making more trips, and were more efficient, as they were moving product, not sitting in line. More efficient use of trucks also enabled CEMEX to negotiate a reduced freight rate. The full roll out of SICADI in Mexico greatly improved the balance of inventory among distribution centers. It also increased the use of rail, and reduced the use of trucks, which are more expensive than rail. Overall, the program provided an estimated $1.5 million/year in savings. It won the CEMEX innovation award. SICADI is also used in South America.

14.7.3   Order-Taking Process

The order-taking process started with a phone call from the customer (usually a CEMEX distributor). During the 4-5 minute call, the order-taker checked the available resources – product availability, dispatch capacity, and transportation capacity. The terms were validated against the buyer's credit status (purchase limits, credit limits, payment terms, etc.) using a customer credit validation program.

The order-taker consulted a screen showing the hourly availability of each product at each plant. The purchase order number was generated in the order entry system and pasted into SICADI. The customer and order information was then transferred automatically to SICADI. The delivery date and time was assigned, and transportation confirmed.

In 2005, CEMEX averaged about 3,000 orders daily. The standard offer was 24 hours delivery for cement, 48 for cement, rebar and steel, and 72 hours for other products. The company tried to schedule within these periods, and work to meet the client needs, using SICADI to give an actual commitment. The standard delivery of cement from competitors was over 72 hours, compared to the CEMEX 24 hour commitment.

About 59 percent of orders were delivered by CEMEX in 2005, the rest picked up by the distributors. This was a substantial increase in deliveries, as in 2002 only 25 percent of orders were delivered. CEMEX deliveries increased as clients (distributors) realized they should use their own trucks to deliver to the end users, and use CEMEX to replenish inventories.

14.7.4   IMATO (Intelligent Multi Agent Transportation Optimization)

IMATO was developed to help execution of the transportation plan. It was developed in-house, and was in the early stages of rollout in mid-2005. The system was used to manage the truck fleet so as to optimize the types of trucks (dump trucks, flat beds, etc.), drivers, and product types, within transport restrictions such as vehicle operating hours. IMATO was a dynamic system. In 2005 it was being used for bulk cement, but there were plans to use it for other products. It was designed to be agent run, checking availability and commitments, and suggesting changes that might be needed to meet customer requirements. A customer representative could advise a client if a delivery is delayed, and make commitments.

IMATO could display the transportation resources available to a plant – the driver, truck type, capacity, required rest times, when the truck is available (working hours), etc. There might be multiple lines on the computer display for a truck in the event that the truck was used by more than one driver. It also could display information related to each truck in transit, at the plant, and not available. In transit information included the truck number, destination, customer order number, when it was loaded, and when it was expected back. Truck information at the plant included truck number, status (loading or waiting), customer name and order number (for waiting trucks, this was highlighted and could be changed by the dispatcher until loading began), and destination. For trucks not available, the displayed information included whether the truck was undergoing maintenance, the driver was unavailable, or if the truck owned by a contractor was used for another of the trucking firm's customers.

The truck schedules could also be displayed graphically, with colored bars showing the status of each truck at each time of the day. This allowed CEMEX to identify slow periods, and when there was peak demand. For instance, most customers wanted their deliveries first thing in the morning (9AM), but there was no work for trucks in the very early morning, or in the afternoon. Using this information, CEMEX could talk to customers about taking deliveries at off-peak times. Drivers had previously changed shifts at the same time – and at rush hour – a practice that turned out to be very inefficient, hurting on-time deliveries. If there was a delay while lots of trucks were waiting to be loaded, the drivers socialized with each other while they waited. Staggering truck schedules greatly increased efficiency. Also, CEMEX had previously washed all its trucks at the same time. Data from IMATO caused the company to change to a staggered schedule so that deliveries could be spaced out and trucks used more efficiently.

IMATO become a best practice driver, highlighting times when trucks were more available, staggering shifts, and staggering planned load times. This had a large impact on delivery commitments. It provided information that allowed CEMEX to have the right vehicle at the right place, improving customer service, proper supply of the distribution centers, and fleet efficiency.

14.7.5 SAPE (Exception Handling Administrative System)

SAPE was introduced in 2004 to identify and manage exceptions in the supply chain. It was a real-time tool that constantly monitored all information in the system and created a hierarchy of information for each person. SAPE took data from SICADI (client order and commitment time, inventory levels and historical demand), and analyzed deliveries, stockouts, and inventory at risk of becoming a shortage. All operations personnel in Mexico had access to SAPE.

SAPE could display the map of Mexico, and the user could select items relevant to him/her by region, product, etc. One of the key measures monitored was stockouts, which were shown as red dots on the map. Potential stockouts were shown in yellow. SAPE could detect a usage trend that suggested a problem. For instance, the historical usage pattern might indicate that the inventory level at a certain distribution center was insufficient, and would result in a shortage if not replenished. SAPE saw that there is no material in transit, so the system signaled an alarm, using a yellow dot. If no action was taken to solve the problem, a stockout would occur and the dot turned red. The system also showed the availability of products that could be substituted for out-of-stock items. Another measure was late customer delivery, where the causes of the late delivery are also tracked.

14.7.6   Vehicle Visibility System

The IMATO system enabled CEMEX to know when a truck left a plant, and when it was due to return, but did not provide a way to determine the vehicle's status while it was gone. In 2005, CEMEX began implementing a tool that provided real time monitoring of the location of each vehicle using GPS and cellular phone technology. It consisted of a catalog of vehicles, locations, and "geofences" (defined areas such as warehouses, customer locations, and toll roads).

Earlier, the company had developed a GPS-based tool to manage its ready-mix fleet. This tool originally used the satellite communications system, although it was later updated to use more efficient and less costly cell technology. The program for managing ready-mix trucks had enabled CEMEX to greatly improve the delivery performance of these trucks.

Once the vehicle visibility system was operating, CEMEX would be able to know if there was a problem with a truck, contact the customer if a truck would be late, and reschedule loading if a truck was expected to return to the plant late. CEMEX planned for the GPS data to be an input to IMATO, so that it could incorporate this information into IMATO's dynamic planning process.

This system was pilot tested in Monterrey in mid-2005, with plans to roll it out to the entire fleet by the end of the year. In order to accomplish this, however, CEMEX had to sell four of its most important trucking companies on the program. The trucking companies would need to buy the GPS, but CEMEX would provide the IT platform for them to use without charge. The system was designed so that trucking companies could use it for non-CEMEX business. The reports from the system could help make their fleets more efficient. For example, they could use geofences to not only identify destinations, but also to establish forbidden areas where vehicles should not go. In some cases, drivers detoured around toll-roads, but submitted the tolls for reimbursement using fake receipts. This wasted both time and money. Geofences could be established around the toll-roads to prevent detouring.

The system could also be used to redirect trucks in the event of a canceled order. Vehicle location could be seen, and the truck sent to another customer.

The system had an event window, which recorded every time that a vehicle entered or left a geofence, and when PDA (or email, voice mail, or cell phone) messages were sent and received. It could be set it up to provide alarms when a specific truck entered or left specific geofences (or whatever other events were defined for notification). Since the GPS position was reported every three minutes, the base could also determine vehicle speed, helping to identify traffic problems and unsafe driving.

In addition to integrating the system with IMATO, there were plans to use an artificial intelligence system to identify problems, and to incorporate traffic condition information.

14.7.7   Delivery Visibility

Another problem facing CEMEX was the timely and accurate preparation of invoices. The system in place in early 2005 required that the bill of lading signed by the customer upon receipt of a delivery be sent back to CEMEX by the contract trucking company. This often took many days. The signed bill of lading started the invoice and bill collection process. The bill of lading went to the invoicing group, who prepared an invoice, which is sent to the sales person, who took it to the client, who challenged one or more items on the invoice, which then needed to get corrected. The payment term did not start until the invoice was accepted by the customer – and often the customer did not accept the initial invoice, claiming that the delivery data was incorrect. If the terms were 30 days, it often took more than 20 days to get the invoice generated, resulting in payment 50 days after delivery.

To address this problem, CEMEX developed a tool to provide a paperless bill of lading and customer receiving document system. The system used the same GPS and PDA technology that was used in the vehicle visibility tool. When the truck went to the plant or distribution center for loading, the loader uploaded an electronic bill of lading to the driver's PDA, showing what was actually loaded and providing customer information. Any information about the customer that the driver should know was also loaded – for instance previous complaints or service issues – so that the driver could follow up if needed.

During transit, the PDA terminal screen could be used for communication with the base. When the truck arrived at the customer's site, both the customer and driver went through the receiving process on the PDA, verifying all the information, such as customer name, billing information, and each line item and quantity. The system was extremely easy to use, which was necessary as most drivers had little formal education.

If there were missing items, or broken cement bags, this could be noted in the electronic receiving documents. The PDA also told the driver if he needed to pick up any papers from the customer. After going through the entire delivery, line-by-line, and making any needed adjustments, the customer signed the PDA, agreeing that he had received the order. This was immediately sent to CEMEX via the cellular network. CEMEX planned to include a printer, so that drivers could leave a hardcopy of the receiving document with the customer. With the electronically accepted bill of lading, CEMEX did not have to wait before generating the invoice, and the collection process was greatly streamlined.

14.7.8 Delivery Metrics

In mid-2005, CEMEX began using a tool to measure the overall effectiveness of the order fulfillment process. The tool introduced in 2005 was designed to measure on-time delivery, but more importantly, to help understand the reasons for late delivery – and to provide information upon which the company could take action. The system provided classifications of reasons for late deliveries. Sometimes, deliveries could not be made on time because of customer problems (such as late payment, in which case the shipment was not even made until the credit problem was resolved). Other problems included lack of trucks to make delivery, stockouts, failure of the customer to pick up an order, or backorders of some line items.

The signed delivery receipt was entered into the CEMEX ERP system. That evening, the metrics system identified all line items that were not delivered on time. The time that the customer accepted the order was used as the delivery time for determining late delivery.

The administrator then reviewed the list, and chose the reason from a menu of 12 causes. If this was not done within 7 days, the system automatically assigned "unknown cause," since by then the administrator probably did not know why the delivery was late. This information went into a data base.

Problems could be analyzed by time window, product type, region, and customer type. Reports showed line items delivered within the time committed to the customer, not delivered on time, total items delivered, and percent delivered on time. Performance could also be analyzed according to items delivered within the standard CEMEX offer (24 hours for cement). The system also provided metrics for the percentage of perfect orders (all line items in an order delivered complete and on time). For line items not delivered on time, the reasons for each line item delay could be reviewed.

This information was used to identify opportunities for improvement. It was considered to be a "thermometer of all tools," since delivering the right products to the customers, on time, was the final result of all the other tools – planning, sales, logistics, and operations

14.8. Results and Lessons from Transformation

The customer-focused programs and the associated information systems deployment delivered great results. As an ultimate measure of performance, delivery to customers within the promised time, which had been just 30 percent in 2000, was greater than 90 percent in 2005. This enabled distributors to keep just 2–3 days of cement inventory in their stores, compared with the more than 30 days of inventory they kept for other products. A delivery metrics program automatically monitored late deliveries (both compared to the standard delivery time, and to customer commitments), and allowed reasons for late deliveries to be categorized and evaluated. Once proven in Mexico, these tools were then rolled out to other CEMEX locations.

We can draw a number of lessons from the supply chain transformation of CEMEX. Despite a very old and traditional industry, it shows that a company that invests in technology and people, can still make significant transformation, leading to admirable business results. The transformation started from Mexico, a country that is not as economically developed as others in North America and Western Europe. Yet Mexico has provided CEMEX as the origin of many of their innovative programs. The new initiatives were tested and refined in Mexico, and they were then successfully rolled out to the other part of the world at CEMEX. The first key lesson is that supply chain innovation does not always have to come from well-developed economies or in fast growing and technology-based industries. Opportunities are everywhere, and there is no boundary for us to explore them.

Why would Mexico be a good testing bed for CEMEX to experiment with new innovations? It was actually not just a choice that CEMEX could make, but a necessity. The Mexico cement market was highly fragmented, with customers having very diverse needs. This meant that CEMEX had to be capable of creating solutions that served a wide range of customers with very different requirements. The Mexican country thus contained segments that were highly sophisticated, as in well developed economies, but also segments that were more primitive, as in developing economies. It was therefore a rich environment for CEMEX to test innovations for diverse global markets. The requirement for highly reliable and timely customer service in Mexico was also among the most stringent in the global market. Hence, it was a good location for CEMEX to engineer the best service solution that they could come up with. Finally, since CEMEX was headquartered in Mexico, it was more effective to use Mexico as a test bed to breed innovations, as close collaborations and communications among the customers, the service providers, the information technology developers, and engineering resources at CEMEX, could be easily facilitated.

Transformation in an emerging economy requires customization of the specific solutions that fit the local needs. The complex distribution system of Mexico, the fragmentation of the channel, and the increasing requirements for delivery reliability by customers, dictated the specific forms of the logistics structure and information systems used by CEMEX. Neither direct copying of industry best practices or readily available commercial solutions are adequate, and CEMEX had to create its unique path that is tailored for the special needs and environments in Mexico.

CEMEX engaged in productivity improvement projects that gave them superior efficiencies and a production cost advantage. But the company recognized that there was a limit to which they could continuously cut cost and increase efficiencies. They correctly recognized that cement was not just a product-based industry, but a service-based one. It was in the provision of customer service that they refocused their company strategy. CEMEX successfully changed the focus of the competition from product to service. As a result, the programs to manage inventory for builders, collaborate with customers by providing construction solutions, give highly precise reliability to deliveries to customers, and provide information visibility to customers, were all service-based strategies.

At the heart of all these innovations lies their information system and associated technologies. CEMEX used a combination of in-house and purchased software solutions to support the planning and coordination of their logistics operations. It stressed giving real-time information visibility to decision makers that matter. Armed with real-time and relevant data, decision makers could dynamically adapt their decisions so that the logistics flows were not disrupted as well as efficient. The integration of information started with the supply side – the plants, through distribution to the customers. Transportation providers and channel partners all had access to the fully integrated information system. CEMEX was, therefore, a data-rich company.

Real-time information availability enabled CEMEX to be a "Sense and respond" company. It could sense the latest heartbeat of every part of the supply chain and detect potential problems early. But it could also respond rapidly by enabling the right parties to act promptly. The culture that the company instilled throughout the supply chain was one of continuously asking questions and seeking answers. An action-oriented culture enabled the company to be agile and respond quickly to potential problems.

The end-to-end transformation at CEMEX started with suppliers, but was extended all the way to the end of the supply chain – the end-customers. The transformation required the full participation of all supply chain partners. The company's strategy changed from a product-based to service-based focus. It involved the creation of a customer-oriented corporate culture and mission, and was fully supported by an extensive deployment of information technologies. The investment efforts were huge, but the payoffs were even greater.

Hau L. Lee and Chung-Yee Lee (Eds.)
*Building Supply Chain Excellence
in Emerging Economies*
© 2007 Springer Science + Business Media, LLC

Chapter 15

THE IDS STORY
Reinventing Distribution Through Value-Chain Logistics

Ben Chang and Joseph Phi
The IDS Group

Abstract: The increasingly cutthroat competitive environment today has led to an unprecedented level of business failures in Asia and around the world. Keeping
one's business model current is paramount to organizational survival. Making
one's service offering relevant is key to gaining competitive advantage.

During the past couple of decades, we have witnessed the rise and fall of the
distribution business in Asia. The major players in the industry have been engaged in head-to-head competition resulting in an industry-wide bloodbath and
leading to margin erosion and profit shrinkage. We happen to be in the thick
of this environment. Upon Li & Fung Group's acquisition of the Asia-Pacific
marketing services business of Inchcape PLC in 1999, we spent the last six
years analyzing the typical relationship between a distributor and its principals,
suppliers and customers. We studied their revenue and profit patterns. We took
note of what worked and what didn't.

Our conclusion suggests the need to develop a non-traditional approach focusing less on direct "buy/sell" competition but more on service "value" generation. Upon working through the various iterations and simulating different business models, we have formulated the framework for Reinventing Distribution
in Asia. We call this Value-Chain Logistics.

Value-Chain Logistics provides an illuminating perspective on how to innova-
tively use Logistics as the fundamental enabler to seamlessly connect with our
two other core businesses of Marketing and Manufacturing. This is the story of
the IDS Group, its creation and our successful launch on the Hong Kong Stock
Exchange in December 2004. And this is a story that's just beginning to unfold.

15.1.   The Li & Fung Story Of Export Trading

15.1.1   Origins And Success

Before the Li & Fung Group's trading subsidiary, Li & Fung Ltd. (a Hong Kong listed company and a main constituent of the Hang Seng Index) perfected its supply chain management process, foreign organizations sourcing goods from Asia had to establish buying offices in the locations they do business. Costly to maintain and cumbersome to operate, such buying offices worked directly with different manufacturers, entered into multiple agreements, and arranged various consignments, as appropriate.

The alternative was to deal with Asian trading companies who operate as middlemen between the buyer and the manufacturers. While this eliminated the need for setting up one's own buying offices, the process was opaque and inflexible, as oftentimes the buyer was kept in the dark on exact order status. Compounding the problem is the fact that most trading companies operate on razor thin margins, thus unable to invest on infrastructure and technology to improve service level.

Founded in 1906, Li & Fung was a traditional Asian trading company until Dr. Victor Fung and his brother Dr. William Fung began executing their supply chain management model in the 1980's. This ultimately led to the total revamp of the export trading business. With around US$6 billion of annual revenues in 2004, Li & Fung is Asia's leading garments and hard goods export trading company, managing the supply chains for retailers and brand owners around the globe. In the next three years, Li & Fung is poised to exceed US$10 billion in annual revenues.

15.1.2   The Li & Fung Supply Chain Management Model

In its supply chain management model, Li & Fung "orchestrates" the supply chain by providing its customers the convenience of a one-stop integrated service through a Total Value-Added Package: from product design and development, through raw material and factory sourcing, production planning and management, quality assurance and export documentation to shipping consolidation. With a proven track record in product quality, quick response capability and cost effectiveness, Li & Fung offers foreign buyers a one-stop, optimal and customized sourcing solution, drawing from its extensive network of 72 offices and over 3,000 supplier factories located in 41 countries throughout the world.

Driven by the success of the Li & Fung supply chain management model in export trading, Dr. Victor Fung together with Ben Chang and their team at Li &

Fung set out to create an entirely new business model for distribution throughout Asia. They set their sights on another antiquated model that once dominated Asian trade – the import distribution of Western branded Fast-Moving Consumer Goods (FMCG) and Healthcare products in Asia.

15.2. The Emergence Of IDS Group In Asian Import Distribution

15.2.1 The Inchcape Opportunity

Import distribution of FMCG and Healthcare products in Asia had been a very profitable business for over a century. Its origin could be traced back to the late 1800's when European distributors facilitated the entry of Western brands into Asia. Soon, a multitude of local distributors mushroomed across Asia with brand agency rights to access the much-coveted Asian consumer base.

The distribution industry flourished. But before long, the world economic order changed. Commencing in the 1980's, distributors saw their margins erode, as they had to grapple with the increasing complexity of the retail trade. Fundamental changes in the industry caused many Asian importers to refocus their businesses and some began to divest their distribution arms.

One such company that decided to sell its Asian distribution business was the marketing services arm of the Inchcape Group, a London PLC company. Founded in 1874 by James Lyle Mackay, the first Earl of Inchcape, the company engaged in various businesses from the trading of tea, sugar and textiles to insurance, banking and shipping. Mackay's grandson, the third Lord Inchcape, rationalized the disparate businesses into a series of complementary activities, and brought them together into a holding company. That company, Inchcape & Co Limited, the forerunner of today's Inchcape Group, was floated on the London Stock Exchange in 1958.

For nearly forty years, the company grew rapidly. However, by mid 1980's, market conditions changed and the yields from distribution began to decline. At that time, Inchcape decided to divest from many of its conventional distribution businesses to focus on international automotive distribution, making available for sale its Asian marketing service companies Inchcape Marketing Services Ltd (IMS), a Singapore public-listed company, and Inchcape Marketing Asia-Pacific (IMAP), a wholly owned Inchcape PLC company. IMS & IMAP were companies with a solid reputation and well known in the trades in which they operated. Collectively they had an illustrious multinational client list and a commendable industry footprint in Asia.

15.2.2   Enter IDS

Dr. Victor Fung is a visionary, a business contrarian who thrives on taking on the most arduous and complex business challenges.

"William and I have always felt that our trading picture is not complete until we really could have both the export trading and import distribution side," Victor had remarked. He could see how the Fungs' invaluable experience of transforming Li & Fung Ltd could be applied to the re-modeling of the Asian distribution industry. He invited Ben Chang to join Li & Fung. Along with Jeremy Hobbins, then CEO of IMS & IMAP, he set out to turn his vision into reality. Ben Chang brought invaluable experience to the team from his 12-year stint with the HAVI Group, the Chicago-based company which had played a crucial role in building the logistics and supply-chain infrastructure for McDonald's in Asia. With Chang on board, Victor then led a group of institutional investors in the buy-out of the Inchcape companies. Chang reminisced: "Victor was driven by the firm belief that we could turn Inchcape's hodgepodge of Asian assets into a regional distribution powerhouse. Nothing could be more exciting than taking an old business that was broken and re-making it in a totally different way." Chang, in the meantime, had convinced Joseph Phi, a leading practitioner in the Asian logistics and supply chain industry, to champion the Logistics business at IDS. Together, the senior management team of Victor, Chang, Hobbins and Phi, aided by the wise counsel of William and backed by the consortium of institutional investors, set the stage for another Li & Fung-led rebirth of a new business model for trade in Asia – the emergence of the IDS Group in Asian import distribution.

15.3.   The Challenge Of Traditional Distribution

15.3.1   Traditional Distribution – A Supply-Driven Model

In Asia, "Distribution" has chiefly come to mean the sales and marketing of brands. The traditional Asian distribution system had started and evolved in a marketplace where demand outstripped supply. During the good old days, all a Western manufacturer had to do was to ascertain origins of supply and extend penetration into targeted localities. This in itself had unfailingly created a market presence. Demand was given. In such a world, both distributor and manufacturer enjoyed hefty profit margins.

In traditional distribution, distributors essentially operate as brand principals in markets where they have distribution rights. They purchase products from the manufacturer/brand owner, undertake all marketing and merchandising functions, organize storage and transportation to retail outlets, and perform credit checks and cash collection. By operating as a brand principal, the local distributor also assumes all the trading risks including product quality, inventory obsolescence, discounts, rebates and credit risks. It is also obliged to finance heavy working capital driven by high inventories and long credit terms. In the past this had not been a problem as margins were high, and competition fairly mild.

The economics of the supply-driven model of traditional distribution worked perfectly as long as consumers had limited choices and supply of Western goods was scarce. In a world where the brand manufacturer was king, distributors developed a tendency to hoard as much inventory as they could, because the more products they got, the more they could sell and the more profit they made. Too bad the rules of engagement changed. The practices of distribution that were the cornerstone of success in a supply-driven world would slowly but surely lead to the demise of many distributors.

15.3.2 Modern Consumerism – A Demand-Driven Model

Today, modern market dynamics are radically different. We live in an over-supplied world where the consumer is king. For every consumer need, there is a multitude of choices. Beginning in the early 1990's, there was an unprecedented proliferation of Asian and Western brands that were of improved quality and increased sophistication. Companies were attracted by the rising tide of consumerism in Asia. Add in the rise of the modern trade, hypermarkets and category killer retail chains who competed head on against incumbents with in-house brands, and you have a completely new game and new rules to deal with.

Increasingly, brand owners were forced to respond to intense competition by lowering their generally generous distribution margins. Meanwhile, distributors were also being squeezed by the power of modern trade retailers who had started to dictate terms to brand principals. Well aware of their power in the market, retail chains demanded listing fees, prominent displays and higher service levels, thus putting pressure on distributors to spend more on promotion and merchandising, and invest more to improve operations effectiveness.

This, of course, didn't happen overnight. Rather, there has been a gradual change where the economics of consumerism evolved from a supply-push to a demand-pull market environment. Today, Asian distributors have become less and less relevant because they have not repositioned themselves to cater to a demand-driven economic model. What was once a thriving business has now degenerated into a state of moribundity.

15.3.3   Asian Supply Chains – A Broken System

To make matters worse, distributors' margins were further eroded by the antiquated and creaky supply-driven infrastructure that still dominates much of Asian distribution today. The supply chain was essentially designed based on a supply-push model where the manufacturer basically "channels" its products to market. This model was characterized by long order lead-times and slow market response. Typically the manufacturer produces in bulk to take advantage of low cost and pushes inventory onto the distributor, unaware and unconcerned about the hidden costs and inefficiencies created in the supply chain.

The inefficiency of the system is exacerbated by retailers who are hesitant to share sales and consumer data with distributors. The dynamics are such that every party works on their own silo and seeks to profit at the expense of everyone else. Without such information, distributors are forced to maintain costly buffer inventory as a cushion or risk their product supply drying up on the store shelves. The lack of visibility about consumer behavior causes distribution inefficiency and lost sales.

As if these challenges weren't enough, distributors seeking economies of scale through Asia-wide distribution face enormous difficulties. Asia is a region of enormously diverse languages, cultures, currencies, regulations, taxes, business practices, organizational forms and economic development levels. Attempting to optimize the supply chain seems to be an effort in futility.

The geopolitical realities of Asia mean that distributors must approach each market separately, and establish a distinct supply chain for each country. Each market features a different infrastructure, language, trade conventions and relationships. It is no wonder that most distributors that exist today are single-country entities. For the big players who have operations in multiple countries, they manage in country silos with little or no coordinated regional strategy.

15.3.4   In Adversity Lies Opportunity

The consequences of the outdated, supply-driven supply chains of Asia are manifold and debilitating for retailers, brand owners and distributors. It is estimated that the supply chain cost of bringing consumer products to market in Asia is between 8% and 12% of product costs, twice that of the US and the Western world. These costs of "un-coordination" stem from huge inventories, high costs of services, and process inefficiencies. Even more costly are product mark-downs, write-offs and product obsolescence. The biggest impact to business is lost sales resulting from out-of-stock situations.

Distribution had become synonymous with being a sunset industry. It was a middleman's role that had become irrelevant characterized by razor-thin margins, high working capital, huge risks and little or no profits.

Déjà vu! This was an all-too-familiar story for Victor and William.

In adversity lies opportunity. Just as Li & Fung perfected the export trading business, IDS was determined to find a new way to skin an old cat – reinventing distribution. The key to the winning formula lies in providing the *right services and solutions* to unlock the huge costs and lost sales opportunity trapped in these complex and un-coordinated Asian supply chains. Fundamentally, IDS needs to find a way to reverse the adversarial relationships that have characterized the industry to a more collaborative platform where distributors act as partners to retailers and brand owners.

The journey began in March 1999. Senior management took a holistic view of the challenge and decided on an expanded definition of distribution to "integrated-distribution" services (IDS), encompassing the end-to-end service proposition of Marketing, Logistics and Manufacturing. A whole new approach to the business was in the making.

15.4.   The IDS Approach – Value-Chain Logistics

15.4.1   Reinventing Distribution Through Value-Chain Logistics

"Our aim is to reinvent the traditional distribution business by introducing Value-Chain Logistics. This means that we position our deep and extensive Asia-Pacific Logistics Network as the fundamental enabler of our business. Coupled with our sophisticated regional IT infrastructure, Logistics connects seamlessly with our other two core businesses of Marketing and Manufacturing to form an end-to-end Value-Chain. We see distribution not just as the marketing and selling of brands but more as a "Menu of Services" comprising a clearly defined, distinct and tangible set of offerings of value-add along the Value-Chain. In working with our customers, we keep a flexible and responsive "Plug and Play" approach, allowing IDS to customize meaningful offerings from our Menu of Services to meet the local, regional or global needs of our customers. We call our holistic approach Value-Chain Logistics."

Ben Chang
Group Managing Director
IDS Group

15.4.2   IDS Business Purpose And Value Proposition

Shortly following the acquisition, IDS began to define its business purpose and value proposition. It articulated its vision by clearly and simply stating its commitment *"To be the Premier Pan Asian provider of Integrated-Distribution Services specializing in Marketing, Logistics and Manufacturing."* The focus was to build an Asia-wide regional business with in-country operations excellence for all its operating units engaged in Marketing, Logistics and Manufacturing. Each should be built on its own strength to compete against best-in-class competition. Yet, as a Group, IDS could see its competitive edge in its ability to integrate the three core businesses and customize its services and solutions across Asia.

Simply put, IDS defined its business purpose as follows:

- IDS is an Integrated-Distribution service provider in three core businesses of Marketing, Logistics and Manufacturing covering the entire value chain.
- IDS operates in the economies of Association of South East Asian Nations (ASEAN) and Greater China specializing in FMCG and Healthcare products as well as consumer durables like footwear, apparel and bedding products.
- IDS focuses on serving brand owners and retailers who wish to penetrate the Asian market. IDS can efficiently and responsively move products from factory floor to the hands of the consumer.

At the onset, IDS made three important decisions that laid the foundation for its repeated pattern of profitable growth:

- Create a Logistics business by transforming it from a back-end support function to a front-end leading business. Build a new and solid Logistics management team and invest in a Pan-Asian Logistics infrastructure.
- Organize across Asia in the three business streams of Marketing, Logistics and Manufacturing. Build expertise in each to ensure in-country flawless execution and augment the capability to provide regional and/or global solutions.
- Invest in regional IT applications to ensure common business systems, practices and processes. Build an integrated and regional IT infrastructure to improve information integrity and visibility thereby enhancing management control and customer relationship management.

15.4.3   Value-Chain Logistics

This reengineering effort was key to launching the concept of Value-Chain Logistics. The goal was ambitious i.e. to establish a new paradigm for distributing FMCG and Healthcare products across Asia, and to eventually dominate the market by offering flexibility and value that could not possibly be matched by operating in the traditional way.

In a nutshell, Value-Chain Logistics takes a holistic view of the supply chain, positioning Logistics as the fundamental enabler to connect Marketing and Manufacturing together to form a complete value chain that covers the entire process from procurement of raw materials to delivery of finished goods to end consumers. Logistics links and drives every part of the distribution process, allowing the entire value-chain to be driven by a closely monitored consumer off-take demand.

Over the years, IDS has progressively developed a *menu* of stand-alone distinctive services under its three core businesses. In this way it can provide one or more services to its customers, and if required provide a one-stop integrated-distribution solution. With Value-Chain Logistics, IDS carves a unique position in the marketplace.

- A strong Logistics offering as the fundamental enabler, integrating the Value-Chain from Manufacturing to Marketing Services

- A Menu of Services along the Value-Chain underpinned by a deep and extensive Asia-Pacific Logistics and Technology Infrastructure

- A One-Stop Solution of Flexible & Responsive "Plug & Play" from our Menu of Services enabling us to customize services for the local, regional or global needs of our customers

15.5. IDS – A Regional Powerhouse

15.5.1 Extensive Asian Network

Today, the IDS Group has an extensive Asian network of offices and operations covering its three core businesses. At the heart is its logistics offering. IDS Logistics has a comprehensive logistics infrastructure which includes more than 50 strategically located distribution centers and depots supported by a deep and extensive transportation network across Asia. IDS Logistics provides a whole range of logistics services including storage and delivery, drayage, bonded warehousing, cross-docking and transportation management, as well as value-added services like export logistics and regional hubbing, import and export customs clearance, labeling and contract packing.

IDS Logistics has an extensive Asia-wide network and provides regional hubbing solutions for multinational customers. Its custom-bonded, temperature-controlled ASRS (Automated Storage and Retrieval System) in Singapore has a capacity of 44,000-pallet positions.

As demand for logistics services continues to grow, IDS Logistics will remain the fastest-growing core business of the IDS Group. The Logistics clientele has become increasingly diversified ranging from consumer and healthcare products to footwear and apparels, retail chain stores, automotive parts and electronics.

IDS Marketing has a well established and extensive in-country distribution networks across Asia. Its wide range of services includes selling, marketing, merchandising, credit and cash management, inventory and working capital management, and various value-added services. Across Asia, IDS Marketing gives brand owners the most mileage for their marketing investment. Through close collaboration, it builds brand equity by leveraging its product knowledge and market intelligence capability, applying key account management tools and deploying traditional trade coverage techniques. It brings products to various outlets including hypermarkets, supermarkets, convenience stores, as well as mom-and-pop stores and corner groceries. Additionally, its healthcare channels cover hospitals, pharmacies and clinics.

IDS Marketing provides a wide range of marketing, selling and credit & cash management services to FMCG and Healthcare multinational brands. Its experience in distribution dates back to the 1800's.

Under its marketing business, IDS also owns the brand rights of Slumberland and its associated brands of mattresses and bedding products in 22 countries across Asia. Other brands in its portfolio include VONO, Sleepmate and Starry Nite. In this instance, IDS controls the entire value-chain by providing all the services from in-house manufacturing to logistics and marketing. Slumberland has five factory locations and its bedding products are distributed all over Asia.

On the manufacturing side, IDS Manufacturing has three centers of excellence in Malaysia, Thailand and Indonesia producing Food & Beverage, Home & Personal care and Health & Beauty products. As a contract manufacturer, IDS Manufacturing does not formulate or own any brands but acts as the manufacturing extension of its customers. It focuses on providing manufacturing know-how, product knowledge, plant efficiency, in-process quality control and assurance, thereby enhancing its customers' competitiveness. It puts strong emphasis on quality. All of the manufacturing plants hold GMP licenses and ISO 9001 certification. Today, IDS Manufacturing is able to offer a wide range of packing options including TetraPak, aerosols, hot PET, tubes, sachets, strip packs, tubs, and more.

IDS Manufacturing produces a wide range of food & beverage, home & personal care and health & beauty products including the regional production of Listerine in its new purpose-built plant in Thailand.

15.5.2   Technology – A Competitive Edge

IDS inherited a variety of stand-alone and out-dated legacy systems that came with the acquisition. This created substantial hindrance on its ability to execute consistently and effectively across the region. Additionally, the visibility of business critical information was lacking. To build a regional business with on-the-ground operations excellence, it was abundantly clear that IDS had to invest in technology to drive visibility, velocity and value in the supply chain. A robust common regional IT system was imperative. IDS needed an integrated regional IT platform designed for scalability and reliability, and capable of supporting rapid business expansion and growth.

Today, the IDS Group has a single, unified technology platform that allows its core business to operate with common business practices and processes. Its systems are also able to interact with each other and ensure free flow of information across all business units.

Based upon this single, integrated platform, each of IDS core businesses adopts best-in-class IT applications in their respective fields. IDS Marketing uses Oracle E1 system. IDS Logistics operates a Warehouse Management System (WMS) called SSA WMS EXceed. IDS Manufacturing runs the QAD MFG/PRO platform. All of these applications are standardized across the region and are fully integrated to enable full visibility of operations on a regional basis. IDS also established its in-house regional technology center and support teams in Malaysia and Hong Kong.

IDS Group is also in the process of refining its cross border Value-Chain management system, Visible IDS Transactional Application Lifeline (ViTAL). This is the core application system used in managing and coordinating regional and global supply chains. Additionally, it is rapidly deploying two down-line application systems – Road Warrior and Tradex. Road Warrior is an application which automates the collection of field data and sales order processing by equipping field sales staff with Personal Digital Assistants (PDA's). It has substantially shortened order cycle time as well as addressed out of stock situations at the retail level. Tradex enables customers to have visibility on sales off-take at the sub-distributor level, which used to be a black box.

To enable IDS customers to access business-critical information and provide visibility of operations and performance, IDS created a web-based portal called Trigantic. Customers can log on to Trigantic anytime, anywhere, and enjoy a custom-designed portal tailored to their requirements. Through Trigantic, customers can view the latest sales information and inventory levels, track order status as well as monitor service level based on agreed Key Performance Indicators (KPI's). All information is drawn from the IDS regional data warehouse which is fully integrated real-time with all its operations throughout Asia.

15.5.3   IDS Customers And Business Partners

IDS partners with an enviable list of blue-chip multinational customers. These include Unilever, Pfizer, Nike, Timberland, Gillette, GSK, Sara Lee, Johnson & Johnson, L'Oreal, Kelloggs and Abbott, to name just a few. Additionally, IDS services retail chains such as Carrefour, Starbucks, Toys "R" Us and Watsons.

IDS has also successfully developed multi-country, multi-business relationships with many major customers. It currently works with over 380 brand owners and retailers. Of these, just under 20% are multi-stream, and/or multi-country customers but they constitute over 65% of its 2004 business. Enormous potential exists for IDS to continue enhancing cross-stream, cross-country relationships building on the strong base of customers it already serves.

15.6. Business Transformation – Change Management And A New Culture

Upon completing the acquisition, IDS was immediately faced with the need to implement widespread change. The most important part of the change process hinged on developing a unique company culture in which Value-Chain Logistics could take root and flourish. Business goals and process can always be imitated, but there is no way to replicate a strong company culture.

To kick off this exercise, senior management went through a lengthy process of brainstorming, consultation and focus group discussion with employees, customers and key stakeholders. The outcome is a set of values that defines what IDS stands for as a company. These values are grounded in respect, teamwork, entrepreneurship, success, ethics, service and loyalty. While management philosophy is anchored upon modern Western management principles, the Group takes pride in its Asian roots embodied by Li & Fung, and in its ability to bring harmony in diversity. Its long-standing Asian heritage helps everyone appreciate the local nuances in each of the markets IDS operates, and to adjust its strategies according to the vastly diverse trade practices, cultures, regulatory requirements and consumer behaviors.

Although it was important to ensure rapid change management and build a new culture, it was even more critical to preserve business continuity and minimize business disruption. To change the rules of the game, it was important to stay in the game. This led IDS to focus its immediate priority on the need to execute the core businesses very well, and to operate flawlessly on the ground in every one of its business units. In the meantime, its regional business development strategy required delivering operations performance and meeting or exceeding service level KPI's. Thus part of the new culture was to change by winning and succeeding. This required IDS to build winning teams by "thinking regionally and acting locally", by combining practical experience with theory, by sharing best practices, and by living the values. This new emphasis forms the cornerstone of its business growth strategy, and allows it to gradually secure cross-stream and cross-country regional accounts.

Very early on, IDS emphasized the need to be thought leaders in the Distribution and Logistics industry in Asia. The leadership team comprises experienced professionals who bring together strong local knowledge, varied cultural background and diverse international perspective. IDS blended the best of management talent from Inchcape, Li & Fung and new management from the outside. To assimilate and help develop best practices, IDS has collaborated with educational institutions, participated in research projects with established universities, joined industry forums and exchanges, and closely monitored industry trends and developments.

One of the most important areas of focus is the high emphasis placed on recruitment, development and retention of staff. When recruiting outside talent, IDS chooses people with a performance track record and potential to grow. In developing its people, IDS works with the concerned staff in jointly assessing competency gaps and developing action plan to address them. As part of the retention program, IDS makes it a point to cascade the vision down to the operating units and spend time listening to and acting on frontline feedback. The company embraces teamwork and celebrates success. On a business unit level, each business unit head acts as entrepreneurs with full authority and accountability. IDS pays for performance.

In evaluating its approach to change management, IDS follows key business fundamentals as follows:

- Live the values and promote the new culture, by leveraging Li & Fung's strong Asian heritage and combining the best of East and West management practices
- Build winning teams by sharing best practices and focusing on flawless execution on the ground; grow by securing cross-stream and cross-country regional accounts

- Champion continuity of employment with no significant retrenchment or downsizing thus fostering strong two-way loyalty between company and employees
- Create a performance-based culture where Asians and Westerners, old and new employees coexist in a meritocracy with no imposition on an individual's background
- Develop a knowledge-based, learning environment to foster career and personal growth; provide opportunities for learning, sharing and relationship building by organizing regional leadership meetings, training workshops, stream and cross-country seminars and team building exercises
- Evolve change through a phased Three-Year Strategic Planning cycles with strong, clear communication of its plan objectives. Subtly embed the new approach and business model by allowing legacy practices, systems and conventions to work alongside new processes, applications and services

15.7. Three-Year Strategic Plan

15.7.1 Phased Aggressive Evolution

Soon after the acquisition, senior management introduced a three-year strategic planning cycle that would align direction and commitment throughout the organization. When formulating a strategic plan, IDS adopted a zero-based approach. Nothing is sacrosanct. Any previous operating assumptions can be put to challenge. However, once the three-year strategic plan is formulated, then it is cast in stone and stays unchanged for three years. Unlike most, it is not a rolling strategic plan that gets updated every year. This Strategic Planning Process is often referred to as an organized and orchestrated methodology of *"Phased Aggressive Evolution"* of IDS Group's business re-make.

To ensure clarity of direction and alignment of thinking, IDS involves management and employees, up and down the organization in an iterative process of Strategic Planning. It develops a clear communication plan of cascading down information. It circulates a one-page summary of its formulated strategic plan to all concerned. It then conducts quarterly reviews to measure progress against the plan. The following chart summarizes the highlights of the three Three-Year Plans IDS Group has developed and implemented so far.

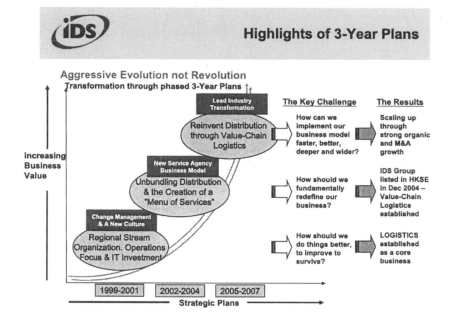

15.7.2   Strategic Plans 1999–2001 & 2002–2004

In the first Three-Year Strategic Plan 1999–2001, the challenge was managing change and creating a new culture. IDS broke down country silo structures and organized itself as a regional company with three core business streams and key support functions across Asia. The primary focus was on operations excellence to retain and build customer accounts and regional relationships. The Group invested heavily in technology. By far the most important outcome of this change process was the establishment of Logistics as a leading core business of the IDS Group. Knitting together the existing Marketing and Manufacturing businesses with Logistics, IDS conceptualized the end-to-end positioning of its core businesses as a regional Asia-wide integrated-distribution business.

The second Three-Year Strategic Plan 2002–2004 was a very involved and extended process. During its course, the competition had recognized the emergence of IDS, and the battle-lines were drawn. It was imperative that the Group needed an extremely robust Strategic Plan that could "make the difference". The first Three-Year Plan was about how to do things better. This second plan addressed the fundamental question of how to do things differently. By the end of 2004, IDS had emerged as a totally different company, having redefined and fundamentally changed its business model.

Way back in December 1998 when the consortium to buy the Inchcape companies was formed, senior management had committed to a listing of IDS at the Hong Kong Stock Exchange "within five years" of acquisition. Everyone at IDS lived by this commitment. In December 2001, IDS announced the breakthrough goal of its second Three-Year Strategic Plan 2002–2004 – IDS Group's Initial Public Offening (IPO) by end 2004, the end of the second planning cycle.

15.7.3 Success As A Public Company & Strategic Plan 2005–2007

On 7 December 2004, this dream became a reality. The listing of the IDS Group on the HKSE was a resounding success. The public-offer shares recorded an over-subscription of 152 times. After 18 months on 30 June 2006, the IDS Group (#2387) touched a high of HK$14 per share, four times the debut price of HK$3.50 per share.

This phenomenal support from institutional and retail investors was not without basis. In the few years of its business re-making, it had moved towards a new Service Agency business model. This model separates out service value-adds from the risks associated with traditional distribution. IDS now represents itself more as an agent to brand owners and works closely in partnership with them to help mitigate risks associated with the selling and marketing process. Furthermore, it also re-positions itself as a service provider to both retailers and brand owners. This contrasts heavily from the traditional distributors whose sole allegiance is to the brand owners. Significantly, IDS is now seen as a company with a regional value proposition.

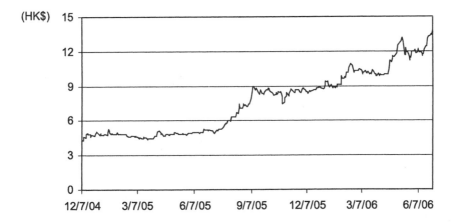

Above all, by the time IDS became a public company, the IDS name and brand had become synonymous with Value-Chain Logistics. Value-Chain Logistics has been established as a holistic approach and compelling alternative to the tired and past-due trading proposition of the traditional distributor. Its financial track record was good, margins better and outlook promising. Looking ahead to the current Three-Year Strategic Plan 2005–2007, much remains to be done if IDS were to achieve the goal of eventually dominating the industry. The major challenges that must be addressed are: How can IDS implement its business model faster, better, deeper and wider? How can IDS lead distribution industry transformation? Can IDS "Reinvent Distribution through Value-Chain Logistics"?

15.8.   IDS – New Business Model

15.8.1   A Scalable And Service-Driven Model

Keeping one's business model current is key to organizational survival. Making one's service offering relevant is key to gaining competitive advantage. At the very onset of the journey to establish the IDS Group, there was an early recognition to develop a non-traditional approach focusing less on the outmoded middleman's *buy & sell* role but more on distinct, value-adding services.

In creating the IDS business model, IDS adhered diligently to key fundamentals observed and determined as being central to reversing the failures that had befallen many distributors:

- Separate services from risks; unwind the many elements of risks associated with an all-in bundled pricing of traditional distribution
- Build scalability with strong operating leverage including a unified and homogeneous business processes and practices, regional organization and common country support function
- Champion an asset-light, knowledge-based business that owns little or no hard assets, declares war on working capital and builds a learning organization of talented people
- Create services in line with the demand-driven economics, dissociating from wasteful economics of uncontrolled supply
- Develop a unified, integrated and regional IT system to drive business value, visibility and velocity of product, work and cash flow, and to provide timely and accurate information and intelligence to key business partners

15.8.2 Unbundling And The Creation Of Menu Of Services

IDS characterized the creation of Menu of Services to the process of separating out or "*unbundling*" the multitude of activities along the distribution chain.

In the past, the distributor sets a single margin for an all-in bundled service including selling, marketing, logistics, credit control, billing and collection, and assumes all trade and principal-related risks. However, a subtle shift has been emerging in customer preferences. More and more customers favor the flexible "menu type" service over the traditional "one-size-fits-all" offering. Most times, customers want only logistics service. Sometimes they require the administrative billing and collection service. More often than not, brand owners do not want to relinquish control of the actual selling and marketing process. Having an unbundled service model provides customers a tailored entry point into Value-Chain Logistics.

The most fundamental step IDS took in unbundling was to take the back-end support function of Logistics and transform it into a front-end distinctive core business of IDS. Progressively over the years, IDS has followed this unbundling concept in creating the Menu of Services.

Today, IDS customers can pick and choose the services they need from the vast array of services available under any of the three core businesses of Marketing, Logistics and Manufacturing. IDS can essentially customize through a

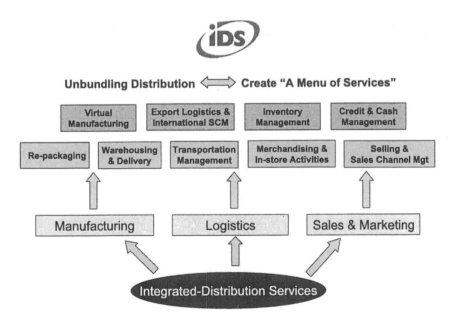

"plug and play" approach to be relevant to any brand owner or retailer, in any locality or across the region. Services can either be stand alone or integrated. Over the course of the next few years, IDS is poised to deliver new and better services that will allow an increasing level of integration along the value-chain. It is this very process of step-by-step creation of new and improved services that will eventually provide the impetus for more collaboration amongst brand owners, retailers and service providers.

15.8.3   The 5 'S' Of Service Growth

The IDS approach to business development is guided by a methodological step-by-step development of the key service offerings based on the 5 'S' of services – Substitution, Scope expansion, Scaling up, Structural change and Simplification. In providing services to the customers, it is important to consider:

- Substitution – this is typically providing a service that the competition can also provide; margins are thin and competition severe
- Scope expansion – on the basis of strong performance, an IDS business unit can then add more service offerings or expand the relationship to another in-country business unit
- Scaling-up – this involves creating a regional relationship and building multiple cross-stream and cross-country relationships, drawn from its menu of services
- Structural change – this entails pursuing innovation and integrating the services provided that will structurally improve the cost-effectiveness and responsiveness of a customer's supply chain
- Simplifying – always strives to make it easy to work with IDS; organize for the customers, locally or regionally, in a single or multiple business relationship

Aptly applying these guidelines of service expansion and differentiation has allowed IDS to retain existing customers, expand regionally with them and more importantly, secure numerous new relationships.

15.9.   The Future – Poised For Growth

15.9.1   Global Outsourcing Trend & Rise Of Consumerism

Asia is home to the world's fastest developing economies. The increased purchasing power of consumers in Asia is irresistible to Western FMCG and

Healthcare brand owners who are faced with mature markets and aging populations at home. As people become more affluent, there is a clear trend towards brand proliferation, including the emergence of Asian brands. The growing consumer market in Asia brings substantial opportunities for professional service providers like IDS.

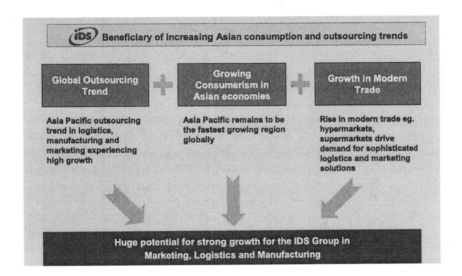

As large retail chains extend their reach and market prowess in Asia, the demand for efficient logistics and increasingly sophisticated marketing services grows. These retail chains require a capable logistics partner to support their growth whilst at the same time, expect brand owners and distributors to meet stringent professional standards before listing their products on shelves.

Additionally, large multinational companies are more and more focusing on their core competencies and outsourcing non-core businesses to service providers. While outsourcing logistics is common amongst companies in the developed economies of the West, businesses in Asia are just catching up with the trend. In Asia, especially the Chinese Mainland, the rate of growth in logistics outsourcing is twice as fast as other parts of the world. Even more encouraging is the extension of outsourcing into areas which brand-owners have shown hesitation in the past. A clear example of this is Contract Manufacturing. As trade barriers gradually ease, more and more multinationals are opting to partner with reputable regional contract manufacturers, instead of setting up their own factories in each market.

15.9.2   Taking Brands To Hearts And Minds In China

China is attracting more investment capital than any other nation in the world. Retailers and owners of consumer and healthcare brands quite rightly see it as the next major market of growth. Today, no multinational business strategy will be complete without an aggressive growth plan for China. However penetrating this market is fraught with difficulty. Business risks abound.

Success in China requires extensive and deep-rooted local knowledge, commitment, tenacity and a long-term view. IDS has had a long and successful on-the-ground experience of doing business in China. It is its fastest growing market today. Through Li & Fung, IDS is inextricably linked to China for over a century. This strong Chinese heritage is key to brand owners and retailers looking for a hassle-free and effective partnership in China.

IDS Logistics now manages an extensive logistics infrastructure of nine major distribution centers, a multitude of depots, and a strong transportation network, making it a partner of choice in China. IDS Marketing is the first distribution company to be awarded a wholly owned national distribution license with import/export rights in April 2004. Its distribution channels are deep and extensive.

By end 2006, IDS will further strengthen its leadership position in China distribution by having branch presence in more than 20 cities with direct selling and key account management capabilities. No other company comes close to its distribution network spread.

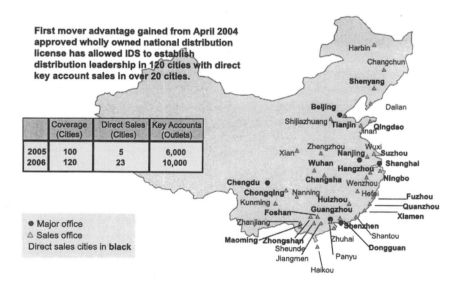

First mover advantage gained from April 2004 approved wholly owned national distribution license has allowed IDS to establish distribution leadership in 120 cities with direct key account sales in over 20 cities.

	Coverage (Cities)	Direct Sales (Cities)	Key Accounts (Outlets)
2005	100	5	6,000
2006	120	23	10,000

● Major office
△ Sales office
Direct sales cities in black

China is a market of limitless opportunity. The IDS network offers an open gateway into China through which multinationals can take their brands into the hearts and minds of consumers of the most populous nation in the world.

15.9.3 Growth Drivers

The third Three-Year Strategic Plan ending in 2007 calls for IDS to lead industry transformation through Value-Chain Logistics. Its financial target is to at least double the net profit by 2007. To make this happen, IDS expects to primarily drive growth via organic expansion and secondarily through selective acquisitions.

(iDS)	**Growth Drivers**

(iDS) expects strong growth in the next 3 years through
Aggressive Organic Expansion and Selective M&A

Selective M & A	Aggressive Organic Expansion

• **Local small and medium size M&A opportunities**	• **New service offerings**
	  ▪ Export / Global Logistics
	  ▪ Regional Hubbing
	  ▪ Credit & Cash Management

• **Fill in the mosaic for critical mass**	• **Brand proliferation**
• **New market entry e.g. Vietnam and Logistics in Indonesia**	• **Outsourcing trend**
	• **Rapid expansion in China**

• **Slumberland brand rights for 22 countries**	• **Growth of existing clients**
	  ▪ Service scope
	  ▪ Regional expansion

15.10. The Journey Continues

The Li & Fung Group journey of business creation continues. IDS is still in the early phase of an evolution into an integrated-distribution services company. While we are excited by the challenge of championing another industry re-make, we are fully cognizant of the roadblocks ahead. Nevertheless, we will be steadfast in constantly updating our business value proposition, in further fine-tuning our Value-Chain Logistics approach, and in eventually extending our expansion beyond Asia and into the global business arena.

This chapter is less about the success we have achieved to date but more about sharing what we have learned. We sincerely hope that you will find valuable insights in these pages. It is in the spirit of sharing and learning that we bring forth to you the IDS Story, a story which is all a part of the Li & Fung Supply Chain Management journey.

Hau L. Lee and Chung-Yee Lee (Eds.)
*Building Supply Chain Excellence
in Emerging Economies*
©2007 Springer Science + Business Media, LLC

Chapter 16

BUILDING A SUSTAINABLE SUPPLY CHAIN
Starbucks' Coffee and Farm Equity Program[1]

Hau L. Lee, Stacy Duda, LaShawn James, Zeryn Mackwani,
Raul Munoz and David Volk
Graduate School of Business, Stanford University, USA

Abstract: As companies increasingly move their supply bases to emerging economies, they maybe exposed to increasing risk of supply disruptions. In particular, they can be subjected to the risk of their suppliers going out of business or changing their lines of business due to the rapid changes in the socio-economic development of these economies. This is especially the case when we are talking about emerging economies in underdeveloped countries such as Africa, parts of East Asia like Indonesia, Central America and parts of South America, In industries requiring raw materials that come from natural resources – mining and agriculture, for example – many companies do not have a choice but to source from the emerging economies endowed with such resources. In such situations, we need to pay attention to the dimension of sustainability of supply as a business objective in managing the supply chain. Efficiency in cost and time is simply not sufficient. What can we do to assure supply in the long term? The key is that we must invest in the suppliers, giving them assistances and incentives, so that they could be successful in their business, and hence, be sustainable suppliers. Otherwise, the cost of not having sufficient supply in the long run, or the cost of having to develop new supply sources, could be excessive. In this chapter, we describe how the world's leading coffee company develops its supply chain to be sustainable. To build a sustainable supply chain, Starbucks helped to enable their suppliers to have equitable returns in their business, be sound global citizens of the environment, and have employees whose welfare is improving over time. The sustainable supply chain has also led to Starbucks having a socially responsible supply chain.

---

[1] This chapter is based on the case "Starbucks: Building a Sustainable Supply Chain," Graduate School of Business, 2006. The authors acknowledge the help of Dub Hay of Starbucks and Eric Poncon of Agroindustrias Unidas de Mexico SA de CV in this study.

16.1.   Sustainable Supply Chain

As companies increasingly move their supply bases to emerging economies, they maybe exposed to increasing risk of supply disruptions. Such disruptions could be the result of natural disasters, but could also be a result of suppliers going out of business or changing their lines of business. As emerging economies are going through rapid changes in their socio-economic development, the risks of such disruptions cannot be underestimated. This is especially the case when we are talking about emerging economies in underdeveloped countries. By now, emerging economies such as China, India, Brazil and Eastern Europe are relatively mature in terms of the economic structure and the social welfare of the people. But emerging economies such as Africa, parts of East Asia like Indonesia, Central America and parts of South America, are still going through significant political and social transformations. Sourcing from these countries could pose a significant risk in the form of stability and sustainability of supply.

In industries requiring raw materials that come from natural resources – mining and agriculture, for example – many companies do not have a choice but to source from the emerging economies endowed with such resources. In such situations, we need to pay attention to the dimension of sustainability of supply as a business objective in managing the supply chain. Efficiency in cost and time is simply not sufficient. What can we do to assure supply in the long term? The key is that we must invest in the suppliers, giving them assistances and incentives, so that they could be successful in their business, and hence, be sustainable suppliers. Otherwise, the cost of not having sufficient supply in the long run, or the cost of having to develop new supply sources, could be excessive.

To build a sustainable supply chain, companies should help to enable their suppliers to have equitable returns in their business, be sound global citizens of the environment, and have employees whose welfare is improving over time. Hence, building sustainable supply chains would also enable a company to be socially responsible. In this chapter, we describe how the world's leading coffee company develops its supply chain to be sustainable, and in the course of doing so, created a socially responsible supply chain.

16.2.   The Starbucks Corporation

Starbucks Corporation is the world's largest specialty coffee retailer, with $6.4 billion in annual revenue for the fiscal year ended Oct 2, 2005. The company has continued to expand the number of retail stores worldwide, and has

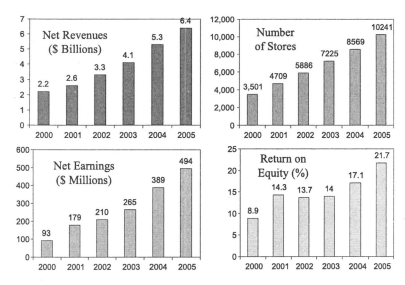

Exhibit 16.1. Starbucks Corporate Performance.

seen strong growth in the sales and net profits consistently (see Exhibit 16.1). Since going public in 1992, its stock has appreciated more than 4,000% after adjusting for stock splits.

In the 90's, the specialty coffee industry experienced gigantic growth, and consumers in this segment came increasingly from the educated class. However, in the past few years a worldwide oversupply of lower-grade coffee has depressed the world's market prices, making it difficult for coffee farmers to earn enough revenue to cover the cost of production. Although Starbucks only purchased the highest quality Arabica coffee and pays premium prices, all farmers have been affected economically by the oversupply of coffee (Exhibit 16.2).

Starbucks was at a challenging point in its history. It boasted more than 10,000 stores – up from 676 a decade ago – and roasted 2.3% of world coffee production. It opened an average of four stores and hired 200 employees each day. To support such a high growth rate, it is clear that an integral part of the company's future success would come from meeting increased demand through a secure supply of high-quality coffee beans. Coffee beans constituted the bread and butter of Starbucks' business, and the company had to ensure a sustainable supply of this key commodity. Consequently, Starbucks partnered with Conservation International, an environmental nonprofit organization, to develop C.A.F.E. Practices (Coffee and Farmer Equity Practices) to help contribute to the livelihood of coffee farmers and to ensure high-quality coffee for

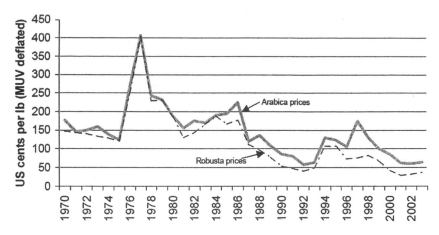

Exhibit 16.2.   Arabica and Robusta prices, 1970–2002.

the long term. This initiative was based on the notion that sustainable supply of high quality coffee beans depends on a stable source of coffee farms with coffee farmers that are not exploited by their trading partners, have lands that are farmed with environmentally sound methods, and have families that live in healthy, secure and supportive societies. Such farmers would be more inclined and able to invest in productivity improvement tools and activities, and in their communities, thereby contributing to being a source of stable and sustainable coffee supply.

In 2005, Starbucks Corp. operated and licensed more than 10,000 coffee shops in more than 30 countries. The shops offered coffee drinks and food items, as well as beans, coffee accessories, teas, and CDs. Starbucks operates more than 5,200 of its shops in five countries (mostly in the US), while licensees operated more than 2,800 units (primarily in shopping centers and airports). The company also owned and franchised the Seattle's Best Coffee and Torrefazione Italia chains in the US (more than 100 shops). In addition, Starbucks marketed its coffee through grocery stores and licenses its brand for other food and beverage products

16.3.   The Specialty Coffee Industry and the Starbucks Coffee Supply Chain

Since the 1980's and especially in the last decade, the specialty coffee industry has grown dramatically. Many experts felt that the differentiated coffees supported by the specialty industry will continue to expand at a much faster rate than conventional coffees. However, the definition of specialty in the United States has continued to be refined. It currently includes coffees that may not necessarily be high-quality and are otherwise only differentiated by being flavored (e.g., chocolate, cinnamon, and hazelnut, etc.) and served as an espresso or milk-based beverage. The industry is beginning to redefine "specialty" to reflect more of a quality orientation. Also called "gourmet" or "premium" coffee, specialty coffee is made from exceptional beans grown only in ideal coffee-producing climates. It tends to feature distinctive flavors, which are shaped by the unique characteristics of the soil that produces it. Specialty coffee has become one of the fastest growing food service markets in the world. The percentage of adults in the U.S. that consumed specialty coffee daily increased from 9% in 2000 to 16% in 2004, and 56% of the adults claimed to be occasional consumers. The total specialty coffee market was estimated to be $8.06 billion in 2004.[1]

In 2004, there were an estimated 17,400 specialty coffee outlets in the United States.[2] Starbucks' success had prompted a number of ambitious rivals to scale up their expansion plans. Observers believed there was room in the category for at least two or three other national players.

Exhibit 16.3 gives a simplified picture of the supply chain of green coffee to Starbucks. In reality, the flows could be much more complicated than as shown in the Exhibit. Coffee beans could come from all over the world. About 50% came from Latin America, 35% from the Pacific Rim, and 15% from East Africa. Most of the coffee producers were small to medium-sized family-owned farms. Some farms were able to process their coffee beans, but most sold their outputs to processors through local markets (mills, exporters or cooperatives). The processors turn coffee "cherry" into parchment or green coffee, and then sold them to suppliers, who were exporters or distributors. These suppliers provided many services to processors and farmers, such as marketing, dry milling, technical coffee expertise, financing, and export logistics.

Coffee farms for Starbucks are often located in high altitudes, and maybe difficult to access. Many farmers are not well educated, and their living conditions are very poor. Farmers may not have the knowledge to get high productivity from their farms, and in some cases, used fertilizers or farming methods that could be very harmful to the environment. The villages or small towns where the farmers live are also in very poor conditions, with the farmers' families sometimes struggling to survive or make ends meet. Some of the farms are in areas where farms for illegal drugs are common, leading to tremendous pressure as well as temptation to the coffee farmers.

---

[1] Specialty Coffee Association of America Web site, http://www.scaa.org

[2] Ibid, http://www.scaa.org

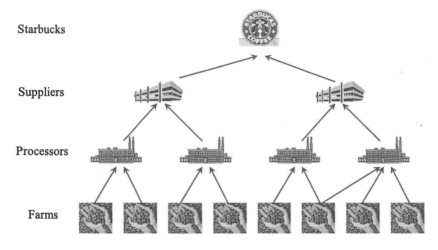

Exhibit 16.3. The Starbucks Coffee Supply Chain.

Starbucks also purchased coffee through agents from individual estates, producer associations, in addition to the suppliers or directly from the processors.

16.4.   C.A.F.E. Practices

Despite its dominance of the specialty coffee industry, Starbucks did not use its purchasing power as a way to squeeze its coffee suppliers in order to improve margins. Instead, Starbucks decided to use its market power as a way to implement social change within its supply chain through the C.A.F.E. Practices initiative. The C.A.F.E. Practices initiative was a way for Starbucks to ensure a sustainable supply of high quality coffee beans, which was an essential component of Starbucks' business. The initiative built mutually beneficial relationships with coffee farmers and their communities. It also helped to counteract the oversupply of low-grade coffee on the world's market, which caused suppressed prices making it difficult for farmers to cover the cost of production. When Starbucks implemented the C.A.F.E. Practices initiative, it had six objectives in mind:

1. Increase economic, social, and environmental sustainability in the specialty coffee industry, including conservation of biodiversity.
2. Encourage Starbucks suppliers to implement C.A.F.E. Practices through economic incentives and preferential buying status.
3. Purchase the majority of Starbucks coffee under C.A.F.E. Practices guidelines by 2007.

4. Negotiate mutually beneficial long-term contracts with suppliers to support Starbucks growth.
5. Build mutually beneficial and increasingly direct relationships with suppliers.
6. Promote transparency and economic fairness within the coffee supply chain.

The C.A.F.E. Practices initiative was a set of coffee buying guidelines designed to support coffee buyers and coffee farmers, ensure high quality coffee and promote equitable relationships with farmers, workers, and communities, as well as to protect the environment (Exhibit 16.4). It was not a code of conduct or a compliance program. Instead, it was a way of doing business that was aimed at ensuring sustainability and fairness in the coffee supply chain. This sustainability and fairness was achieved through a set of global guidelines for Starbucks suppliers and a set of incentives to reward farmers and suppliers who followed those guidelines. The guidelines consisted first of a set of prerequisites that must be met in order to be considered for the C.A.F.E. Practices initiative. These prerequisites set a minimum standard for Starbucks suppliers, including coffee quality and economic transparency. The transparency prerequisite meant that suppliers were expected to illustrate economic transparency on the amount of money that was ultimately paid to farmers.

Exhibit 16.4
C.A.F.E. Practices Self-Evaluation Checklist

Product Quality – Required
- Green Preparation – Prerequisite
- Cup Quality – Prerequisite

Economic Accountability – Required
- Demonstration of Economic Transparency
- Equity of Financial Reward
- Financial Viability

Social Responsibility
Hiring Practices and Employment Policies:
- Minimum/Living Wage/Overtime Regulation*
- Freedom of Association/Collective Bargaining
- Vacation/Sick Leave Regulation
- Child Labor/Discrimination/Forced Labor*

Worker Conditions:
- Access to Housing, Water and Sanitary Facilities
- Access to Education
- Access to Medical Care
- Access to Training, Health & Safety

Coffee Growing – Environmental Leadership
Protecting Water Resources:
- Watercourse Protection
- Water Quality Protection

Protecting Soil Resources:
- Controlling Surface Erosion
- Improving Soil Quality

Conserving Biodiversity:
- Maintaining Coffee Shade Canopy and Natural Vegetation
- Protecting Wildlife
- Conservation Areas and Ecological Reserves

Environmental Management and Monitoring:
- Ecological Pests and Disease Management and Reducing Agrochemical Use
- Farm Management and Monitoring Practices

Coffee Processing – Environmental Leadership
Wet Milling
Water Conservation:
- Minimizing Water Consumption
- Reducing Wastewater Impacts

Water Management:
- Waste Management Operations/Beneficial Reuse

Energy Use:
- Energy Conservation/Impacts

Dry Milling
Waste Management:
- Waste Management Operations/Beneficial Reuse

Energy Use:
- Energy Conservation/Impacts

After the initial prerequisites had been met, suppliers were graded based on a set of environmental and social criteria. All suppliers were evaluated not just on their performance, but also their supply networks of farms. Farmers were rewarded for coffee growing and processing practices that contributed positively to the conservation of soil, water, energy, and biological diversity, and had minimal impact on the environment. Also, workers' wages should meet or exceed the minimum requirements under local and national laws. Effective measures should be taken to ensure workers' health and safety, and provide them with adequate living conditions. Based on their performance, as measured along the environmental and social criteria, suppliers might earn up to 100 percentage points in C.A.F.E. Practices.

Under C.A.F.E. Practices, farms, mills, and suppliers must illustrate equitable payments to those who work for them or sell to them. They must demonstrate economic accountability and document their hiring and employment practices. Scores were audited by an independent verifier, licensed by Scientific Certification Systems. Since the verifier was independent of Starbucks, the cost of the verification must be negotiated between the supplier and the verifier. However, there was no cost to the supplier to submit a C.A.F.E. Practices application to Starbucks.

In order to qualify for C.A.F.E. Practices supplier status, suppliers must be independently verified and meet minimum Social Responsibility criteria. Points above 60% increased the status of the supplier. For scores above 60%, the supplier qualified as a Preferred supplier and would gain preference in future Starbucks coffee purchases. Additionally, suppliers who earned scores above 80% would qualify as strategic suppliers and would earn a Sustainability Conversion Premium of $0.05 per pound of coffee for one year.[3] In order to encourage continued improvement, Starbucks also offered an additional Sustainability Performance Premium of $0.05 per pound of coffee to suppliers who were able to achieve a 10-point increase above 80% over the course of a year.

Besides the price premium for Strategic Suppliers, the C.A.F.E. Practices initiative allowed Starbucks to buy from preferred suppliers first, paying high prices and offering preferential contract terms to those with the highest scores. The premium prices helped coffee farmers make profits and support their families, despite a global glut in the coffee bean industry. Additionally, Starbucks provided access to affordable credit to coffee farmers through various loan funds. They invested in social development in coffee producing countries and collaborated with farmers through the Farmer Support Center in Costa Rica to provide technical support and training. If a supplier failed to meet C.A.F.E. Practices criteria, Starbucks sponsored information sessions in coffee growing regions for farmers.

16.5. Success Story of a Coffee Farmer – Investment Payoff for CAFÉ Standards[4]

For years, the Santa Teresa farm did well enough by producing regular extra-prime coffee rather than higher quality specialty grade. But that changed when world coffee prices hit rock bottom several years ago. Ervin Pohlenz

---

[3] On average, Starbucks pays about $1.20 per pound of coffee. (FY04 CSR Report)

[4] This section is taken from Starbucks FY04 CSR Report Coffee Section.

Cordova, the son of the farm's owner, wasn't earning enough for his crops to cover the farm's expenses. The farm nearly went bankrupt.

Cordova was introduced to Starbucks through his exporter and discovered he could earn more by producing higher-quality coffee. Santa Teresa is located in Chiapas, Mexico, an area known for its optimal altitude, fertile soil and shade trees – perfect coffee-growing conditions. Some investments were needed to improve quality and implement sustainable farming practices, a commitment Cordova was willing to make, despite initial resistance from his elderly father.

It took three years before the coffee grown on Santa Teresa reached Starbucks quality standards. Along the way, the exporter worked with Cordova on implementing quality improvements. In 2003, Starbucks signed a three-year contract to buy all of Santa Teresa's high-quality coffee at premium prices and added a provision that earmarks funds for social improvement and environmental protection projects to benefit the farm. Cordova's accomplishment is now the pride of his father.

The Starbucks contract gives Cordova security in knowing he has a buyer for his future crops and one that contributes to the quality of life on Santa Teresa. "Now I feel that I will work my entire life as a coffee producer because my farm is sustainable," he said. Clearly, Cordova's investment is paying off. And for Starbucks, we gain a wonderful source of high quality coffee grown under sustainable conditions. Cordova is a firm believer in C.A.F.E. Practices, and his goal is to become a Starbucks preferred supplier in 2005.

16.6.   Benefits to Starbucks

Even though the direct benefits of the C.A.F.E. Practices initiative helped suppliers and farmers, Starbucks received significant indirect benefits from the program. The program strengthened Starbucks' supply base, improved their marketing ability, and increased their visibility into the supply chain. Therefore, the benefits of the C.A.F.E. Practices initiative extended all the way through the supply chain, from the farm to the end consumer.

16.6.1   Supply Base

On the supply base side, the program served to lock in strategic and high quality suppliers. This consistent, quality supply could provide Starbucks with a competitive advantage over other coffee roasters in the industry. Since suppliers would have invested resources in complying with Starbucks programs, they would have an incentive to remain with Starbucks and would face switching costs should they try to demonstrate their excellence to another coffee roaster.

The large pool of high quality suppliers would also smooth supply fluctuations by providing a base supply of high quality growers. Since Starbucks' long purchase cycle included signing purchase agreements before the crop had even been harvested, any reduction in supply uncertainties and fluctuations could lead to better planning of future supply in the form of faster procurement. The benefits of the C.A.F.E. Practices initiative could also improve Starbucks' reputation among suppliers, which would make it easier to expand into purchasing in different countries or locations.

In the long run, the C.A.F.E. Practices initiative also sought to buffer against a form of Bullwhip effect that existed in the coffee industry supply chain. As coffee sales increased during the 1990's with the growth of Starbucks and the specialty coffee industry, suppliers and farmers began to respond with a huge increase in the amount of land dedicated to coffee farming. This glut of coffee beans on the market led to decreased prices and a shortage of high quality coffee. This type of fluctuations and swings in price and supply was common in commodity products that faced very long supply response times. In order to combat this price and supply volatility, the C.A.F.E. Practices initiative induced longer-term supply relationships with a consistent set of suppliers. Starbucks wished that this program would reduce its susceptibility to such price and supply volatility in the global coffee market.

16.6.2 Marketing

On the marketing side, the C.A.F.E. Practices initiative supported Starbucks' goal to be a socially responsible company. While C.A.F.E. Practices were not yet widespread and were not directly marketed to customers, an increased awareness of Starbucks corporate social responsibility (CSR) practices could help justify the premium prices that Starbucks charged. The C.A.F.E. Practices initiative would allow Starbucks to market its coffee as procured through a highly selective process that ensured only the highest quality bean. Awareness of this program might encourage other coffee roasters to join in the C.A.F.E. Practices program; however, Starbucks would be known as the inventor of the program. They might also be able to brand their practices and sell the know-how to other roasters that were looking to implement similar initiatives. Such wider-spread expansion of the program would simply serve to expand the benefits of the program towards creating a base of high quality coffee beans. With each improvement in the supply of beans, Starbucks achieved more flexibility in being able to charge premium prices at its stores. It also improved employee morale by creating an atmosphere of working at a socially responsible corporation.

16.6.3 Supply Chain Visibility

Finally, the C.A.F.E. Practices initiative increased Starbucks visibility of its supply chain by demanding documented and verified product and financial flows through their suppliers' supply chains. In the past, Starbucks had very poor visibility into their supply base, as coffee farmers and processors were not very technologically sophisticated or mature in their business processes. By increasing visibility into their supply base, Starbucks would be able to gain a better understanding of the needs and the conditions of their suppliers. The increased visibility would also allow Starbucks to improve its relationships with growers, who until now had been isolated from them in the supply chain due to the existence of intermediaries – coffee exporters and distributors – between the two sides.

On a more practical side, increased visibility in the supply chain could allow Starbucks to better predict supply shortages as they arose. Since the majority of Starbucks coffee was grown in relatively unstable countries in Latin America, Africa, South America, and Southeast Asia, Starbucks had a significant risk of supply shortage due to regional instability. Without visibility into the supply base, Starbucks did not have a good way to predict the impact of regional instability on its supply base. With increased visibility, an outbreak of regional instability could be linked to a particular quantity of expected coffee supply, giving Starbucks advance notice of the need to find alternate sources of coffee supply. This could allow Starbucks to be proactive in managing supply disruptions even before they arose.

16.7. Corporate Social Responsibility

Starbucks provided various resources to promote and help farmers comply with the guidelines of C.A.F.E. Practices and ensure sustainability. In January 2004, the company opened a farmer support center called the Starbucks Coffee Agronomy Company in Costa Rica that contained a team of experts in soil management and field-crop production (agronomists), and in coffee quality and sustainable practices. These experts collaborated directly with farmers and suppliers in Central America and provided services to farmers and suppliers in Mexico and South America. This helped build long-term and strategic relationships with members in the supply chain who were committed to the sustainable production of high-quality coffee. They also administered C.A.F.E. practices, oversaw regional social programs, and engaged with local government on sustainability issues.

Starbucks also bought certified or eco-labeled coffees that had been grown and sold in ways that helped preserve the natural environment and/or promote economic sustainability. There were three such types of environmentally sustainable coffee purchased by Starbucks:

Conservation Coffee (shade-grown): Starbucks, with its partnership with CI (a nonprofit organization dedicated to protecting global biodiversity), encouraged coffee farmers to use traditional and sustainable cultivation methods. The basic aim was to protect shade trees, which were often stripped away and replaced with tight rows of coffee trees on large coffee plantations. This not only destroyed the habitats of numerous species but also results in lower producing coffee.

Certified Organic Coffee: This coffee is grown without the use of synthetic pesticides, herbicides, or chemical fertilizers to help maintain healthy soil and groundwater.

Fair Trade Certified Coffee: With a licensing agreement with TransFair USA, Starbucks tried to ensure that coffee farmers were fairly compensated for their crops. The Fair Trade Certified Coffee label certified that the coffee met Fair Trade criteria. These criteria focused primarily on price and other sustainable needs. Fair Trade Certified coffees only came from democratically owned cooperatives, not large farms or coffee pulled across supply channels.

In order to improve farmers' access to financing, Starbucks provided loan funds to several organizations to ensure that farmers could obtain affordable loans and to help them gain some financial ability to improve their agriculture techniques. In 2004, Starbucks committed \$6 million to several loan programs. The importance and alignment of this upstream support component was highlighted in a quote from Shari Berenbach of Calvert Foundation: "Starbucks has taken a leadership position by aligning its investment capital with the company's mission and products to create more sustainable coffee growing communities."

Finally, Starbucks worked with local farmers to understand the greatest needs of their rural communities, which often lacked basic necessities such as adequate housing, health clinics, schools, good roads, and fresh drinking water. Starbucks worked together with these farmers to develop projects that helped meet their needs, especially in areas where the company bought large volumes of coffee. In fiscal 2004, the company contributed nearly \$1.8 million for 35 social programs.

16.8. C.A.F.E. Practices Implementation

There were two main challenges of the C.A.F.E. Practices implementation that could potentially be addressed with better integrated-information technologies. First, since some members of the supply chain had very poor information systems, it could be very difficult to gain economic transparency –

a key goal of C.A.F.E. Practices – from these members. Second, as C.A.F.E. Practices were updated and refined, it became a daunting job to effectively communicate the revised requirements and practices to farmers, suppliers as well as other members of the industry.

In addition, it had been a very labor-intensive and slow process to evaluate farmers for scores in the C.A.F.E. program. There did not seem to be any means to avoid the need of the auditors to travel to the farms, which were often located in difficult to access geographies.

The company was in the process of developing an internal system to track compliance with C.A.F.E. practices, and link such data to support procurement. The plan was to integrate the C.A.F.E. data that were currently stored in spreadsheets with the more versatile Oracle data-base, and to then link the data with its procurement system, together with other information systems on quality data.

To Starbucks, it seemed that a more comprehensive information system was needed to support a large-scale implementation of C.A.F.E. Practices.

16.9. Concluding Remarks

Starbucks has continued to aggressively expand the C.A.F.E. Practices initiative towards meeting its goal of supplying the majority of its coffee through the program by 2007. The C.A.F.E. Practices initiative has gone a long way towards improving the sustainability of its coffee supply chain while at the same time improve Starbucks' image as a socially responsible corporation.

The Starbucks case provided us with a few lessons. First, the company clearly saw that the usual objectives in supply chain management – cost, speed, quality and flexibility, etc. – were not sufficient. Given the strategic importance of coffee as the raw material to its supply chain, and the nature of the supply bases in underdeveloped economies, Starbucks recognized that sustainability was a key objective that it has to focus on.

Second, sustainable supply chains in underdeveloped economies and social responsibility have to go hand in hand. In fact, both are synonymous in this case. Starbucks has given us an example of focusing on the three dimensions of social responsibility – economic transparency, environmental soundness, and social welfare. Together, a sustainable supply chain can be built.

Third, building sustainable supply chains could mean a short-term reduction of profit. Starbucks has to pay a premium on its coffee, eroding part of its margin. Hence, it is important that senior management and even board level executives recognize that we are trading short term profit with long term sustainable supply which would lead to long term profits.

Fourth, as an industry leader, Starbucks has to take the lead in defining what sustainability means, and helps the supply chain members to master the concepts and make progress. Starbucks has to provide the necessary assistance and training, especially to supply chain members in underdeveloped countries.

Fifth, Starbucks has to create the right incentives for the farmers and others to be willing to participate. Such incentives were provided in the form of premium prices given by Starbucks. Economic incentives are necessary, again, in emerging economies due to the economic need of the people there. But the objective is to use economic incentives to induce social transformation.

Finally, sustainability and social responsibility are not just something that companies do to please their corporate conscience. They are the key to long-term business success.

Hau L. Lee and Chung-Yee Lee (Eds.)
*Building Supply Chain Excellence
in Emerging Economies*
© 2007 Springer Science + Business Media, LLC

Chapter 17

BUILDING A DISTRIBUTION SYSTEM IN EASTERN EUROPE
Organic Growth in the Czech Republic

M. Eric Johnson[1]
Center for Digital Strategies
Tuck School of Business
Dartmouth College, USA

Abstract: In an environment of changing customer expectations and evolving distribution infrastructure, firms can develop new business models that generate rapid growth. In this article, we examine the organic growth of a Czech office supply firm, Papirius. By introducing express delivery with service guarantees, Papirius was able to quickly grow and solidify its based. However, to defend its market against expanding multinationals, firms like Papiris must exploit their unique customer knowledge while driving efficiencies to world-class levels.

17.1.   Introduction

Building distributions systems in emerging markets like those of Eastern Europe offer many opportunities and challenges. In an environment of changing customer expectations and evolving distribution infrastructure, firms can develop new business models that generate rapid growth. In this paper, we examine the organic growth of an office supply firm in Eastern Europe. The distribution firm, Papirius, developed a strong business over its twelve-year history tracked in this article (1993–2005). This article first chapters the company through 2000 where it faced some major decisions about its growth and

---

[1] This article was written with research assistance from Daniel R. Justicz and Jay A. Altizer.

the defendability of its markets. It then examines the outcome of those decisions through 2005. The success of this firm provides many important lessons for distributors in emerging markets.

17.2. Early Company History

In the spring of 1993, two 18-year-old college students (Petr Sýkora and Jan Èerný) wondered how they might make some money between classes. One of them had a friend that worked for a paper wholesaler, and so they decided to try their luck as sales representatives. After a week of work, they finally landed their first client, for a toner cartridge. When they tried to fill the order, however, they realized that the wholesaler didn't have the cartridge in stock. Afraid to let down their very first client, they searched around Prague and found the product elsewhere. As they had very little money, they bought the product on credit and hopped on the metro, stopping only to buy some snowdrops as a gift for the customer. Smiling at the box of snowdrops, their customer thanked them heartily for their on-time delivery.

With this experience in hand, the two of them fronted 10,000 Kc (cca. $275) of capital and started their own business called Papirius. After some time, they were able to borrow a car, and a few months later they bought their first fax machine. Shortly thereafter they bought an old Skoda for 20,000 Kc ($550) and hired their first employee.[2] Their "customer first" approach, while commonplace in Western Europe and the USA, struck a chord with Czech consumers. Both their overall sales and customer base grew at compound annual rates in excess of 100%,[3] with 2000 sales exceeding 800M Kc ($22M). By 2000 they were one of the largest players in the office supply market competing with many small local players with limited offerings.

The company's growth was financed entirely by internal cash flows (see Exhibit A1) and the initial founders' equity contribution of 10,000 Kc until early 2000, when the company opened a line of credit with a local bank (Ceska Sporitelna). Jan and Petr had repeatedly declined offers of equity investment from foreign and domestic investors, choosing instead to retain their independence and financial control. While there was a cultural bias against the use of debt in the country, their decision to wait so long before seeking financial support from a bank was in large part due to the inability of local banks to understand the operations of a rapidly growing distribution business. The economic climate of the Czech Republic throughout the late 1990s made investment capital scarce and expensive.

---

[2] Adapted from the "How it all began" section at www.papirius.cz
[3] Ibid.

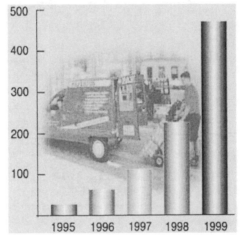

Customers

Net Sales
in Million
Kc

Figure 17.1. Papirius Growth in the first five years.

17.2.1 The Czech Economic Climate

Initially the darling of Eastern Europe, the Czech Republic began a period of economic stagnation in 1996. Most economic observers attributed the economy's sluggish performance in the late 1990s to structural issues that have impeded the country's transformation from the centrally planned economy that existed prior to 1989. As noted in 1999 by *World Trade Magazine,*

> Even three or four years ago the Czech Republic still looked like a bright spot, but now it's clear that privatization has been mishandled and that banking laws and creditor protection remain week. The political logjam means none of these

problems is being addressed. The only encouraging sign is last year's increase in FDI to $1.8 billion, which should at least underpin some productivity growth.[4]

The initial period following the 1989 "Velvet Revolution" was dominated by two issues: the country's separation from their sister state Slovakia after more than 70 years of joint statehood and the coupon privatization scheme championed by the conservative government of (Economic and then) Prime Minister Václav Klaus.

Coupon privatization involved distributing shares in non-strategic state-owned companies directly to individual citizens. While seemingly a capitalist's dream, the plan had two major shortcomings. First, the existing laws formulated under a totalitarian regime were inadequate to protect small shareholder rights. Second, individual investors seeking financial security either sold their shares to or placed the shares certificates and voting rights in the hands of banks that remained majority state owned. These facts, coupled with inadequate "Chinese Wall" provisions, often meant that banks with significant shareholdings in money-losing companies often loaned money to these companies in an effort to support their equity investments. The results were disastrous for the economy. Not only was scarce capital loaned to companies that would never repay the money, but massive bureaucratic enterprises were also effectively given stays of execution, delaying much needed reform and redistribution of human capital in the Czech economy. The backdrop of a massive state-sponsored bailout of all the major commercial banks clouded the start of the new millennium in the Czech Republic. The minister of trade was trying to get a restructuring program going, but it was not clear how it would be done or financed. Many wondered if the government would be willing to endure the pain.[5] Throughout 2000, a major effort was underway to dress up the banks for privatization. But this process was riddled with revelations of scandal. The headline on the July 4, 2000 *Prague Post* was like many others that summer declaring "Discovery of massive-debts and a criminal investigation follow high-level housecleaning." This painful economic climate made it difficult for a young business, like Papirius, to access the capital needed to fund growth. And banking was not the only institution going through upheaval. The same day's paper also headlined turmoil in the state-controlled television network, where independent thinking journalists who pressured Prime Minister Klaus for openness, quickly lost their jobs. Even with these challenges, many felt optimistic about positive economic changes ahead as the Czech Republic worked toward joining the E.U. in 2004.

---

[4] *World Trade Magazine*, Andrea Knox, July 1999.

[5] Ibid.

17.2.2   Expanding Outward from Prague

With poor opportunities in the capital market, Jan and Petr had to carefully choose their internally funded growth initiatives. By the summer of 2000, Papirius supported a catalog of over 3,500 office-related supplies and delivered anywhere in the Czech Republic within 24 hours. While the item selection was modest by western standards (an typical Office Depot store stocked over 8000 skus with over 14,000 skus stocked in fulfillment centers for business delivery), it surpassed private stationary stories and the local wholesale competitors who focused on specific supply segments (like paper). Moreover, the overnight delivery concept was new in the Czech Republic. The relatively large item selection and the fast, reliable service was a winning combination for Papirius' core small-business customers.

The firm started their operations by servicing only the Prague market. By early 1995, they expanded into Brno, the Czech Republic's second largest city, and by the end of 1998, they covered the entire country. They had outlined plans to enter some of the larger Slovak cities by spring 2001, and were considering strategies for entering Poland. Their bias was towards expanding into countries that had similar cultures and lacked established competitors.

The relative geographic centrality and market size of Prague led to a hub-and-spoke distribution system (described below). While they felt like they had only recently moved into their $3,300m^2$ customized facility in Dolni Chabry, their rapid growth and the lack of adjacent warehouse space meant that another move was a necessity. They had contracted with a developer who was willing to build out a $7,800m^2$ facility with options for another $13,000m^2$ – more than enough room to support an operation eight times the company's 2000 size. While intuitively this felt like ample space, they were anxious to plan far enough ahead to accommodate at least three years' growth.

17.3.   Papirius' Strategy

Papirius sold office products directly to business customers. Papirius was a pure distributor, in that it did not manufacture any of its products. All products were purchased from domestic wholesalers or a few international suppliers. The product line included assorted office products (paper, pens, pencils, toner cartridges, etc.), food and beverage products for at-work consumption, and a small number of specialty items stocked specifically for key customers. This was similar to the model followed in the U.S. by companies like Corporate Express and B.T. Office Products.

17.3.1   The Papirius Approach: Customer Service

The typical Papirius customer was a small firm of approximately 50 employees. Larger firms often bought directly from some of the same wholesalers who supplied Papirius. The orderer and recipient of the Papirius delivery were usually the same person, typically filling the role of an office manager or a similar position. The customer would place her order before 5:00 P.M. and expect delivery before noon the next day. Her interface with Papirius would typically be a detailed glossy catalog that was available upon request and automatically sent annually to existing customers. About 65% of the customers placed their orders by filling in the requisite information on the order request form and faxing it via an 0800 toll-free fax number to Papirius' Distribution Center. Papirius launched a web site in the late 1990s, quickly capturing 5% of the sales. The remainder used the phone to order. Petr estimated that internet access among their fax ordering accounts was as high as 95% and believed that it would rapidly replace the fax orders as customers became more comfortable with the web. However, he felt that the fraction of phone orders, originating from old-fashion buyers and those without internet access, would likely remain stable. The next interaction a customer would have with Papirius would usually be with the driver/delivery person the next morning when products were delivered.

The Papirius strategy was to offer its customers excellent service and value as defined by on-time free delivery (for orders larger than $40 US), competitive pricing, ease of ordering, and a large, complete selection of goods. The history of price consciousness among Czech consumers stemming from 50 years of economic hardship continued to exert a subtle, but important impact on Papirius' business. Even when purchasing for a relatively large business, Papirius' typical customer was extremely price conscious, and as a result, Papirius was hesitant to exert too much upward pricing pressure on its customers. For their small business customers, Papirius had to keep prices of many items below those found in small stationary shops. Given that the Papirius did not have to maintain expensive store-front real estate and quickly had larger volumes than small stores, they could easily compete with the stationary stores – even burdened with the cost of free delivery. Competing on price with the wholesalers for high-volume customers (like larger businesses) and products (like copy paper) was more difficult. Here Papirius emphasized the value of their wide range of products and fast service. Much of their advertising was used to educate customers about the value proposition that Papirius provided by saving their customers' employees' time. In tandem, however, they were pursuing a customer acquisition and retention campaign centered around a "Price Jackhammer" mini-catalog. Papirius viewed customer confidence in their delivered value to be of paramount importance.

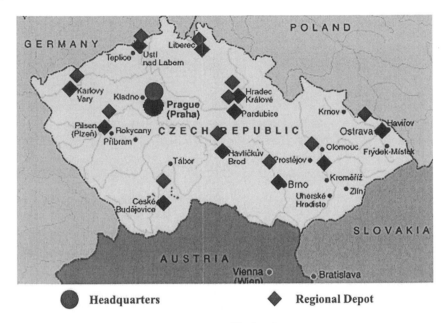

● **Headquarters**   ◆ **Regional Depot**

Figure 17.2. Facility locations.

17.3.2   Overnight Delivery Model

Papirius used an overnight delivery model to fulfill customer orders. The quick response and large SKU selection were the primary reasons large customers would choose Papirius over existing wholesalers. The order-processing department received customer orders throughout the day, and then the warehouse filled the orders in the evening. Each order was picked and then placed on a "spur" according to its final destination. Orders bound for Prague were queued according to their delivery route around the city. In addition to the Dolni Chabry distribution center, Papirius had 11 regional depots situated around the Czech Republic. Orders bound for different regions were queued up for the "line-haul" trucks that carry orders to depots around the Czech Republic. At these depots, orders were then reloaded into delivery vans according to their specific regional delivery route.[6] The regional depots were simply transfer stations and did not hold significant inventory.

By the early morning hours, order picking was complete and delivery drivers loaded their trucks. Orders for regional customer would first be trucked

---

[6] Map of Czech Republic from Maps.com

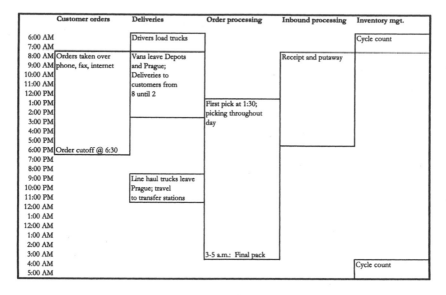

Figure 17.3. Operations Schedule.

to transfer terminals close to the customer and then transferred to delivery trucks by mid-morning. The order cycle was complete when drivers finished their delivery routes by around 2:00 P.M. The breakdown of a typical day at Papirius is shown in the Figure 17.3.

In 2000, Papirius used a conventional set of information systems to manage its processes. There were four important components in their suite of systems:[7]

Web Applications	Order Management System	Warehouse Management System	Accounting System
• Web front end for porder management • Online catalog • Online brochure	• Order processing • Purchasing • Inventory management • Customer invoicing	• Manages order fulfillment • Pick/pack/ship • Slotting • Receipt & put away • Queues orders but outbound spur • Manages material handling equipment in warehouse	• General ledger • Accounts payable • Accounts Receivable • Supplier invoicing

Figure 17.4. Key Systems.

---

[7] Source: Interview with Papirius' senior management.

17.3.3 Papirius' Organization

Papirius' business processes were straightforward. There were eight groups within the company:[8]

Order processing	Order entry and customer service
Marketing	Prepare and distribute catalogs; choose products for inclusion in catalog; promotion; Sales
Product group	Purchase products from suppliers; establish service levels for various products; establish and maintain vendor relationships
Warehousing	Receive inbound purchase orders from vendors; Pick/pack/ship customer orders
Distribution	Deliver products to end customers; manage delivery fleet
Information technology	Systems development and maintenance
Accounting	Accounts payable, accounts receivable, management reporting
Personnel	Human resource management

Figure 17.5. Business Processes.

All of these functions were housed at corporate headquarters in the central Dolni Chabry distribution facility. Petr and Jan split responsibilities, each taking four groups, and managed the business on a day-to-day basis:[9]

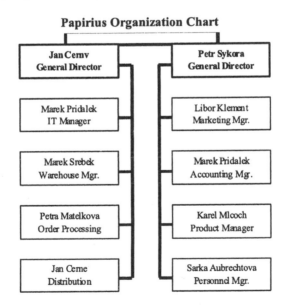

Figure 17.6. Organizational Structure.

---

[8] Ibid.

[9] Source: Internal Papirus documents.

17.3.4   Capabilities

Papirius served a base of over 16,000 customers, delivering approximately 1000 orders daily. While an average day generated about 3M Kc ($82,500), swings of 30% were common with Monday, and Fridays typically slower than midweek. The sample monthly operations report, shown below, provided a set of sales and fulfillment statistics:[10]

Date:	27/04/00
Turnover	
Today's gross revenues	2,578,024
Month-to-date revenues	77,340,720
Lines per day	
Items ordered	5,367
Items fulfilled	5,351
Items unfulfilled	16
Percentage	99.70%
Invoices per day	
Total invoices	743
Invoices completely filled	728
Invoices not completely filled	15
Percentage	97.98%

Figure 17.7. Sample monthly operations report.

Exhibit A2 contains other statistics about the Papirius operation. Papirius strove to provide very high levels of order fulfillment with fill rates of greater than 99.5% for each item they stocked. To showcase its high service, Papirius offered customers the guarantee, "Tomorrow or It's Free." Items that were not in stock would be delivered later at no charge. Of course there were some limits on the types of products covered under the guarantee and customers could not order multiple units of an out-of-stock item and receive them all free. Petr estimated that the promotion cost less than $200/day and thought the publicity was well worth it. In fact, he felt that it was good that some customers received free merchandise – he argued that it was a positive thing to give something to the customer from time to time to reinforce the value of their service and ability to meet promises. Providing high levels of customer service did cost Papirius precious capital. With tight payment terms from many of their suppliers, inventory investment was a key consideration in growth –

---

[10] Source: Internal Papirius reports, interviews with Papirius management.

particularly as Papirius added more items to its offerings. Papirius worked hard to keep inventories low – holding about one month of demand in inventory, which is less than half of Office Depot. The company did this by carefully monitoring its customers' interests and the sales data to determine which items to add and maintain in the catalog and which items to prune.

17.4. Superstores Expand Globally

Over the past decade, U.S. and European superstores, like Wal-Mart, Carrefour, Toys 'R' Us and Office Depot, have taken their concepts abroad, attempting to gain market share and economies of scale on a global basis. Wal-Mart, for example, opened its first international supercenter in Mexico in 1991.[11] By 2000, Wal-Mart's international division operated in nine countries, with a European presence in Germany and the U.K.[12] Underscoring the increasing importance of international expansion, Wal-Mart expected 30% of its profit growth to come from abroad by 2005.[13] So-called "Category Killer" retailers had headed abroad as well. Toys R Us International opened its first stores in 1984, in Singapore and Canada. By 2000, Toys R Us maintained over 450 international stores in 27 countries, including franchise operations.

17.4.1 International Struggles

While international expansion had provided these retailers with a good avenue for growth, many superstores and category killers encountered difficulties abroad. Long known for its flawless execution stateside, even Wal-Mart stumbled numerous times in entering overseas markets.

Through a joint venture with C.P. Pokphand Co., Wal-Mart opened three Value Club discount stores in Hong Kong in late 1994. Joe Hatfield, who headed Wal-Mart's China unit, acknowledged that those stores stocked some merchandise that was more appropriate for Westerners than Hong Kongers. One of the outlets prominently displayed a basketball hoop – an unlikely adornment for Hong Kong's high-rise apartments.[14]

Likewise, in Japan, retailers met with mixed success. The office superstores were among a big crowd of American retailers that had run into trouble there.

---

[11] Source: Company information posted at http://www.walmartstores.com/corporate/coinfo_international.html

[12] Source: Company information posted at http://www.walmartstores.com/corporate/coinfo_international.html

[13] "Wal-Mart Goes Shopping In Europe," Jeremy Kahn, *Fortune*, 6/7/1999.

[14] "Wal-Mart Expands Cautiously in Asia," *Wall Street Journal*, Bob Hagerty and Peter Wonacott, 8/12/1996.

During Japan's recession of the 1990s, U.S. companies stormed in, aiming to make a killing by revolutionizing a tradition-bound retail industry. They sparked profound changes with their new products, offering better value and wider selections. But only a few, including Toys R Us Inc. and Gap Inc., truly succeeded. Many others were scrambling to revise their strategies, and some were even giving up.[15]

Though it was still early in Wal-Mart's Brazilian foray, the Bentonville, Arkansas retailer had been posting wider-than-expected losses, resulting from a ruthless price war with well-established local competitors.[16]

Other retailers felt the pain along with Wal-Mart as they adapted to new markets. Some retailers failed to merchandise properly for the new markets they entered. Other firms were unable to circumvent local middlemen in purchasing. All firms had to deal with hostile responses from home-market competitors. These expansion difficulties all came at a time when growth in the home markets was drying up, making growth abroad even more imperative.

17.4.2   Office Depot Follows Suit

Wrangling with all of these competitive pressures, Office Depot had also chosen to go abroad for growth. Office Depot claimed to be among the world's largest office products retailer, operating a total of 989 stores operating in 10 countries worldwide by mid 2000.[17] Having begun operations in Delray, Florida, Office Depot ran operations in Canada, France, Japan, Mexico, Israel, Poland, Hungary, Columbia and Thailand.[18] These international expansion plans were in part enabled by the acquisition of Viking Office Products. Viking, an office products cataloger, had already developed a strong international presence in both catalog and web sales. Office Depot had made it clear that expansion in Europe was coming and that it planned to rely on Viking's knowledge of the marketplace to succeed.[19]

Viking had an established record in cracking new European markets. It reportedly broke even in the U.K. during its first year; when it launched in Germany a few years later, it reportedly showed a profit after 18 months. As in the U.S., Viking was not the cheapest option: customers just thought it was. After

---

[15] "U.S. Superstores Find Japanese Are A Hard Sell," *Wall Street Journal*, Yumiko Ono, 2/15/2000.

[16] "Wal-Mart Won't Discount Its Prospects in Brazil, Though Its Losses Pile Up," *Wall Street Journal*, Matt Moffett and Jonathan Friedland, 6/4/1996.

[17] Source: Company information posted at www.officedepot.com

[18] Source: Company information posted at www.officedepot.com

[19] "Viking Purchase Readies Chain for Overseas Expansion," *Discount Store News*, 10/26/1998, Anonymous.

years of lackluster service, U.K. customers appreciated the high levels of service offered and the convenience of buying from a catalog.[20] Office Depot was using this platform to enter new markets via the web. Country-specific web sites had been launched in three countries, with at least three more to come.[21]

As Office Depot continued to integrate Viking into their operations, they increasingly endeavored to gain scale globally in their operations. Office Depot was just beginning to realize benefits from global purchasing. It had deals with well over 65 major manufacturers on a global basis and they planned to go through each category of products to identify cost savings and/or private label opportunities. Company executives and analysts both agreed that the biggest opportunities for the future growth were the internet and overseas sales channels. The company's international division, which included operations in 17 countries, continued to report the biggest increases in existing-store sales of any area outside the internet.[22]

17.4.3 Office Depot in Japan: Not-So-Smooth Sailing

Just as with other superstores and category killers, Office Depot had encountered bumpy roads abroad, including Japan. Like many foreign merchants, Office Depot had linked up with local partners that knew the lay of the land, a strategy that permitted quick expansion but limited operational control. After a promising start, "the Americans soon hit a wall." Japanese office products were very different from those in the U.S. For example, loose-leaf binders had two rings instead of three, requiring Office Depot to buy most products from traditional local suppliers.

In the U.S., Office Depot was able to operate each store profitably by selling a portion of the products from each store via retail and then delivering the rest directly to consumers. They presumed that this model would also work in Japan. Instead, the Japanese stores had been plagued with problems where the challenge was delivery and acceptance. In Japan, Office Depot struggled to shrink the store footprint to make the economics work. These profitability problems were compounded by the actions of local competitor Plus Corporation, the number two competitor in Japan. Struggling to lift its sagging sales, Plus created a small division in 1993 called Askul to sell discounted stationery by catalog. Askul targeted exactly the same customers as its U.S. rivals: small

---

[20] "Business-to-business Overseas Success Stories," Catalog Age, 3/1/1999, Mike McKenna.

[21] "Office Depot: Gravitating Towards Its Strengths," JP Morgan Equity Research, 4/12/2000, Danielle Turnof.

[22] "Office Depot Clicking Online, But Not In Stores," Portland Oregonian, 5/14/2000, Elaine Walker.

business owners that were not getting the discounts that big companies buying in bulk received. In the face of operational difficulties and competitive pressures, Office Depot was forced to make drastic changes, closing its large store in Hiroshima and focusing on Tokyo.

17.5. A Lurking Threat

By 1998, Office Depot had established a sizeable presence in Poland and Hungary through a series of retail super centers and a growing business-to-business direct delivery business. The Viking acquisition gave Office Depot a large presence in Europe and a platform from which they would be able to accelerate expansion across Central and Eastern Europe. Papirius viewed Office Depot's strategy as a serious threat to their business and believed they would enter Prague before the end of 2001. Jan and Petr estimated Office Depot's current Polish turnover (including their supercenter sales) to be around $100 million, or about five times Papirius' current level. But Office Depot was not the only threat. European office supply companies, such as French Lyreco, had already entered Poland and were eyeing other countries in Eastern Europe.

As they began formulating their potential responses to an Office Depot entrance into the Czech market, Jan and Petr thought of several different responses. After seven years of aggressively growing their business under adverse conditions, they were very focused on not squandering their incumbent position. Among the alternatives they discussed were:

- Developing a competing chain of superstores.
- Aggressively growing their current delivery business by adding SKUs to lock-in the best customers.
- Expanding into the home delivery market.
- Opening their distribution network to other retailers.
- Partnering with a retail operation that could benefit from Papirius' distribution network.
- Acquiring domestic competitors to gain market share.
- Competing creatively, not just on price.

Many of the proposed strategies would involve a movement away from the "organic" internally funded growth that had characterized their business to date, and could quite possibly require dilution of Jan and Petr's economic interests and operating control of the business.

They believed that the Czech business environment, despite the country's accession into the E.U. in 2004, was a hostile environment for any new business. They knew that their strong presence in the Czech market could serve as a deterrent to new entrants, and that a new entrant would have to endure a lot of costs as they learned the local landscape.

17.5.1 Plotting Their Future

Although Petr and Jan had traditionally made it their first priority to focus on growth within the Czech market, two things were becoming clear to them. First, Poland offered a market potentially four times larger than the Czech market with similar cultural and operating conditions, and Slovakia, the Czech Republic's former sister state, stood barren with no major competitors. Second, they wondered about their decision to provide coverage to the entire Czech Republic. From a marketing perspective, it was a huge advantage summed up by the tag line "Anywhere within the Republic within 24 hours." Operationally and financially, however, they knew it was a different story.

Petr and Jan were also trying to realistically assess the threat posed by a potential entry by Office Depot into the Czech marketplace. Clearly, Office Depot had deep pockets, and viewed their entry as a long-term investment. Additionally, Office Depot would have the ability to source their product internationally and undercut Papirius on many commodity-type products. Office Depot would also have a broader strategy, opening retail super centers alongside their business-to-business delivery services.

On the other hand, however, they would be entering the country with relatively little experience in the Eastern European market, and far less knowledge of the Czech customers. Petr and Jan were determined to make Office Depot's entry into the Czech market as unattractive and unprofitable as possible, but without resorting to a price war.

17.6. Update and Learnings

Papirius continued its rapid growth, doubling 1999 sales by 2002 (Figure 17.8) while moderately increasing the SKU offerings (from 3000 to 4500). Petr and Jan extended their distribution network into neighboring Slovakia in 2001 and then into Hungary in 2002. Given the formidable competition in Poland, they postponed entering that market. Rather, in 2005, they entered Lithuania through the acquisition of Mabivil – a local office supply firm.

Fortunately, Petr and Jan's fear of foreign competition was not immediately realized. All three major US office supply firms were busy with their own domestic problems, giving Papirius four more years to expand and solidify their base market. However, with the introduction of both Czech Republic and Hungary into the EU in 2004, these markets became even more attractive to foreign competition. U.S.-based Staples made the first move in the Czech

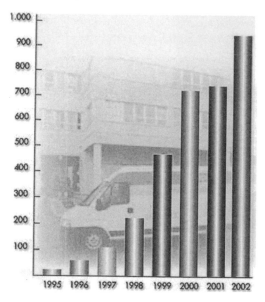

Figure 17.8.   Sales in M Kc.

Republic acquiring Austria-based Pressel Versand International and the office products division of Denmark-based Malling Beck A/S. These acquisitions extended Staples' European catalog operations into both Hungary and the Czech Republic among several other European markets. Also in 2004, Office Depot bought Elso Iroda Superstore Kft in Hungary that operated four Office Depot stores under a license agreement and announced plans to open 17 additional locations by the end of 2006. The arrangement marked the first time Office Depot directly owned international retail stores operating under its name.

Papirius' success in organically developing a distribution business in Eastern Europe illuminates several important lessons:

1. As emerging markets mature, customers' expectations for service increases. Distribution firms can use service to quickly capture these more demanding customers. Focusing on a service based strategy, Papirius was able to leapfrog the competition.
2. Distribution in emerging markets requires more infrastructure development and closer attention to execution. Papirius had to develop and extend is distribution capabilities, without the help of many other outsourced partners. By starting in a major city and building out from an area of strength, Papirius could extend its reach while maintaining high levels of service and efficiency.

3. Organic growth in an emerging market requires careful inventory management to conserve capital. Increasing the product offerings in any distribution business makes it harder to achieve perfect orders without increasing inventory.[23] Papirius carefully managed the SKU growth to maintain high service (high order fill rates) while avoiding rapid increases in working capital.

4. Defending local markets from multinational competition requires a competitive supply chain that is tailored to local tastes. By focusing on the needs of it core customers, Papirius was able to quickly grow and solidify its based while driving efficiencies to make it competitive with world-class firms.

5. Some of the best opportunities to develop a western style distribution business occur before an emerging market ascends onto the world economic stage. Building their business during the decade before the Czech Republic and Hungary entered the E.U. in 2004 gave Papirius the opportunity to solidify their business without multinational competition. Clearly, the competition will now become more intense.

For firms in an emerging market where the competition is not mature, there are always opportunities to exploit some dimensions of service that have not yet been satisfied. One way to gain entry in such markets is to look for such gaps and build a solution to fulfill it. This is what Papirius did. They identified the lack of rich variety and long response time as the gaps of the current market, and created a solution to address them. However, to be successful in the long run such firms must find true competitive advantage to maintain a defense against larger foreign competitors. This requires being efficient and effective so that the solution is sustainable and profitable. For Papirius, growth in their markets while replicating large scale distribution practices enabled them to be efficient enough to withstand global competition. However, a key part of their success was building an understanding and relationship with their customers that could not easily be replicated.

---

[23] "Improving Supply Chain Performance Using Order Fulfillment Metrics," *National Productivity Review*, Johnson, M. Eric and Tom Davis, Summer, 3–16, 1998.

Income Statement (in 000s)

			1997	1998	1999
Net Revenues			108,530	221,514	465,739
COGS			79,285	165,490	325,039
Material/Energy Cons.			4,384	7,015	13,968
Services (leases, rent, etc.)			10,614	22,293	61,896
Wages, Social Security			8,969	21,244	52,893
Fines and Penalties			9	43	93
Depreciation			1,045	2,935	4,528
Taxes			-	-	-
Other Expenses			2,935	1,971	4,902
Net Income			1,289	523	2,420

			1997	1998	1999
Net Revenues			100.00%	100.00%	100.00%
COGS			73.05%	74.71%	69.79%
Material/Energy Cons.			4.04%	3.17%	3.00%
Services (leases, rent, etc.)			9.78%	10.06%	13.29%
Wages, Social Security			8.26%	9.59%	11.36%
Fines and Penalties			0.01%	0.02%	0.02%
Depreciation			0.96%	1.32%	0.97%
Taxes			0.00%	0.00%	0.00%
Other Expenses			2.70%	0.89%	1.05%
Net Income			1.19%	0.24%	0.52%

Balance Sheet (in 000s)

			1997	1998	1999
Assets					
	Fixed Assets		1,263	6,125	11,644
	Current Assets				
		Cash	2,582	5,531	5,615
		Inventory	5,926	15,811	25,834
		A/R	7,719	14,954	36,993
		Accruals	961	3,185	2,409
			18,451	45,606	82,495
Liabilities					
	Line of Credit		6,000		13,000
	Required Reserve		1,148	570	952
	A/P		12,655	31,577	54,153
	Other Payables		3,244	4,766	8,377
			17,047	42,913	76,482
Equity					
	Paid-in Capital		100	100	1,000
	Retained Earnings		1,304	2,593	5,013
			1,404	2,693	6,013

Exhibit A1. Selected Financial Data[24]

Papirius Operations Information			
Delivery Vans		85	
Line Haul Trucks		5	
Total Number of Drivers		90	
Cost/driver/day (wages + benefits)		800	
Cost/truck/day (daily lease expense)		450	
Daily turnover		3,000,000.00	Kc
SKU base (2000)		3500	
SKU base (1999)		3000	
SKU base (1998)		1500	
Time between deliveries (Prague)		10	min.
Time between deliveries (Regional Depots)		20	min.
Time required to complete a delivery		5	min.
Orders processed/regional depot/day		25	
Total number of regional depots		11	
Average order size		3000	Kc
Number of suppliers		80	
Customer credit terms			
	10 days	70%	
	<45 days	20%	
	Cash on Delivery	10%	
Source: Papirius internal reports, interviews with management			

Exhibit A2. Selected Operations Data

---

[24] 1USD = 36.34 Kc (7/2000). The Czech Republic in now part of the EU.

Hau L. Lee and Chung-Yee Lee (Eds.)
Building Supply Chain Excellence
in Emerging Economies
©2007 Springer Science + Business Media, LLC

Chapter 18

A PATH TO LOW COST MANUFACTURING FOR INTEGRATED GLOBAL SUPPLY CHAIN SOLUTIONS

Wesley Chen
Crimson Shanghai (Formerly, with Solectron)

Abstract: Many multinational corporations are seeking low cost labor and better sourcing to improve their competitiveness. China as the largest emerging market provides a great stage for this. The labor cost advantage achieved from operating in China is obvious. However, finding the right sourcing partners and developing the local management talent are equally important and much more difficult to achieve. Any company planning to source goods and services from China must be aware and ready for the challenges China poses, such as the unique regulatory environment, IP protection concern and surging turn over rate. Selecting the right location is the first and the most important step which should be based on the maturity of the local supply chain, customer proximity, pro business level of local government, labor availability, facility cost, logistic convenience, and level of tax intensives. A company must achieve these production and sourcing efficiencies without compromising productivity and quality. In addition, one should implement six sigma lean disciplines and apply "design for China supply chain" to ensure production efficiencies and to fully leverage the competitive advantages that China can provide.

18.1. Background

I appreciate the opportunity to share my Solectron China experience from an integrated global supply chain of electronics manufacturing service perspective.

Solectron provides a full range of global manufacturing and integrated supply chain management services to the world's premier high-tech electronics companies. Solectron offers new product design and introduction services, materials management, product manufacturing, product warranty and end-of-life support. Solectron based in Silicon Valley, California had sales of $11.6 billion in fiscal year 2004.

Solectron was initially established as a solar electronics company in late 70's. Unfortunately, solar electronics never really took off. In 80's, Solectron changed its company value proposition to contract manufacturing which was not yet a popular concept. In 90's, Solectron was recognized twice as Malcom Baldridge National Quality Award winner, focuses on quality of service and positions as supply chain facilitator. Solectron became one of the leading Electronics Manufacturing Service providers in the world.

At 1996, Solectron established its China operations in Suzhou to provide low cost manufacturing for international customers. Solectron first entered China to leverage on low cost labor and most business was transfer-in from high cost regions. When turning into the new millennium, after the "Dot Com bubble" burst, cost reduction was the hottest topic among all companies during the economy recession. At the mean while, China local supply chain for electronics manufacturing is getting very completed even the semiconductor foundries. "Localize supply chain" becomes a very important subject of further cost reduction. Also, an enormous domestic market is emerging whereas the outsourcing concept becoming increasing popular in Asia. Therefore, Solectron China's businesses are no longer limited to transfer-in. Mainland Chinese, Taiwanese and Japanese OEM's/ODM's outsourcings take shares rapidly.

18.2. Why Manufacture Electronics in China?

Around last Christmas my nine year-old daughter came home with a question from school, "Daddy, where does Santa live?" Of course, I answered "North Pole" intuitively. She laughed at me and said, "Daddy, it is China now, because we find all the Christmas gifts are printed Made in China." Yes. China is now becoming the factory for the whole world, not only on toy, but also on high tech devices, automotives, and more. Some said, "Whoever missed China will miss the future." While all companies are looking ways to cut cost, China has the advantage not only in lower wage but also in the lowest total supply chain cost.

China's economy growth was mainly along the coastal belt. They are Pearl River Delta, Yangtze River Delta, and Great Beijing Area. If we draw a diagonal line on the China map, from northeastern to southwestern corner, 96% of

- City per capita annual income at $1,250 and rural per capita income reached $363 (2004).
- GDP $1.68 trillion, with a growth rate of 9.5% (2004).
- Post-WTO, China emerged as the favorite destination for foreign direct investment ($153 billion in contract, 2004).
- During 2004, on average real urban disposable incomes rose by 7.7% and urban consumption rose by 7.5%.

Source: China Statistical Yearbook

population would fall on the right and that leaves two high potential areas yet to be developed. That's Chongqing / Sichuan area & North East area. When the logistic infrastructure gets more completed in the future, these two areas would be riding on the next wave of economy boom in China. With uninterrupted GDP growth in the past decade, even during Asia financial crisis and SARS, China again proves its economic growth momentum.

There are two main reasons why you are here in China; first, cost advantage. Secondly, because you can't afford not to be in China.

18.3. Cost Advantage

China's hourly labor wage is low and stagnant due to large population whereas most low cost countries have relatively shorter low cost span. In other words, low cost advantage of China's non-skill labor will last many more years until there is noticeable change. However, the same rule may not apply to skill labor and white-collar level managers. Although the hourly wage is relatively stable, the wage of white collar is moving up fast. This situation is especially severe for those foreign-language-capable with international business exposure expertise in the booming economy area.

For manufacturing operations at China, initially, the cost reductions are from the wage differences. However, supply chain cost reduction plays an increasing important role for continuous cost reduction efforts. For EMS business, value-added cost is about 10% to 20% and the remaining are mostly material cost and slim profit. Therefore, it is especially important to find ways to reduce cost in material supply chain through localization.

Worldwide Hourly Wage Rates

Source: <<Electronics Manufacturing in China Second Edition 2003>>

18.4.   You Can't Afford Not To Be in China

There are tremendous opportunities in the domestic market. Although there may be only a small fraction of people can afford modern consumer electronics, home appliances, and cars. Let's say only top 5% of China population have the buying power. 5% of 1.3 B is about 65 millions which is more than three times of Australia's total population.

Many MNCs in China heavily focus on export. There is more to be considered beyond production cost. Freight, import / export tax and duty, and lead time increase need to be added in the equation.

China has a total landed cost advantage, even for complex products.

Through a case study of total landed cost comparison, a complex product was manufactured in selective locations and shipped to multiple destinations. Of course, labor cost was not the only consideration. Other variables includes freight cost, import tax and duty, supply chain cost, etc... Not a surprise, China has the lowest total landed cost in many different scenarios.

Customer Sites		Manufacturing Sites							
		Singapore	China		United States		Mexico	Western Europe	
			Shanghai	Shenzen	Oregon	Columbia	Guadalajara	Budapest	Scotland
United States	West	5	2	1	6	8	4	3	7
	Central South	5	1	2	8	6	3	4	7
	Midwest	5	1	2	8	6	3	4	7
	East	5	1	2	8	6	3	4	7
Mexico		5	2	1	6	8	3	4	7
China	East	3	1	2	7	8	5	4	6
	South	4	2	1	6	8	5	3	7
Japan		4	1	2	7	8	5	3	6
Malaysia		3	2	1	7	8	5	4	6
Hungary		4	1	2	8	7	5	3	6
Netherlands		5	1	2	8	7	4	3	6
Scotland		5	2	1	8	7	4	3	6

Notes: Ranking explanation—"1" the lowest landed cost site, "8" the highest landed cost site
= Lowest-cost sites

Source: Solectron study

18.5. Challenges

Labor rate advantage is obvious, however, the key cost reduction should be in production efficiency improvement and supply chain cost optimization. After all, finding a world class manufacturing partner is the key to ensure the quality will not be compromised due to low cost. Some challenges should be aware of while conducting business in China.

18.5.1 Unique China Regulatory Environment

Government regulatory and policies are volatile even after WTO. Of curse, it may be unrealistic to expect government to consult with you before making any changes. For example, the 17% VAT rebate policy for export changed in the beginning of last year which caused us a hard time to communicate with overseas customers.

18.5.2 Intellectual Property Protection

Intellectual Property protection will remain as an ongoing concern before China truly migrates from "Made in China" to "Innovate in China". Although there have been some government efforts to strengthen the IP awareness, there is still some distance to be pirate-free. Most MNCs in China are finding their own ways to protect IP.

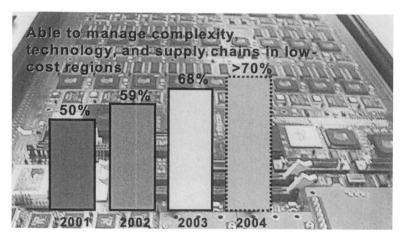

Source: Solectron study

18.5.3   Resource Constraint

Western trained middle-class management shortage, insufficient electricity, and congested transportation on some areas are challenges. I believe they will get worse before getting better.

18.6.   Solectron China Experience

The success of Solectron China Operations should be attributed to not only our own efforts, but also the collaboration with local government and custom to improve the speed and flexibility of supply chain logistics.

Solectron built sophisticated capabilities in low cost regions. It has been an ongoing effort to optimize our low cost capabilities.

By the end of year 2004, more than 70% of capacity will be in low cost regions. Solectron China is able to manage not only the high volume, but also the high complexity and low volume productions.

18.6.1   Right Location Is the First Step

At the initial stage of building up China capability, Solectron selected Suzhou as its China base in 1995. During the site selection process, logistics convenience, manpower availability, local government support, facility cost, supply chain proximity, and customer locations were major considerations. Among them, I believe a right first step is to find the right location.

Some said: China is hot, Yangtze River Delta is hotter, and Suzhou is the hottest.

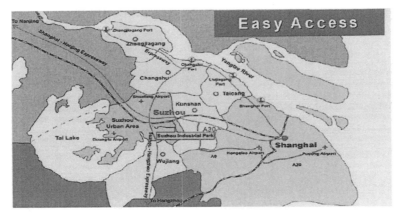

Source: Suzhou Industrial Park Administrative Committee

A small area with only 0.1% of land, 0.5% population, created 2.5% of China's GDP, 8.6% of Export, and captured almost 10% of foreign investments last year. As a matter of fact, Solectron China picked the right location for success.

18.7.   Proximity to Local Supply Chain

To have suppliers close by is one of very important considerations to reduce shipping cost and lead-time. Solectron Suzhou has most local suppliers within 1 day trucking distance and most of them within 2 hours.

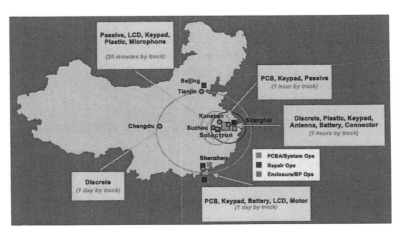

Source: Solectron Suzhou study

18.8.   Government Support and Cooperation

Solectron has been working with Suzhou-Singapore Industrial Park and Suzhou Custom as a team to improve material inbound and outbound logistics.

Manual process trade log is deemed as a very cumbersome formality especially for EMS. Due to large number of different raw material part numbers, Solectron spearheaded the initiative of Electronic Process Trade Log (e-PTL) project with Suzhou Custom. Solectron linked (EDI) its MRP system with custom to make custom clearance process become paperless, mistake-proof, (avoid wrong HS code), short lead-time, and easily reconciled. Solectron was the first to establish e-PTL process with custom in China and now this concept is widely implemented to other customs in China.

The cooperation between local government and Solectron has being a great experience. Beside e-PTL, Virtual Hong Kong and Virtual Suzhou International Airport are some other great on going initiatives. Several initiatives sponsored by Suzhou Industrial Park and Suzhou custom are listed here for your reference:

1. EDI Linkage/EIEMS (EDI 联网 / 电子手册管理系统)
– Ensure the smooth and fast customs clearance, paperless management, effective customs supervision, minimize risk

2. Weekend OT per Request (周末预约加班)
– Ensure consistency of customs clearance, reduce inventory level and improve materials turnover

3. SLC/EBW/EPZ Operation (苏州物流中心/ 出口监管库/ 出口加工区)
Suzhou Logistics Center (SLC) – One-stop service for customs clearance
Export Bonded Warehouse (EBW) – Provide virtual overseas HUB
Export Process Zone (EPZ) – Especially set up for 100% export business

4. SZV Model (虚拟的苏州机场)
– Virtual Suzhou airport, avoid customs formality in Shanghai, improve customs clearance cycle time

5. Bonded Logistics Park (BLP) – Coming in near future (保税物流园–近期l内成立), enjoy immediate Value Added Tax (VAT) rebate, best place for Vendor Managed Inventory (VMI) hub

These initiatives changed many people's bureaucratic perception about China. These ongoing efforts foster an ideal environment for industries to grow their businesses in China.

18.9. Six Sigma Lean Manufacturing

Solectron manufacturing system uses Six Sigma Lean to ensure a high-quality and cost efficient environment in China. We benchmarked the world-class best practice for manufacturing excellence in China Operations.

Six Sigma is a statistical and data driven approach resulting in reducing the process variance. Lean focuses on value stream mapping, value chain flow, pull system, and Kaizen to reduce waste. These continuous improvement initiatives are not limited to production floor, but also the business and logistics processes.

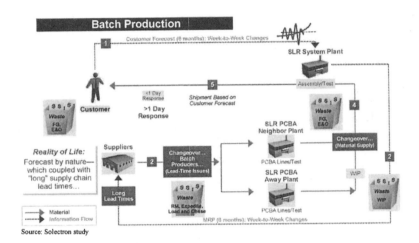

Source: Solectron study

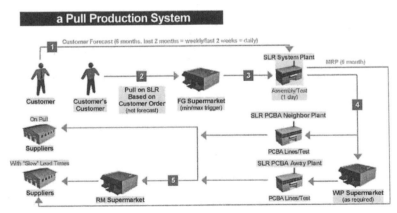

Source: Solectron study

From batch production to a pull production system. Solectron China went a long way to differentiate itself from competitors.

Under ideal case and ideal supply chain, all material should arrive on JIT basis to meet production needs. However, in reality, forecast by nature which coupled with "long" supply chain lead time items, when demand fluctuates, the WIP, excess and obsolete material become the cash cycle killer – inventory.

In a pull production system, forecast is used only to "capacitize" Solectron and supply chain. In other words, there is less waste and more value throughout the production. However, this system requires close participation of both customers and suppliers. The whole supply chain concept is now a "Pull" system using the material supermarket to cope with demand fluctuations.

18.10. Solectron Has Advantages in Logistics in China

China plays a very important role in Solectron's global supply chain. With many years' endeavors in creating a win-win scenario between Solectron China and local government, Solectron Suzhou has earned its creditability in logistics in China. However, some processes listed are unique to Solectron Suzhou operations.

Custom Bonded Factory – Awarded by China Custom

Bonded factory license allows importing and keeping material as bonded goods until the real domestic sales happen.

"AA Class Company"–Certified by China Customs

No import duty/VAT required when importing bonded material
Lowest inspection rate: <1%

"Fast Customs Process Company"–Certified by China Customs

Pre-Customs clearance
Critical shipment release process prior to formal Customs entry

Electronic Process Trade Log (e-PTL)

e-PTL: Electronic tool to manage materials and clearance through Customs
First company in China to deploy e-PTL Customs process (July 2000), which reduces Customs clearance from 7–8 days to 1.5 days

Logistic Hub in Suzhou Industrial Park (Virtual Hong Kong)

VMI model with full range of services from import/transportation/customs clearance to specialized third-party logistics provider.

With the above mentioned prestigious logistics convenience, Solectron Suzhou inbound supply chain enables the shortest logistics lead time. Ship-

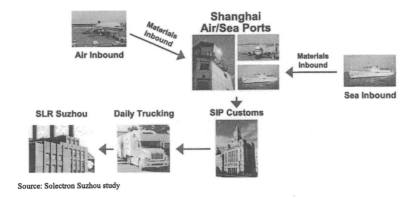

Source: Solectron Suzhou study

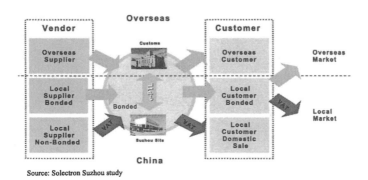

Source: Solectron Suzhou study

ments are shipped through Shanghai air/sea port. Bonded transfer to SIP customs and receive customs clearance at SIP customs.

All raw materials can be imported through e-PTL as bonded material. 1.5 working days from shipment-landed airport shipments delivered to Solectron Suzhou by trucks.

It is imperative to comply with government regulation and have the right logistics model to achieve a frictionless inbound and outbound process.

There are two major process differences compared with most competitors at China. First, "e-PTL" and secondly, "Bonded". e-PTL exempted us from cumbersome manual trade log process and Bonded enable Solectron to delay the payment of VAT to the point of sale.

With these processes, we are able to bring in raw materials and ship out finished goods in a much more efficient way than our competitors do.

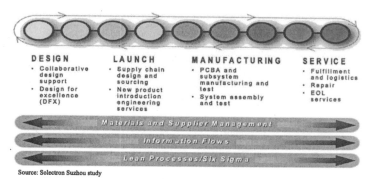

DESIGN
- Collaborative design support
- Design for excellence (DFX)

LAUNCH
- Supply chain design and sourcing
- New product introduction engineering services

MANUFACTURING
- PCBA and subsystem manufacturing and test
- System assembly and test

SERVICE
- Fulfillment and logistics
- Repair
- EOL services

Materials and Supplier Management

Information Flows

Lean Processes/Six Sigma

Source: Solectron Suzhou study

18.11. Design for China Supply Chain

One very important concept needs to be highlighted here is "design for China supply chain." We are in the process of working with our customers' designers to ensure the approved vender list includes the availability of component sourcing in China, therefore, the costs of bring in parts all the way from overseas to China for manufacturing can be saved. At the mean while, minimize the sourcing change during volume production to ensure the consistence of product quality.

Solectron China is an integral part of Solectron's powerful integrated global network. Regional design and New Product Introduction (NPI) centers close to customers R&D teams. Manufacturing focuses on optimizing value and risk. In addition, service localized for end users in each market.

18.12. Conclusion

While China becoming the factory for the whole world, you can't afford not to be in China no matter from cost advantage point of view or the attraction of its fast growing, dynamic and large emerging market.

For those already in China, while enjoying the low cost advantages, there are still some challenges to be addressed. Supply chain localization is the necessary next step for further cost reduction. Above all, the best strategy is to begin with "design for China supply chain". In this unique China regulatory environment, it is imperative to proactively work with government to stay in tune with the changes and to improve the industrial environment. When China becomes such an attractive hot land for foreign players, resource constraints in English-speaking skills will be inevitable.

Time to Accelerate Your Business Partner with Solectron

A Chinese old saying about three pillars for success: "Right Timing, at the Right Location, with the Right Partner!!" –. Wish you all have a successful and fruitful journey in China.

Hau L. Lee and Chung-Yee Lee (Eds.)
*Building Supply Chain Excellence
in Emerging Economies*
©2007 Springer Science + Business Media, LLC

Chapter 19

TRANSFORMING AN INDIAN MANUFACTURING COMPANY
The Rane Brake Linings Case

Ananth V. Iyer
*Krannert School of Management
Purdue University, India*

Sridhar Seshadri
*Stern School of Business
New York University, USA*

Abstract: Rane Brake Linings (RBL) is a manufacturing company, located in India, that epitomizes the dramatic transformation in manufacturing that is currently occurring in the Indian industry. RBL's journey to win the Deming prize in 2003 required organizational and procedural changes whose impact is seen in the way they deal with both managing supply costs and managing customer service upstream and downstream in the supply chain. Today, RBL is one of thirteen Deming prize winning Indian manufacturing companies. RBL's transformation, in the Indian context, suggests an approach to develop a globally competitive manufacturing firm by developing superior management processes.

19.1. Trade Liberalization and its Impact in India

Recent years have seen significant changes in the Indian manufacturing environment in response to changes introduced by successive Indian governments. The key changes, which started in 1991, included progress towards the elimination of quotas and liberalization of imports, reduction of tariffs for imports and liberalization of Foreign Direct Investments (Sivadasan 2003). The impact on the Indian industry has been significant, with projections for GDP growth improving from the 2–3% historic range to the 6–8% range since the 1991 reforms.

We focus on how these macro shifts affect one specific Indian company that had to transform itself to compete. While care is warranted in generalizing from one specific case, the case provides a microcosm of the changes that are necessary, down to the individual level, in order for a manufacturing company to become globally competitive. Harnessing these individual capabilities provided the strategy for Rane Brake Linings – the company we focus on in this chapter.

19.2.   What Does It Take To Be a Global Supplier?

How does a manufacturing company in India, that believes it can compete globally, transform itself? We believe, as do many others, that attaining global competence requires significant changes "back at the ranch" which can lead to a significant improvement in the domestic cost/quality frontier. In other words, thinking globally is good for the local market.

We illustrate this point by focusing on one company in India, Rane Brake Linings (RBL). We use RBL as an example to show the significant transformation in business processes and thus outcomes in one organization and the top to bottom impact of such changes on the company. The examples from this company can serve as a template for managing change that is mission driven in many developing countries. The success of RBL also provides some insights into the manufacturing challenges in India and suggests that it is possible to develop an approach for manufacturing excellence that can work effectively in the Indian environment.

In 2003, RBL won the prestigious Deming prize and joined an elite group of 13 Indian companies that have won the Deming prize (see Table 19.1 for the entire list as of 2004). The Deming prize, awarded by the Japan Union of Scientists and Engineers (JUSE), was the culmination of a three year journey for RBL, which began with a visit by Prof Yasutoshi Washio from Japan. The prize and its citation are proudly displayed in the front lobby of RBL's offices in the outskirts of Chennai. But what was the three year transformation that culminated in the Deming prize? The company won the Deming award in 2003, has ISO 9001 and 14001 certification and has accumulated over 200 man years of R&D experience.

Table 19.1. Indian Deming Award winners list (1998–2004).

DEMING APPLICATION PRIZE	
Sundaram-Clayton Limited, Brakes Division (India)	1998
Sundaram Brake Linings Ltd. (India)	2001
TVS Motor Company Ltd. (India)	2002
Brakes India Ltd., Foundry Division (India)	2003
Mahindra and Mahindra Ltd., Farm Equipment Sector (India)	2003
Rane Brake Linings Ltd. (India)	2003
Sona Koyo Steering Systems Ltd. (India)	2003
SRF Limited, Industrial Synthetics Business	2004
Lucas-TVS Limited	2004
Indo Gulf Fertilisers Limited	2004
QUALITY CONTROL AWARD FOR OPERATIONS BUSINESS UNITS	
Hi-Tech Carbon GMPD (India)	2002
Birla Cellousic, Kharach-A Unit of Grasim Industries Ltd. (India)	2003
JAPAN QUALITY MEDAL	
Sundaram-Clayton Ltd., Brakes Division (India)	2002

Source: JUSE website: www.juse.or.jp

19.3. RBL Before the Transformation

Before its transformational journey, RBL was like any other Indian auto company. The Indian business environment was characterized (before 1991) by significant government control of the industry termed in the press as "License Raj" (Sivadasan, 2003). Individual firms were provided production quotas that governed the quantity produced. There were tight price controls, stringent costs for layoffs, strong union power all of which worked towards producing a stable business environment with low levels of growth and low levels of new product introduction and manufacturing innovation. The goal was to produce parts to specifications and quality was ensured by inspection.

RBL was organized along hierarchical lines with all key decisions being made by top management – even transactional decisions. Individual employees focused primarily on adhering to the rules of operation rather than solving customer problems. Contacts with suppliers were managed by sales managers who were the only conduit between the company and its customers. Manufacturing had no mechanism to contact customers and thus lead times to solve customer problems were long. Defect rates were high with ppm defect levels of 16000 ppm in 1999 and plant in process rejections of 2.1% of total pieces being common. However, RBL did have long term contracts with the Indian

Railways that provided a stable business base, with low levels of competition, based on the history of the company its beginnings as a supplier of components to the Indian railways.

The opening up of the Indian industry provided a new economic environment within which RBL had to compete. It also coincided with a downturn in the Indian auto industry and thus the auto component industry. Existing businesses had to find a way to offer globally competitive products even to survive in the Indian market. Companies had to decide how to transform themselves – many looked eastward to Japan and decided that the quality movement and associated processes used by Japanese companies offered a viable template. Why Japan? Possible similarities between Japanese and Indian companies included the role of a coach, which meshed with the Indian culture's inherent respect for the opinions of a "guru", a historical tradition (often lost in industrialized manufacturing) in the Indian culture for excellence in work (the elaborate temple and cave décor as testimony to this tradition) as being a reward unto itself, an inherent disapproval of "firing" employees and thus a premium for job security and finally the use of age as a proxy for wisdom. Some of these features are at odds with each other, thus they had to be meshed to generate value for RBL as a company. It is interesting to ponder the alternative approach, frequently followed in the US environment, where the idea of a company coach is replaced by the view of the CEO as a charismatic leader. Which of these choices is ideal for a manufacturing company in a developing country? What are the inherent risks associated with making the wrong choice?

What had to be developed was the enthusiasm for individual process ownership and thus the proffer of individual suggestions for improvement, best expressed by the title of the book "40 years, 20 million ideas" at Toyota [Yasuda, 1991]. In addition, the movement to a learning organization required careful documentation of the processes being followed, transparent measurement and use of data, and clear links between performance and payments.

19.4.   New Goals

Mr. Sundarram, President of RBL, explains it as a transformation that begins with changing the goal of the company. He describes the purpose of RBL is to "create a customer". He explains that while there are many possible objectives to be met, there is only one that needs to be maximized and that is to "maximize customer satisfaction". He believes that TQM provides a philosophy that aims to define a methodology that creates Quality that delights the customer while satisfying all other stakeholders. He views this as a competitive imperative because he thinks that for a business to succeed, it must create quality that is superior to the competition "in the customer's eyes." He goes even further than applying TQM for improving quality of products and services and expects to adopt the technique to almost every facet of decision making at RBL.

Is this a significant choice? Yes, as it seems to fly in the face of conventional wisdom that the only goal of a firm is to "maximize shareholder value." But, RBL treats stock returns as a constraint – steady returns as promised are considered a target that has to be achieved, not maximized. We believe that this is a key choice for the company. In other work, we have examined whether Indian companies that have won the Deming prize have delivered better metrics than comparable companies in the Indian marketplace. Our results suggest that Deming companies seem to have chosen a deliberate strategy not to "milk the cash cow" (that is not to drain the firm of resources that fuel the generation of ideas for the future – killing the golden goose through a low carb diet?) but instead focus on improving quality subject to financial returns targets. This resulted in a deliberate choice to focus on the Deming prize – a deliberate strategic choice regarding how to compete.

19.5.   History

Rane Brake Linings (RBL) is a division of the Rane group, an automotive components company with a sales turnover of $131 million and 4600 employees. The Rane group consists of Rane (Madras) focused on steering and suspension, Rane TRW focused on power steering and seat belt, Rane engine valves focused on valves, valve guides and tappets, Rane Brake Linings focused on brake linings, disc pads, composite brake blocks and clutch facings and Rane Nastech focused on energy absorbing steering columns.

RBL started in 1967 in Chennai, with the second plant in Hyderabad in 1991 and a third plant in 1997 in Pondicherry. Sales turnover for RBL was $70 million in 2003–04. RBL has a technical collaboration with Nisshinbo, a Japanese company. Until the 70's, Quality meant fitness for use of product sold at an affordable price. Thus, businesses differentiated themselves on this basis and superior quality in the customer's eyes, fewer defects, less irritants, lower purchase price, etc. Total Quality Control focused on reducing defect rates and reducing the cost in the factory.

But in the new business environment, companies needed to differentiate their offering in the eyes of the customer comparatively, not necessarily superlatively; by making it lighter, faster, safer, etc.; or differentiate in a manner that may be subtle: by improving service, reliability, etc. The TQM journey began in 1999 at RBL. Becoming more profitable required finding new markets, but that required first managing the company better so that costs decreased as quality improved. RBL chose to aim for the loftiest goal, winning the Deming prize, in order to contain costs and improve quality. We believe apart from the fact that this decision was momentous and led to impressive results, the rapidity of the transformation provides further food for thought. How is it that RBL was able to achieve so much within such a small time span? We discuss this issue at the end of the case as being reflective of the potential of engineering and related services available in the subcontinent, as well as, how we believe these capabilities came about.

19.6.   The Japanese Professors at RBL

Rane was ready to announce its commitment to quality by submitting to rigorous examination by their Japanese professors. This represented a movement that had been termed by a local business magazine as "India endeavors to become an economical Japan to the world." Indian manufacturers were eager to become suppliers and collaborators to larger European, Japanese, and US companies. This also reveals their (initial) reluctance to take business risk inherent in product design and direct marketing to end users. It reveals their preference for competing through engineering and technical innovations.

RBL's TQM journey began with the choice of Prof Yasutoshi Washio as their coach. He established the following criteria for excellence: 1. Develop a business model to generate business for long period through Uniqueness in products & technology and Uniqueness in achieving certain excellence 2. Reduce technological dependence on another company 3. Create new market(s) or achieve drastic expansion of existing market 4. Unique and enhanced utilization of manpower/human resources that drives a company to supply excellent people/manpower.

As an example of the criteria used by the examiners for the Deming Prize, Prof Iizuka (Lead examiner for Diagnosis) focused on RBL's answer to the following question: "What is your approach to competition?". Prof Shiba (Lead examiner for final examination) focused on "How do you deal with change?" (Technology, Legislation, Competition, etc.). In other words, winning the Deming prize provided an interesting challenge whereby RBL had to justify not just use of TQM tools but its justification in managing change and managing against competitive forces shaping the business.

19.7. Executing TQM at RBL

RBL decided to focus on Policy Deployment and Daily Routine Management (DRM) to achieve their TQM implementation. As a result, RBL redefined its management of processes for New Product Development System, Manufacturing Quality, Supplier Quality and Customer Quality. In the new system, each manager was required to define his role, his metrics, his measurement of performance to date and the steps being taken to improve performance. DRM deployment meant that (a) Each function will have unique purpose, (b) Each purpose will have role, (c) Each role will have managing points to achieve the purpose, (d) All managing points have measure of performance, (e) All managing points have metrics, (f) All management points have either graphs or vital activity monitoring chart, (g) Development of metrics or indices is important, specifically for non-manufacturing areas. The impact of such measurement and associated financial incentives is such that the discussions of year end bonus is now a short discussion, i.e., merely verifying the numbers and thus the payments.

TQM implementation created tangible and intangible benefits for RBL. Intangible benefits included role clarity so that each person understood their role in the organization, their suppliers and customers, and their metrics. The focus on competency and involvement resulted in a different approach to managing people. The focus on management points and check points and the systematic approach to planning (including catch balling or adjusting plans across roles) all resulted in a management system where charts, goals and current performance relative to plan all became commonplace throughout the company. Also, common for every problem that came up was a systematic analysis of the problem, steps taken to resolve it, impact of the steps and learning from each observation.

How does all this affect execution of specific tasks at RBL? Customer line rejections dropped from 16000 ppm in '99 to 1750 in '03. Plant in process rejections decreased from 2.1% of total pieces to 0.85% of total pieces produced. Sales per employee went up from $22000 to $40000. Number of employee suggestions went from 280 to 7500 during the period. In other words, TQM represented a dramatic and measurable improvement across many specific metrics that would impact the company.

But, the key benefit was the continual improvement potential unleashed by TQM. When a potential buyer contracts to RBL, they become part of the TQM processes for improvement. This means that the specific business processes affect the cost reduction rate or the quality improvement rate for the buyer. The speed of response to engineering specification changes, customer requests, etc., all affect the overall cost to do business.

In short, how RBL deals with cost increases, performance issues, etc., provides important insights into their potential for transforming into a first class global supplier. To illustrate this, we discuss specific examples provided to us by managers at RBL. The purpose of these examples is to provide a glimpse of the processes that would be experienced by a buyer in his interaction with specific functions at RBL.

19.8.   Materials Management at RBL

A significant issue at RBL is that raw material contributes to 39.0% of Total product cost, i.e., 65% of the Total Variable cost (TVC), and has an appreciable impact on product contribution. RBL defined a Material Price index (MPI) to measure the variance in the prices of key raw materials with reference to the base year 2000-01. They also focus attention on the group of high value materials that contribute to 80% of the total purchase value. Their plan was to take the then (2000) current MPI index of 100 and work to decrease it to 96 by March 2003. Of the materials consumed, asbestos constituted 60% of the total raw material cost and was imported in a fiber form from North America. Data regarding potential world sources of asbestos showed that North America provided 30.0% of world volume, CIS (Commonwealth of Independent States represented) provided 46.0%, South America 13.0%, South Africa 5.0%, Zimbabwe 3.0%, China 2.5% and India 0.5%. The decision was made to focus on CIS as a potential source replacing the North American source. In addition, the quality was expected to improve with the new fiber with the potential for reduction of in process rejections. Potential savings were expected to come from alternate source price reductions over time, savings through negotiations, savings through government incentives (to reduce foreign exchange use), etc. The implementation of the plan with the new source material was coordinated with a new product introduction at the higher level of quality, but required the new product to be tested and approved by the user, the chemical composition (called homologation) to be registered with the appropriate authorities, the sources established and deliveries streamlined, planned production coordinated with purchases to reduce air freight (and associated costs), etc. The overall impact was a delivered reduction of the MPI as planned.

How does this help a buyer making a decision regarding RBL as a potential supplier? It is clear that material costs are a large portion of overall costs for many automotive products. A study by Balakrishnan, Iyer, Seshadri and Sheopuri (2004) finds material costs to account for 60 to 70% of the total costs for many auto components and related products. Thus, material cost fluctuations will require creative engineering solutions to maintain or reduce product costs. RBL's carefully documented example of cost reductions shows that the buyer can reasonably expect similar actions for their products over time, thus maintaining price competitiveness without the need for repeated supplier changes. We believe that RBL is thus selling their process management capability in addition to their products. The careful attention to programmed details brings confidence in not only the stability of the process quality but also the buyer-supplier relationship.

19.9. Customer and Supplier Coordination to Improve Product Specification and Performance

One of RBL's customers introduced a new two-wheeler disc pad in the Indian market. While RBL produced the product to specifications, the pads were found to stick during use by the end customer. The customer reported the problem to RBL on 14 April 2004. The two-wheeler manufacturer claimed that the parallelism of the installed pads was not up to standard, and the flatness and surface finish were not acceptable. The possible causes could be attributed to (a) the supplier of some components of the disc brakes to RBL, (b) RBL's manufacturing of the brake linings, (c) the customer (two-wheeler manufacturer) installation of the disc brakes in the two wheeler, or (d) its use by the end customer. However, given RBL's stated goal to maximize customer satisfaction, they decided to solve the problem for the two-wheeler manufacturer.

Should RBL choose to solve the problem on behalf of the supply chain? Historically, employees in the company would have focused on whether this is a service everyone will require, who would pay the costs, what the impact of a failure to solve the problem would have been, whether the customer would even appreciate the effort etc. – all risks associated with taking this action. But in the post TQM world, the decision was easy – since customer satisfaction was important, RBL had to offer to solve the problem and believe that its superior processes were capable of managing the execution. In addition, the execution of processes had to be documented so that similar problems could be mitigated in the future.

The first step was to devise a measurement gauge that would be used by all three companies – the supplier to RBL, RBL's manufacturing personnel and the customer. This gauge was designed to measure thickness all around the pads. RBL stationed its engineers at the supplier and the two-wheeler manufacturing sites and proceeded to use these (now standard) gauges to measure the pads. This step alone decreased error rate from 25% to 3%. The next step was to work on correcting the plate manufacturing process at the supplier's end. The original process at the supplier had a 0.2 mm gap between the rollers, the direction of pass of the roller (the side which faced the roller) was not specified, as well as, the number of pieces per pass was not specified. In the modified process, developed jointly by RBL and the supplier, the machine was set to have a 0.1mm gap between the rollers, the direction of pass was specified as the Lug Side, and the number of pieces per pass was set to one. These changes increased the acceptance rate from 75% to 98%. At RBL's manufacturing, the grinding wheel was changed from a Diamond wheel to Aluminum Oxide (60 Grits). In addition, buffing was done to clean dust. This decreased productivity at RBL but increased roughness necessary to deliver the required performance. To improve productivity, RBL changed to a 120 Grits diamond wheel. To further improve productivity, RBL switched to a three step wheel. The end point was an improved disc pad in which the sticking problem was completely eliminated. The completion date of this project was May 10, 2004!

What does this example show? It demonstrates that RBL is capable of not only understanding how they fit into the supply chain but also how the product is used by the customer. They are interested and capable of both managing process improvement across the supply chain, as well as, completing this process in a short time frame. The top management goal of maximizing customer satisfaction means that employees at RBL and managers do not need to spend time authorizing engineers and other personnel to tackle such problems. Suppliers and customers do not have to worry about paying for such service. This provides RBL with an edge over companies without such top management commitment.

How does it help a potential buyer of RBL's products? The buyer can now potentially decrease the overhead (engineering and procurement staff) that would otherwise be required to play the coordination role described above. This reduction in overhead is an added reduction in direct item related costs that can make RBL more competitive overall even if their product prices are higher. In fact, one of RBL's overseas customers, who chose them over stiff competition from a supplier in another country, claimed he did it so that he can potentially take a vacation as planned, knowing that potential problems would be resolved by RBL within the supply chain. This capability may be worth a lot to a global buyer.

19.10. How Has RBL Evolved as a Result of Its Deming Journey?

Sundarram, the president of RBL describes "the state of being" of an organization as the attitude towards customers, employees and improvements. He describes RBL as having evolved from basic to systematic to strategic in its "state of being" and thus its approach to business.

Prior to the '90s the organization could be described as "basic," i.e., it was focused on survival. The top management and a few employees dealt with the customer. For most employees, there was a distinct lack of awareness of the customer needs. The approach was top-down with instructions provided by top management that were executed by the rest of the company. There was no approach to improvement, the goal was to maintain status quo.

In the early '90s, RBL moved to a "systematic" mode of operation which was methodical but bureaucratic in its approach. The approach to the customer was contractual in nature, thus mechanical. Employees had role clarity, but conformance, not creativity, was rewarded. Small improvements were permitted but large changes required top management intervention.

The end of the '90s through today, RBL is described as having moved to a "strategic" mode of business operation. In such a mode, leadership focuses on strategic direction and uses cross-functional interfaces at all levels. Customer contacts are viewed as opportunities and closeness with customers is encouraged. Employees are regarded as key resources with large investments in human capital. RBL proudly describes their investment in employees, which comprises of sending them for training several times over the course of just a few years to receive advanced technical training in Japan to develop their R&D capability. Finally, the company is focused on making big improvements, market expansion, and achieving its global ambitions. The organizations is now focused on relationship quality with the introduction of (a) customer satisfaction index (CSI) and customer relationship index (CRI) and (b) direct contact between new product development and quality assurance and the customer.

19.11. Discussion

RBL's story of quality improvement, strategic focus and creation of a modern management team empowered by a modern philosophy might seem to be an exception. However, the mantra of success at RBL is being repeated in many manufacturing firms across India today. The rapidity with which the transformation is taking place can be assessed by the table given below comparing quality and delivery related performance metrics for auto component firms.

There are several contributing factors for the improvement in quality and global competitiveness of the Indian manufacturing sector and the auto component sector in particular. We classify these into macro-economic factors, industry-related factors, and infrastructure related factors, human factors and intangibles. Amongst the macro-economic factors are the liberalization strategy pursued by India since 1990 as well as the growing affluence of the middle class, thus, increase in purchasing power that has led to demand for more sophisticated products and services. Amongst the industry-related factors, the establishment of Maurti Udyog in collaboration with Suzuki in the 1980s created a nucleus for propagating Japanese manufacturing best practices amongst suppliers. It is perhaps surprising how large Japanese firms ventured to enter the then small and underdeveloped (in several dimension) Indian market! The industry also saw the entry of major auto OEMs in the nineties. Survival was an issue for auto component suppliers in many ways: meeting quality, productivity, and delivery standards became essential to stay in business. The infrastructure-related factors are better communication facilities and better roads, as well as, some critical urban services. It is easy to document changes in the three areas mentioned above.

The last two factors are difficult to quantify. Amongst human factors is the fact that the Indian engineering schools have kept pace with growing demand and changing technology. Even medium sized engineering schools teach sophisticated and current topics. Moreover, there has been more and more exposure to modern design and engineering ideas due to easier exchange of ideas. Among intangibles, we list the IT revolution and its impact on several aspects: ability to handle large complex projects, communication skills, professionalism, systematic thinking, etc.

Among intangibles is also the ability of CEOs and Presidents, like Sundarram and the Chairman of the Rane group, to take a long term view towards improvement. The change to professional management at all levels that emphasizes detailed planning and customer focused execution is an important factor. RBL has invested in process improvement and capability development almost equally with a view to gaining competitive ability. It is also different because it has adopted TQM to managing non-manufacturing processes. Finally, for RBL, TQM provided a mechanism to compete in the new Indian manufacturing environment by harnessing the individual capabilities of its employees using a systematic process improvement mechanism.

Table 19.2.   Quality Performance of the Indian auto-component industry.

2001	2003
Process conformance through Quality Certifications	Process Improvements through Quality Initiatives like TQM, TPM, Six Sigma
Customer (OE) Line Rejections	Customer (OE) Line Rejections
1000 plus ppm	100–400 ppm
Rework 3–5%	Rework < 1%
First pass yield < 80%	First pass yield 95 to 97%
OEE 70 to 80%	OEE 90 to 95%
Warranty > 95%	Warranty 500–2000 ppm

Source: Customer Satisfaction Tracking Surveys

Table 19.3. Delivery Performance of the Indian auto-component industry.

2001	2003
Functionally oriented delivery mechanisms	Integrated Supply chain Systems
OEMs maintained raw material & components inventory at their end	Stocks maintained by suppliers to service OEMs Just In Time (JIT) systems
Component suppliers used "push" systems – minimum batch quantity	Component suppliers use Kanban, Bin Systems – "pull" system
Key Delivery Metrics:	Key Delivery Metrics:
OTD - OEMs: 70 to 80%	OTD - OEMs: 90 to 100%
JIT Adherence: 80–90%	JIT Adherence: > 95%
Milk Van Residence Time: 60 mins	Milk Van Residence Time: 30–45 mins

Source: Customer Satisfaction Tracking Surveys

The specific motivations for TQM at RBL are thus different from reports in academic studies. There have been studies published that suggest that TQM implementation has not really helped firms improve their financial performance. For example, Ernst and Young (1991) found that firms did not benefit from TQM. Many have blamed the failure of TQM due to poor implementation (see discussion in Sterman, Repenning, and Kofman, 1997). However, Hendricks and Singhal (1999) provide a study that suggests that firms with effective TQM implementation "do better in terms of stock price performance when compared to appropriate benchmarks." A study by the National Institute of Standards and Technology, which administers the US Malcolm Baldrige award, found that "through December 1, 1998, the publicly traded Baldrige award winners from 1988 to 1997 outperformed the S&P 500 by 2.6 to 1 (460 percent return of the winners against a 175 percent return for the S&P 500)." (NIST 1999). Academic studies of TQM impact thus seem to vary from a study of its impact on individual companies (Sterman et al. 1999) to studies of average performance of specific winners of the award in the stock market (Hendricks and Singhal (1999) and NIST (1999)). At this point, there are no similar studies of the Indian Deming award winning companies. Perhaps winning the Deming was an important component of "getting to becoming globally competitive." We expect to continue our research efforts into the Indian manufacturing environment in the future to provide more systematic evidence.

Are there some reasons why a large number of the Deming award winners are located in southern India, in a region close to Chennai? Perhaps the industry associations and specific leadership provided by early Deming award winners, might account for such a spatially concentrated similarity in firm strategies. In addition, the availability of capable managers, who moved between firms as they went through the Deming award quest might have played the role of a catalyst. In addition, the family business units that own the auto ancillary clusters could have helped spread the TQM philosophy.

19.12. Conclusions

How does a manufacturing company in India develop a capability to compete in global markets? We suggest that the strategy used by Rane Brake Linings provides one approach to the development of such capability that focuses on both policy deployment and daily routine management and thus moves a firm from a "basic" state of operation to a "strategic" mode of operation. The end result is an ability to engage with a potential buyer both to deliver current product designs as well as to evolve new designs and products over time. Discussions with senior management suggests that there were some culturally relevant similarities between the Indian and the Japanese environments that enabled TQM implementation as part of the quest for the Deming prize. In addition, as most global buyers would attest, learning about a potential supplier's "mindset" is a crucial step in the final choice – RBL's journey to adjust and document their new "mindset" offers them significant future potential to engage with the global buyer.

References

Balakrishnan, K., Iyer, A., Seshadri, S., Sheopuri, A. (2004). "Indian Auto Supply Chains at the Crossroads". Purdue CIBER Working Paper, Krannert School of Management, Purdue University, West Lafayette, IN 47907.

Ernst and Young (1991). "International Quality Study – Top Line Findings," and "International Quality Study – Best Practices Report." Ernst and Young American Quality Foundation.

Hendricks, K. and Singhal, V. (1999). "The Financial Justification of TQM". *Center for Quality Management Journal, 8* (1).

National Institute of Standards and Technology (1999). "Baldrige Index Outperforms S&P 500 for Fifth Year". Press Release NIST 99-02, February 4, Washington DC.

Sivadasan J. (2003). Barriers to Entry and Productivity: Micro-Evidence from Indian Manufacturing Sector Reforms* Graduate School of Business, University of Chicago, http://bpp.wharton.upenn.edu/Acrobat/Sivadasan_AEW_paper_1_16_04.pdf#search='Barriers%20to%20Entry%20and%20Productivity:%20MicroEvidence', last cited on 24th March 2005.

Sterman, J. D., N. P. Repenning, and F. Kofman (1997). "Unanticipated Side Effects of Successful Quality Programs: Exploring a Paradox of Organizational Improvement." *Management Science, 43* (4), 503–21.

Yasuda, Yuzo (1991). "40 Years, 20 million Ideas – The Toyota Suggestion System". Productivity Press, Cambridge, MA.

Index

Early Titles in the
INTERNATIONAL SERIES IN
OPERATIONS RESEARCH & MANAGEMENT SCIENCE
Frederick S. Hillier, Series Editor, *Stanford University*

Printed in the United States